D1499220

# NUCLEAR MAGNETIC RESONANCE SPECTROSCOPY OF NUCLEI OTHER THAN PROTONS

# Nuclear Magnetic Resonance Spectroscopy of Nuclei Other Than Protons

*Edited by*

## T. AXENROD

*City College of the City University of New York*

and

## G. A. WEBB

*University of Surrey, England*

A WILEY-INTERSCIENCE PUBLICATION

**JOHN WILEY & SONS**

New York · London · Sydney · Toronto

Copyright © 1974, by John Wiley & Sons, Inc.

All rights reserved. Published simultaneously in Canada.

No part of this book may be reproduced by any means, nor transmitted, nor translated into a machine language without the written permission of the publisher.

*Library of Congress Cataloging in Publication Data:*

Axenrod, T. and Webb, G. A.

Main entry under title:

Nuclear magnetic resonance spectroscopy of nuclei
   other than protons.

   "A Wiley-Interscience publication."
   "[Papers] taken in large part from the Advanced
Study Institute on 'nuclear magnetic resonance
spectroscopy of nuclei other than protons' held
in Tirrenia (Pisa), Italy during September 1972."
   Includes bibliographical references.

   1. Nuclear magnetic resonance spectroscopy.
I. Axenrod, T., ed. II. Webb, Graham Alan, ed.

QD95.N8          541.3          73-17377
ISBN 0-471-03847-4

Printed in the United States of America

10 9 8 7 6 5 4 3 2 1

CONTRIBUTORS

Abraham M. Achlama (Chmelnick), *Department of Structural Chemistry, The Weizmann Institute of Science, Rehovot, Israel*

I. R. Ager, *Organic Chemistry Department, Imperial College of Science, University of London, London, England*

J. W. Akitt, *The School of Chemistry, The University of Leeds, Leeds, England*

Joseph R. Autera, *Explosives Division, Feltman Research Laboratory, Picatinny Arsenal, Dover, New Jersey, U.S.A.*

Theodore Axenrod, *Department of Chemistry, The City College of The City University of New York, New York, U.S.A.*

Edwin D. Becker, *National Institutes of Health, Bethesda, Maryland, U.S.A.*

E. G. Brame, Jr., *E. I. du Pont de Nemours & Company, Elastomer Chemicals Department, Experimental Station, Wilmington, Delaware, U.S.A.*

E. Breitmaier, *Chemisches Institut der Universität Tübingen, Tübingen, Germany*

Suryanarayana Bulusu, *Explosives Division, Feltman Research Laboratory, Picatinny Arsenal, Dover, New Jersey, U.S.A.*

Joseph D. Cargioli, *General Electric Corporate Research and Development, Schenectady, New York, U.S.A.*

L. Cavalli, *Montedison, Centro Ricerche, Bollate, Italy*

P. Diehl, *Department of Physics, University of Basel, Basel, Switzerland*

Daniel Fiat, *Department of Structural Chemistry, The Weizmann Institute of Science, Rehovot, Israel*

Carlos F. G. C. Geraldes, *The Chemical Laboratory, University of Coimbra, Portugal*

Victor M. S. Gil, *The Chemical Laboratory, University of Coimbra, Portugal*

George C. Levy, *General Electric Corporate Research and Development, Schenectady, New York, U.S.A.*

Gary E. Maciel, *Department of Chemistry, Colorado State University, Fort Collins, Colorado, U.S.A.*

B. E. Mann, *The School of Chemistry, The University of Leeds, Leeds, England*

W. McFarlane, *Chemistry Department, Sir John Cass School of Science and Technology, City of London Polytechnic, London, England*

Joachim Müller, *Department of Chemistry, University of Marburg, Marburg, Germany*

John F. Nixon, *School of Molecular Sciences, University of Sussex, Brighton, Sussex, England*

O. Oster, *Chemisches Institut der Universität Tübingen, Tübingen, Germany*

L. Phillips, *Organic Chemistry Department, Imperial College of Science, University of London, London, England*

P. S. Pregosin, *Department of Chemistry, University of Delaware, Newark, Delaware, U.S.A.*

E. W. Randall, *Department of Chemistry, Queen Mary College, London, England*

W. Voelter, *Chemisches Institut der Universität Tübingen, Tübingen, Germany*

Rod Wasylishen, *Department of Chemistry, University of Manitoba, Winnipeg, Manitoba, Canada*

G. A. Webb, *Department of Chemical Physics, University of Surrey, Guildford, Surrey, England*

Felix W. Wehrli, *Varian AG, NMR Laboratory, Zug, Switzerland*

## PREFACE

The contents of this volume are taken in large part from the Advanced Study Institute on "Nuclear Magnetic Resonance Spectroscopy of Nuclei Other Than Protons" held in Tirrenia (Pisa), Italy during September 1972.

Although the potential of NMR studies of nuclei other than protons was recognized very early, experimental difficulties rather than any fundamental limitations were responsible for the initial slow developments in the field. The recent advent of commercial pulse Fourier transform NMR spectrometers has revolutionized the field, and non-proton NMR is now one of the most rapidly expanding branches of chemistry. In a few short years NMR studies of other nuclei have become an essential part of molecular spectroscopy and an indispensable aid in the elucidation of molecular structures.

In contrast to conventional proton spectroscopy, NMR of such important nuclei as $^2$H, $^{11}$B, $^{13}$C, $^{14}$N, $^{15}$N, $^{17}$O, $^{19}$F, $^{29}$Si, and $^{31}$P share certain characteristics that render the NMR experiment difficult; these include unfavorable natural abundance and/or low inherent sensitivity arising from their small magnetogyric ratios. The dramatic increase in sensitivity afforded by the pulse Fourier transform technique through multichannel excitation has made the study of these nuclei almost routine and has simultaneously accelerated progress in theory, applications, and instrumental development. One of the objectives of the institute, and of this book, is to bring together an authoritative and up-to-date treatment of the recent advances, problems, and new techniques involved in the study of the NMR parameters of the more important nuclei.

Grateful acknowledgment is made to the NATO Scientific Affairs Division for their generous support of the Advanced Study Institute.  We wish to express our thanks to Professor G. Berti for his advice and assistance in organizing the institute and to the contributors for their cooperation in the production of this book.

Theodore Axenrod
Graham A. Webb

*New York, New York*
*Guildford, Surrey, England*
*June 1973*

# CONTENTS

# Contents

# NUCLEAR MAGNETIC RESONANCE SPECTROSCOPY
## OF NUCLEI OTHER THAN PROTONS

# CHAPTER 1

# NUCLEI OTHER THAN HYDROGEN: A REVIEW OF SOME NUCLEAR PROPERTIES AND A DISCUSSION OF THEIR RELAXATION MECHANISMS

Edwin D. Becker

*National Institutes of Health*
*Bethesda, Maryland*

If we adopt the familiar $^1$H resonance as a standard for comparison, then all other magnetic nuclei (except $^2$H and $^3$H) are set apart by their enormously greater range of chemical shifts—usually hundreds of ppm, rather than the 10 ppm or so for hydrogen. In sensitivity, too, most nuclei are at least an order of magnitude (and often several orders of magnitude) lower than $^1$H. Thus both the problems of studying "other nuclei" and the information available are often quite different from those pertaining to proton NMR. The first portion of this chapter reviews briefly the pertinent nuclear properties and includes a discussion of such topics as chemical shift scales and references, isotope effects, and the consequences of negative magnetogyric ratios. The major portion of the chapter is devoted to an examination of how and why nuclei relax, since relaxation behavior governs many facets of the experimental methods we can apply. Moreover, among "other" nuclei, studies of relaxation can often provide valuable information on chemical bonding and molecular dynamics.

Table 1-1 lists the nuclear spin and relative sensitivity

TABLE 1-1.  Some nuclear properties

| Nucleus | Spin | Relative Sensitivity ($S$) | % Natural Abundance ($A$) | $S \times A$ |
|---------|------|---------------------------|---------------------------|--------------|
| $^1$H   | $\frac{1}{2}$ | 1.00 | 100. | 100. |
| $^2$H   | 1 | $9.65 \times 10^{-3}$ | 0.015 | $1.4 \times 10^{-4}$ |
| $^3$H   | $\frac{1}{2}$ | 1.21 | — | — |
| $^{10}$B | 3 | $1.99 \times 10^{-2}$ | 20. | 0.40 |
| $^{11}$B | $\frac{3}{2}$ | 0.17 | 80. | 13.6 |
| $^{13}$C | $\frac{1}{2}$ | $1.59 \times 10^{-2}$ | 1.1 | 0.017 |
| $^{14}$N | 1 | $1.01 \times 10^{-3}$ | 99.6 | 0.10 |
| $^{15}$N | $\frac{1}{2}$ | $1.04 \times 10^{-3}$ | 0.37 | $3.8 \times 10^{-4}$ |
| $^{17}$O | $\frac{5}{2}$ | $2.91 \times 10^{-2}$ | 0.04 | $1.2 \times 10^{-3}$ |
| $^{19}$F | $\frac{1}{2}$ | 0.83 | 100. | 83. |
| $^{23}$Na | $\frac{3}{2}$ | $9.25 \times 10^{-2}$ | 100. | 9.3 |
| $^{29}$Si | $\frac{1}{2}$ | $7.84 \times 10^{-3}$ | 4.7 | 0.037 |
| $^{31}$P | $\frac{1}{2}$ | $6.63 \times 10^{-2}$ | 100. | 6.6 |
| $^{199}$Hg | $\frac{1}{2}$ | $5.67 \times 10^{-3}$ | 16.8 | 0.095 |
| $^{207}$Pb | $\frac{1}{2}$ | $9.16 \times 10^{-3}$ | 22.6 | 0.21 |

for many of the nuclei that are discussed in this volume. Except for $^{19}$F and the very infrequently studied $^3$H, the inherent sensitivities of all the nuclei are far below that of $^1$H. In addition, many magnetic nuclei exist at only low natural abundance, so that the sensitivity problem is compounded. Only within the last few years has it been possible to make studies of many of these nuclei in a wide variety of compounds, usually by pulse Fourier transform methods (see Chapter 2) for direct observation or by double resonance methods (see Chapters 3 and 23) for indirect study. Nuclei with spin > $\frac{1}{2}$ usually relax rapidly by quadrupole interactions, as we shall see, with concomitant line broadening and reduction of the peak intensity even further

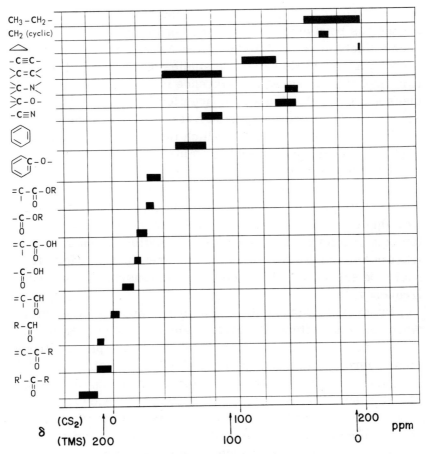

Figure 1-1.   Correlation of $^{13}$C chemical shifts with functional groups.

than would be suggested by the figures in Table 1-1.

For nuclei other than hydrogen the dominant effect in determining chemical shifts is the temperature-independent paramagnetic term, in contrast to hydrogen, where this term is almost negligible.[1] One result is that the range of chemical shifts commonly found for "other" nuclei is at least an order of magnitude greater than that found for hydrogen, as exemplified in Fig. 1-1, an approximate correlation chart for $^{13}$C.

Figure 1-1 gives two scales for $^{13}$C chemical shifts, one relative to $^{13}$CS$_2$, which has been used widely in the past, and one relative to $^{13}C$ in tetramethylsilane (TMS). The latter reference has been employed in three books[2] on $^{13}C$ NMR and seems likely to be accepted by most NMR spectroscopists. In general, as we begin extensive studies of a variety of nuclei, it is important to agree on reference compounds for each nucleus and to establish a sign convention for chemical shifts. The vast body of data for proton NMR is based on the internationally accepted δ scale in which the reference, TMS, occurs at lower frequency (higher field) than most other resonances.[3] To provide mostly positive numbers for chemical shifts, then, the scale has been chosen to increase with increasing frequency. The TMS-based scale for $^{13}C$ is in accord with this convention, and it seems highly desirable that data for other nuclei be expressed in a compatible manner. (At least one major NMR publication requires that data for all nuclei be expressed with values of δ increasing with increasing frequency.[4])

In several instances it is possible to observe NMR with more than one isotope of a given element; often different isotopes have specific advantages. One good example is nitrogen, where there is an advantage of narrow lines arising from $^{15}N$, compared with the generally broad lines from rapidly relaxing $^{14}N$. On the other hand, the almost 300-fold advantage in natural abundance of $^{14}N$ relative to $^{15}N$ makes it the isotope of choice in many cases. It is important to know to what accuracy data obtained for different isotopes can be compared. Some early NMR data suggested that chemical shifts for $^{14}N$ and $^{15}N$ in the same compound might differ substantially—a "zero-bond" isotope effect analogous to well-known effects on the chemical shift of a nucleus on isotopic substitution of nearby atoms.[5] We can write the Larmor equation, including the chemical shielding σ, for the two isotopes:

$$^{15}\nu = \frac{^{15}\gamma}{2\pi} B_0 (1 - {}^{15}\sigma)$$

$$^{14}\nu = \frac{^{14}\gamma}{2\pi} B_0 (1 - {}^{14}\sigma) \ .$$

TABLE 1-2.   Ratio of $^{15}N$ and $^{14}N$ resonance
frequencies

| Compound | $\delta_N$ (ppm) | $R = {}^{15}\nu/{}^{14}\nu$ |
|---|---|---|
| $(CH_3)_4N^+I^-$ | 0 | 1.402 757 06 |
| $C_6H_5CH_2NC$ | 125.5 | 1.402 756 83 |
| $CH_3ONO_2$ | 298.6 | 1.402 756 98 |
| $(CH_3)_2NNO_2$ | 311.7 | 1.402 756 91 |

If the magnetic field $B_0$ is held constant, the ratio $R = {}^{15}\nu/{}^{14}\nu$ can be found for several compounds with very different chemical shifts, and checked for constancy. Several years ago we carried out this determination by double resonance methods[6] with a series of compounds chosen for their (a) narrow $^{14}N$ resonance lines, (b) ease of synthesis of the $^{15}N$ derivatives, and (c) coverage of 300 ppm in chemical shift range. The results, summarized in Table 1-2, show that $R$ is very accurately constant over this range of chemical shifts and strongly indicate that there is no significant isotope effect on the shielding. The ratio $|{}^{15}\gamma/{}^{14}\gamma|$, obtained for the bare nuclei by atomic beam methods,[7] is 1.4027, in good agreement with the value found for $R$. (In fact, $R$ probably represents a more accurate value for the ratio $|{}^{15}\gamma/{}^{14}\gamma|$.) It certainly seems probable that for isotopes of other elements also there is no significant difference in shielding.

According to the generally accepted theory of spin-spin coupling, coupling constants are directly proportional to the magnetogyric ratios of the coupled nuclei, so that predictable (and often large) isotope effects occur. Often coupling constants are expressed in reduced form as

$$K_{ij} = 2\pi \frac{J_{ij}}{\gamma_i \gamma_j}$$

in order to facilitate comparison of the magnitudes of couplings involving various nuclei. Some recent reports of $^1J(P-H)$ and

$^{1}J$(P–D) and of $^{2}J$(H–H) and $^{2}J$(H–D) in selected compounds indicate that $K_{ij}$ is not independent of isotopic substitution, as expected.[8,9] The discrepancies are rather small and difficult to measure accurately in the deuterated compounds because of line-width effects, but they appear to be real. The explanation for the isotope effect is not entirely clear.[8] It is worth noting that $\gamma$ is negative for some nuclei, including $^{15}$N and $^{29}$Si. For these nuclei particular caution must be exercised in discussion of the signs of coupling constants; usually it is least confusing to cast the discussion in terms of $K$'s rather than $J$'s.

Another important consequence of the existence of negative $\gamma$'s is found in the nuclear Overhauser enhancement (NOE), which occurs when protons are decoupled while another nucleus is observed. The maximum NOE possible in the observation of a nucleus X while protons are irradiated is given by

$$\eta = \frac{\gamma_H}{2\gamma_x} \, .$$

The intensity ratio decoupled/undecoupled is, then, $1 + \eta$. In many cases, as we shall see, relaxation proceeds by paths that lead to a smaller value of $\eta$ than given above. If $\gamma_X > 0$, then $(1 + \eta) > 1$, and the signal observed with decoupling is at least somewhat larger than without decoupling. If $\gamma_X < 0$, the signal can be rather large (but negative) if the full NOE is obtained; however, it can also be close to zero if a partial NOE occurs. This effect is of importance in studies of $^{15}$N (see Chapter 7) and $^{29}$Si (see Chapter 17).

An understanding of the origin of the nuclear Overhauser effect and of its magnitude in various compounds requires some familiarity with the mechanisms by which nuclei relax. Let us now turn to a brief review of this important topic. We do not attempt a rigorous treatment, such as that given in various monographs[10]; instead we attempt to indicate qualitatively why nuclei relax and to point out via some simple equations the important factors in the relaxation processes. This treatment is largely abstracted from the book by Farrar and Becker.[11]

Consider the classical picture of a precessing nucleus. An ensemble of identical nuclei all precessing at the same fre-

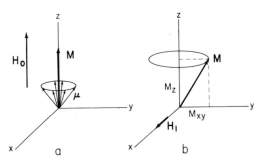

Figure 1-2.   (a) Ensemble of identical nuclei precessing about $H_0$, and the resultant macroscopic magnetization **M**.   (b) M tipped in the $yz$ plane by interaction with radio-frequency field $H_1$.

quency about an applied field $B_0$ is depicted in Fig. 1-2.  Since the nuclei have no preferred orientation in the $x$ or $y$ directions, their components of magnetization along these axes average to zero, and the net macroscopic magnetization **M** is at equilibrium directed along the $z$ axis parallel to $B_0$.  It is helpful to treat the motion of **M** not in the laboratory coordinate system, but in a coordinate system or frame of reference that rotates about $B_0$ at the Larmor frequency.  In this rotating frame, with axes denoted by primes, a radio-frequency field, $B_1$, that is rotating at the Larmor frequency appears to be stationary along the $x'$ axis.  It can thus interact with **M** and cause a torque that leads to precession of **M** in the $y'z'$ plane of the rotating frame. This tipping of **M** away from the $z'$ axis results in a decrease in $M_{z'}$, which is eventually restored to its equilibrium value by spin-lattice relaxation in a time characterized by $T_1$, the longitudinal relaxation time.  The components $M_{x'}$ and $M_{y'}$ relax to their equilibrium values of zero by spin-spin relaxation in a time characterized by $T_2$, the transverse relaxation time.

Nuclear relaxation is caused by an interaction between the nuclear magnetization **M** and small incoherent magnetic fields, **h**, that arise from random Brownian motions of the molecules. (For quadrupole relaxation it is electrical, rather than magnetic, fields that interact with the electric quadrupole moment of the nucleus, but the results are in many ways similar to

those given here for magnetic fields.) We can denote a typical
microscopic magnetic field in vector notation:

$$\mathbf{h} = \mathbf{i}h_{x'} + \mathbf{j}h_{y'} + \mathbf{k}h_{z'} \ ,$$

where $\mathbf{i}$, $\mathbf{j}$, and $\mathbf{k}$ are as usual unit vectors, but in this case
taken along the axes of the rotating frame. The torque that $\mathbf{h}$
exerts on $\mathbf{M}$ in attempting to return $\mathbf{M}$ to its equilibrium position is, as measured in the rotating frame,

$$(\mathbf{h} \times \mathbf{M})_{\text{rot}} = (\mathbf{i}h_{x'} + \mathbf{j}h_{y'} + \mathbf{k}h_{z'}) \times (\mathbf{i}M_{x'} + \mathbf{j}M_{y'} + \mathbf{k}M_{z'})$$

$$= \mathbf{i}(h_{y'}M_{z'} - h_{z'}M_{y'}) + \mathbf{j}(h_{z'}M_{x'} - h_{x'}M_{z'}) + \mathbf{k}(h_{x'}M_{y'} - h_{y'}M_{x'}).$$

(In the expansion of the vector cross product some terms are
zero since $\mathbf{i} \times \mathbf{i} = \mathbf{j} \times \mathbf{j} = \mathbf{k} \times \mathbf{k} = 0$.) Note now that the only components of $\mathbf{h}$ that multiply $M_{z'}$ (the component of $\mathbf{M}$ that is involved
in spin-lattice relaxation) are $h_{x'}$ and $h_{y'}$, not $h_{z'}$. There is a
fundamental difference between $h_{x'}$ and $h_{y'}$ on one hand and $h_{z'}$
on the other: since the frame rotates about the $z$ axis of the
laboratory, a vector component along $z'$ is also a static component along $z$ in the laboratory, whereas a component fixed
along $x'$ or $y'$ is actually moving relative to the laboratory at
the rotation rate of the frame—in this case, at the Larmor frequency of the nuclei. Thus it is only high-frequency fluctuating
magnetic field components (those at approximately the precession frequency of the nuclei) that lead to spin-lattice relaxation.
$M_{x'}$ and $M_{y'}$, the components of $\mathbf{M}$ important in spin-spin relaxation, are each multiplied by one of the components of $\mathbf{h}$ that
has the same high-frequency behavior we have just described,
but are also multiplied by $h_{z'}$, which is static in the laboratory
frame. Thus spin-spin relaxation can be induced by processes
with a frequency near the nuclear precession frequency, just
as spin-lattice relaxation, but can also be brought about by
processes near zero frequency.

To treat relaxation quantitatively we must define a correlation function $K(\tau)$ that measures the extent to which a molecule alters its angular position by tumbling motions in the time
$\tau$. $K(\tau)$ is defined as follows:

$$K(\tau) = K(0) \exp\left(\frac{-|\tau|}{\tau_c}\right) .$$

The time constant in the exponential, $\tau_c$, is called the correlation time. A detailed treatment requires construction of $K(0)$ in terms of spherical harmonic functions and an examination of the Fourier components of $K(\tau)$. The result, in general terms, is that the component of the random motion at angular frequency $\omega$ is proportional to

$$\frac{\tau_c}{1 + \omega^2 \tau_c^2} .$$

For spin-lattice relaxation it is, as we have seen, the Larmor frequency $\omega_0$ that is of interest. (For some processes $2\omega_0$ also turns out to be important, but we shall not discuss this point.)

We now ask what causes the fluctuating fields h. There are several distinct processes that can cause fluctuating fields; it is convenient to list here the important processes that lead to relaxation, then discuss them separately:

1. Magnetic dipole-dipole interaction
2. Electric quadrupole interaction
3. Scalar coupling
4. Spin-rotation interaction
5. Chemical shift anisotropy

Magnetic dipolar fields arise from the magnetic nuclei within a molecule. Thus as a molecule tumbles the fluctuating magnetic field from one nucleus can relax another nucleus in the same molecule or in another molecule. With the formalism described previously to handle the average molecular motion, it is not difficult to derive an expression for the spin-lattice relaxation rate $R_1$ $(\equiv 1/T_1)$ and the spin-spin relaxation rate $R_2$ $(\equiv 1/T_2)$ for a nucleus relaxed by magnetic dipolar interaction with other nuclei of the same type within the molecule.

$$(R_1)_{\text{rot}} = \frac{2\gamma^4 \hbar^2 I(I+1)}{5r^6} \left[\frac{\tau_c}{1 + \omega^2 \tau_c^2} + \frac{4\tau_c}{1 + 4\omega^2 \tau_c^2}\right] ,$$

$$(R_2)_{\text{rot}} = \frac{\gamma^4 \hbar^2 I(I+1)}{5r^6} \left[3\tau_c + \frac{5\tau_c}{1 + \omega^2 \tau_c^2} + \frac{2\tau_c}{1 + 4\omega^2 \tau_c^2}\right] .$$

Edwin D. Becker

Here $r$ is the distance between the two nuclei and $\tau_c$ is the rotational correlation time, as described previously. Note that
the relaxation rate depends on the fourth power of $\gamma$. If the
relaxing nucleus is a different species, a similar but more
complex equation applies which depends on the squares of both
magnetogyric ratios, so that nuclei with large magnetic moments play the dominant role. The inverse sixth power dependence on $r$ means that nearby nuclei are by far the most
significant. For example, a proton is normally relaxed by
other nearby protons, but a $^{13}C$ (even in an enriched molecule
with other nearby $^{13}C$'s) is primarily relaxed by nearby protons,
not other $^{13}C$'s. A plot of $T_1$ and $T_2$ versus $\tau_c$ (Fig. 1-3) shows
the behavior predicted by our qualitative discussion of the frequency requirements for spin-lattice and spin-spin relaxation.

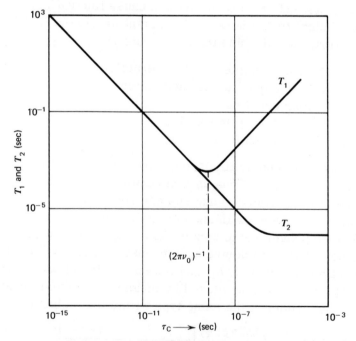

Figure 1-3. Dependence of $T_1$ and $T_2$ for magnetic dipole–dipole relaxation on the correlation time $\tau_c$ (after Bloembergen, Purcell, and
Pound[12]).

Spin-lattice relaxation is most efficient ($T_1$ shortest) when molecular tumbling occurs with a correlation time $\tau_c = 1/\omega_0$, while spin-spin relaxation becomes more efficient with increasing $\tau_c$ (eventually approaching a limit determined by a rigid crystal lattice). For sufficiently rapid molecular tumbling, such as that experienced by most small molecules in the liquid or gas phase, $T_1 = T_2$.

Intermolecular dipole-dipole relaxation can also occur. The expressions for $T_1$ and $T_2$ in this case are quite similar except that the correlation time is that for translational, rather than rotational, motion.

The nuclear Overhauser enhancement that occurs when a nucleus $I$ is observed while one or more nuclei $S$ are irradiated has its origin in magnetic dipole-dipole relaxation processes. The expression we gave above for the maximum NOE is based on nucleus $I$ being relaxed *entirely* through dipolar relaxation by the nuclei $S$ that are irradiated. To the extent that any of the processes described below occur, the NOE is reduced.

The dipolar fields required for magnetic dipole-dipole relaxation can also arise from unpaired electrons if they are present. Since the electron magnetic moment is about 658 times as great as the proton moment, and since the magnetic moment of the relaxing species enters the equation as the square, such relaxation can be very effective provided $r$ is small enough. Paramagnetic species present in low concentration, such as atmospheric oxygen, are rather effective in relaxing hydrogen nuclei in many compounds but much less effective in relaxing nuclei such as $^{13}C$ that lie in less exposed positions. Paramagnetic ions that can be complexed directly to a molecule are often very efficient; their use in increasing the relaxation rate of $^{13}C$ is discussed in Chapter 9. On the other hand, some paramagnetic ions, notably certain rare earths such as $Eu^{3+}$, have a $T_1$ for the unpaired electron that is much smaller than $\tau_c$. In this case the nuclear relaxation rate depends on $T_1$ (electron), rather than $\tau_c$, and the relaxation is rendered much less effective. Such ions are useful as shift reagents since they induce pseudocontact shifts but do not lead to appreciable decrease in $T_2$ with concomitant line broadening.

A nucleus with spin $> \frac{1}{2}$ possesses an electric quadrupole moment. If the electron environment around the nucleus is asymmetric, then the fluctuating electric field gradients at the nucleus induced by molecular tumbling permit nuclear relaxation. We shall not go into the details of quadrupole relaxation, but qualitatively the situation is much like that for dipolar relaxation in terms of the dependence on correlation time. In particular, for the situation where molecular tumbling is rapid enough that $\tau_c \ll 1/\omega_0$,

$$R_1 = R_2 = \frac{3}{40} \frac{2I+3}{I^2(2I-1)} \left(1 + \frac{\eta^2}{3}\right) \left(\frac{e^2 Qq}{\hbar}\right)^2 \tau_c \, ,$$

where $\eta$ is an asymmetry parameter that measures the departure of the nuclear environment from cylindrical symmetry and $(e^2 Qq/\hbar)$ is the quadrupole coupling constant, which depends on the nuclear quadrupole moment $Q$ and the electric field gradient $q$. Most quadrupolar nuclei relax predominantly by this means, since $T_1$'s are often measured only in milliseconds, or even microseconds in some cases. The only exception is the situation where the electron distribution around the nucleus has tetrahedral or higher symmetry.

The ordinary spin-spin coupling (scalar coupling) that exists between two nuclei $I$ and $S$ can furnish a mechanism of relaxation for $I$ if $S$ relaxes at an appropriate rate, since relaxation by $S$ alters the magnetic field experienced by $I$ through the spin coupling. If the spin-lattice relaxation time of $S$ is much shorter than $1/A$ (where $A$ is the coupling constant in rad/sec), then no splitting is observed in the resonance of $I$, and the relaxation rates of $I$ are given by

$$R_1^I = \frac{2A^2}{3} S(S+1) \frac{\tau_S}{1 + (\omega_I - \omega_S)^2 \tau_S^2}$$

$$R_2^I = \frac{A^2}{3} S(S+1) \left\{\tau_S + \frac{\tau_S}{1 + (\omega_I - \omega_S)^2 \tau_S^2}\right\} \, ,$$

where $\tau_S$, the correlation time for this process, is the spin-lattice relaxation time of $S$. Scalar relaxation is most commonly found when $S$ is a quadrupolar nucleus that relaxes rapidly. Even in this case, however, it is rare that scalar relaxation has much effect on $T_1{}^I$ unless the resonance frequencies $\omega_I$ and $\omega_S$ are nearly equal. $R_2{}^I$ is very commonly affected, however; line broadening (i.e., increase in $R_2{}^I$) in the $^1$H spectrum of nuclei coupled to $^{14}$N is commonly observed, for example. $T_2{}^I$ may also be affected in cases where the spin-lattice relaxation of $S$ is much greater than $1/A$. For example, in molecules with $^{13}$C coupled to $^1$H, the $^1$H $T_1$ is normally long enough so that spin-spin splitting is observed, but $T_2$ ($^{13}$C) is nonetheless shortened by scalar coupling to the proton.[13] Specific examples are given in Chapter 12.

Spin-rotation relaxation arises from magnetic fields generated at the nucleus by the fluctuating motion of the *molecular* magnetic moment that is dependent on the overall electron distribution in the molecule. This effect is most significant for small, symmetric molecules that rotate rapidly or for similar portions of molecules, such as methyl groups. One important point to note regarding spin-rotation interaction is that it becomes more significant as the molecule increases its rotational rate; hence it is most effective for molecules in the gas phase or in liquids at high temperature. Spin-rotation relaxation is found to compete effectively with magnetic dipolar relaxation for such nuclei as $^{13}$C, $^{15}$N, and $^{29}$Si in many small molecules (see Chapters 12 and 17).

Finally we mention relaxation by chemical shift anisotropy. In general the shielding factor $\sigma$ is anisotropic, so that the local magnetic field at the nucleus varies as the molecule tumbles. The molecular tumbling rate in general is so rapid that only a single resonance line is seen, and the ordinary NMR spectrum measures the average value of the chemical shift. The fluctuating local field, however, may serve as a relaxation mechanism. For an axially symmetric situation,

$$R_1 = \tfrac{1}{15}\, \gamma^2 B_0{}^2 (\sigma_\| - \sigma_\perp)^2\, \frac{2\tau_c}{1 + \omega^2 \tau_c{}^2} \,,$$

$$R_2 = \tfrac{1}{90}\, \gamma^2 B_0{}^2 (\sigma_\| - \sigma_\perp)^2 \left\{ \frac{6\tau_c}{1 + \omega^2 \tau_c{}^2} + 8\tau_c \right\} \,,$$

TABLE 1-3. Summary of relaxation mechanisms

| Interaction | $T_1$ range[a] (sec) |
|---|---|
| Dipole–dipole | $1 \leq T_1 \leq 100$ |
| Quadrupole | $10^{-7} \leq T_1 \leq 10^2$ |
| Chemical shift anisotropy | $10 \leq T_1 \leq 100$ |
| Scalar coupling | $0.1 \leq T \leq 100$ |
| Spin–rotation | $10^{-2} \leq T_1 \leq 100$ |

[a]For $\tau_c \sim 10^{-11}$ sec.

where $\sigma_\parallel$ and $\sigma_\perp$ are the values of $\sigma$ along and perpendicular to the symmetry axis, respectively. Note that this relaxation mechanism is explicitly dependent on the value of $B_0$, since it is $B_0$ that induces the electron currents leading to the chemical shift. The quadratic dependence on $B_0$ suggests that this mechanism will be of importance principally at high fields, and that is indeed the case. In general, other mechanisms lead to much more rapid relaxation, so that chemical shift anisotropy is seldom of much importance.

Table 1-3 summarizes the five relaxation mechanisms we have described and indicates the common ranges of $T_1$ found for each (with the assumption that $\tau_c \approx 10^{-11}$ sec, a fairly typical value for many small molecules in the liquid phase). Many examples are discussed in later chapters, as are the consequences of particular relaxation mechanisms insofar as they affect the design of experimental methods or the interpretation of data.

## References

1. See, for example, J. W. Emsley, J. Feeney, and L. H. Sutcliffe, *High Resolution Nuclear Magnetic Resonance Spectroscopy*, Pergamon Press, Oxford, 1965, pp. 151–157.

2.  (a) J. B. Stothers, *Carbon-13 NMR Spectroscopy*, Academic Press, New York, 1972; (b) G. C. Levy and G. L. Nelson, *Carbon-13 Nuclear Magnetic Resonance for Organic Chemists*, Wiley-Interscience, New York, 1972; (c) L. F. Johnson and W. C. Jankowski, *Carbon-13 NMR Spectra*, Wiley, New York, 1972.

3.  International Union of Pure and Applied Chemistry, "Recommendations for the Presentation of NMR Data for Publication in Chemical Journals," *Pure Appl. Chem.*, **29**, 627 (1972).

4.  E. F. Mooney, *Ann. Rep. NMR Spectros.*, **3**, xi (1970).

5.  For a good discussion of this question, see E. W. Randall and D. G. Gillies, *Prog. NMR Spectros.*, **6**, 119 (1971).

6.  E. D. Becker, R. B. Bradley, and T. Axenrod, *J. Magn. Resonance*, **4**, 136 (1971).

7.  See, for example, E. D. Becker, *High Resolution NMR*, Academic Press, New York, 1969, Appendix A.

8.  A. A. Borisenko, N. M. Sergeyev, and Yu. A. Ustynyuk, *Mol. Phys.*, **22**, 715 (1971).

9.  R. R. Fraser, M. A. Petit, and M. Miskow, *J. Am. Chem. Soc.*, **94**, 3253 (1972).

10. A. Abragam, *The Principles of Nuclear Magnetism*, Oxford University Press, London, 1961, Chap. 8; see also Ref. 1, pp. 20–33; T. C. Farrar, A. A. Maryott, and M. S. Malmberg, *Ann. Rev. Phys. Chem.*, **23**, 193 (1972).

11. T. C. Farrar and E. D. Becker, *Pulse and Fourier Transform NMR*, Academic Press, New York, 1971, Chap. 4.

12. N. Bloembergen, E. M. Purcell, and R. V. Pound, *Phys. Rev.*, **73**, 679 (1948).

13. R. R. Shoup and D. L. VanderHart, *J. Am. Chem. Soc.*, **93**, 2053 (1971).

# CHAPTER 2

# INTRODUCTION TO PULSE AND FOURIER
# TRANSFORM METHODS

Edwin D. Becker

*National Institutes of Health*
*Bethesda, Maryland*

The use of short radio-frequency (rf) pulses, rather than continuous wave (cw) radio frequency, provides great flexibility in NMR studies. With the advent of rapid Fourier transform (FT) methods for analyzing the data obtained in pulse experiments it has become feasible to apply pulse techniques to complex molecules and often to obtain spectra in a small fraction of the time formerly required. We first discuss the basic features of pulse methods, then explore the requirements for efficient utilization of FT techniques, and finally touch on the pulse *sequences* used to measure $T_1$ and $T_2$ and those used for other purposes.

As in Chapter 1, we use the classical picture of an ensemble of identical precessing nuclei, with a macroscopic magnetization **M**. We shall again find it helpful to consider the motion of **M** in a frame (coordinate system) that is rotating at or near the nuclear precession frequency. If the frame rotates at exactly the nuclear precession frequency, then relative to the rotating frame the nuclei appear stationary. Thus to an observer in the rotating frame it would appear that *no* magnetic field is acting on the nuclei. A quantitative treatment of this

effect of change in coordinate system[1] shows that the apparent
magnetic field acting on the nuclei, as viewed in a frame ro-
tating at $\omega$ rad/sec, is $B_0 - \omega/\gamma$, where $\gamma$ is the magnetogyric
ratio of the nucleus.

Consider a radio-frequency field applied to the nuclear
spin system with $B_1$ as the rotating magnetic vector of the rf
field. If the frequency of rotation of the rotating frame is taken
to be the radio frequency, then $B_1$ is fixed in the rotating frame,
say along the $x'$ axis. The net, or effective, magnetic field
acting on the magnetization in the rotating frame is then given
by the vector sum

$$B_{eff} = B_0 - \frac{\omega}{\gamma} + B_1 . \tag{1}$$

Fig. 2-1 shows this relation pictorially. In the rotating frame
M always precesses about $B_{eff}$.

If $B_1$ is applied along the $x'$ axis in the frame rotating at
the resonance frequency of a nucleus, then

$$B_{eff} = B_1 \tag{2}$$

and M precesses in the $y'z'$ plane. In a time $t_p$ the angle $\theta$
through which M precesses is

$$\theta = \gamma B_1 t_p (\text{rad}) . \tag{3}$$

If $\theta = 90°$, then M ends up along $y'$ and induces a signal in the
receiver coil placed along that axis. This signal decays as nuclei lose phase coherence and $M_{y'}$ decreases in magnitude (the

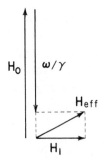

Figure 2-1. Formation of effective magnetic field in
a frame rotating at $\omega$ rad/sec. $H_0$ is the static applied field, $\omega/\gamma$ the "fictitious" field arising from the
rotating frame, and $H_1$ an rf field at $\omega$ rad/sec.

*free induction decay*, FID). If $B_1$ is large enough, nuclei off resonance may also precess in the $y'z'$ plane provided $B_{eff} \approx B_1$. This requires that for each nuclear precession frequency the vertical component in Fig. 2-1 be negligible compared with $B_1$. If $\Delta$ is the range of nuclear frequencies to be affected by the pulse (in general, the possible chemical shift range for the particular nuclear species), then

$$\gamma B_1 \gg 2\pi\Delta \ . \tag{4}$$

For a $90°$ pulse, equations 3 and 4 tell us that

$$t_p(90°) \ll \frac{1}{4\Delta} \ . \tag{5}$$

The FID for a single resonance frequency, where the pulse is applied exactly on resonance, is (as shown in Fig. 2-2$a$) an exponentially decaying function with a time constant $T_2^*$ related to the transverse relaxation time $T_2$ and the magnetic field inhomogeneity $\Delta B_0$:

$$\frac{1}{T_2^*} = \frac{1}{T_2} + \frac{\gamma\Delta B_0}{2} \ . \tag{6}$$

Because of field inhomogeneity $T_2^*$ is seldom more than a few seconds. The line width observed in an ordinary spectrum would be

$$\nu_{1/2} = \frac{1}{\pi T_2^*} \ . \tag{7}$$

If, again, the spectrum consists of only a single frequency, but the rf pulse is applied somewhat off resonance, then the signal from the precessing nuclear magnetization interferes with the rf signal from the oscillator that is used as a reference for the phase sensitive detector. The result is a FID of the sort shown in Fig. 2-2$b$, where the difference in frequencies between the radio frequency and the nuclear precession appears as an oscillation superimposed on the exponential decay. Now if there are several precession frequencies due to chemical shifts and spin-spin coupling, then the FID is a complex interference pattern. The FID contains

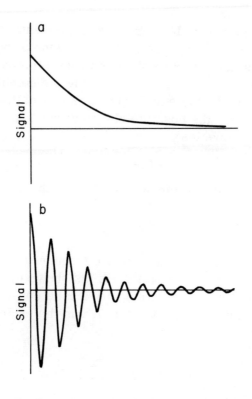

TIME

Figure 2-2.    Free induction decay (FID) following a 90° pulse.    (*a*)
Pulse applied exactly at the resonance frequency of the nucleus.    (*b*)
Pulse applied off resonance.

all the information about the frequencies and their relative intensities, but it is present in a form that is not directly interpretable. This information can, however, be extracted by
Fourier transformation:

$$F(\omega) = \int_0^\infty f(t)\, e^{i\omega t}\, dt \,, \tag{8}$$

where $f(t)$ is the FID and $F(\omega)$ is the resultant spectrum. By
this procedure the spectral information may be acquired in a
second or so, rather than in the hundreds of seconds it normally takes for a spectral scan with conventional techniques.

Thus rapid processes may be followed by NMR; or by repetitive pulsing and coherent addition of the FID following each pulse, signal/noise may be improved without the expenditure of inordinate amounts of time.

For nuclei other than protons the FT method is especially valuable, since signals are generally very weak, as indicated in Chapter 1, and a large number of repetitive scans would be needed. Furthermore, chemical shift ranges tend to be quite large, so that extremely long times might be required for a conventional scan. With the pulse FT method the time needed for data acquisition is independent of the spectral range. On the other hand, there are some important factors that must be taken into account for proper utilization of FT NMR.

For coherent addition of FID's, as with any time averaging process, the magnetic field and radio frequency must be well locked to each other to prevent field drift. For FT work the long-term stability must be better than for time averaging of conventional cw scans, since it is not possible to correct for drift by "triggering" on a particular spectral line in each scan. Most commercial spectrometers have adequate field/frequency lock systems for FT applications.

We already pointed out in equations 4 and 5 that $B_1$ must be large enough to ensure that nuclei with different chemical shifts experience a 90° pulse. Often it is difficult to achieve sufficiently large $B_1$ to meet this criterion strictly, but if some distortion in relative intensities of lines is permissible (as is often the case for $^{13}C$ NMR, for example), then the weaker condition $\gamma B_1 \geq 2\pi\Delta$ suffices. In this situation phase distortions are also introduced but can be corrected along with other, unavoidable phase errors, as we shall see later.

One factor that exaggerates the problem of rf power is that normally only half the power is effective, since the pulse frequency is usually set so as to be completely outside the range of chemical shifts expected. If the radio frequency is set in the middle of the spectrum, the significance of positive and negative frequencies is lost in the Fourier transformation, and "foldover" of half the spectrum occurs, as shown in Fig. 2-3. Some instruments now operate with two phase sensitive

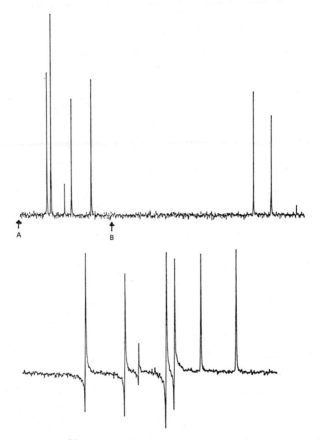

Figure 2-3.    (*Top*) $^{13}$C FT spectrum (proton decoupled) of 3-ethylpyridine, with pulse applied at frequency A.    (*Bottom*) Same sample, with pulse applied near frequency B.    Note that all lines to the left of the pulse frequency have folded over with phase distortion.    (Spectra courtesy T. C. Farrar, JEOL, Inc.)

detectors and a computer programmed to carry out a complex Fourier transform.    In this way the rf power can be used more effectively by being applied in the middle of the spectrum, and signal/noise also is improved by $2^{1/2}$.

Another type of foldover occurs in the process of digitizing the FID for computer calculations if the rate at which the digital data points are taken is not sufficiently rapid.    A theorem from information theory states that a sine wave can be reproduced faithfully if it is sampled two times per cycle. Thus if data from the FID are taken at $2f$ points/sec, the maximum frequency (measured relative to the radio frequency as zero) that can be displayed in the transformed spectrum is $f$. Any spectral lines and any noise of higher frequency will be folded back into the region $O-f$. To prevent such foldback a filter with a sharp high-frequency cutoff near $f$ is applied in the data acquisition circuit.

The time during which data are acquired for each FID is also important, since it may govern the resolution obtained in the Fourier transformed spectrum. Simple FT theory shows that a FID sampled for $T$ sec leads to lines of width $1/T$ Hz in the transformed spectrum. Although resolution better than that dictated by $T_2^*$ (equations 6 and 7) cannot be obtained by long acquisition times, early truncation of the FID can lead to line broadening. Another factor that must be taken into account is the loss in signal/noise as sampling time is increased, and the free induction signal decays into noise. There is no point in increasing data acquisition time beyond $3T_2^*$ from the standpoint of resolution, and often from a signal/noise (S/N) standpoint it may be wise to truncate earlier. Often the best compromise is to acquire data for about $3T_2^*$, then in the computer before Fourier transformation to multiply the FID data by an exponential function with a smaller time constant than $T_2^*$. The result is to discriminate against the data at longer time with lower S/N. On Fourier transformation, as shown in Fig. 2-4, a considerable improvement in S/N can be obtained, with only modest loss in resolution. Most FT programs now include a routine for such "exponential filtering." In cases where S/N is sufficiently high, a negative time constant in the exponential function can be used to enhance resolution above that naturally obtainable with the sample and magnet, at the expense of S/N. The latter application is only beginning to come into use.

When $T_1 \gg T_2^*$, some thought must be given to the optimum rate of repetition of pulses, for all FID signals after the first might be reduced substantially in magnitude if spin-lattice relaxation has not occurred sufficiently during the interval be-

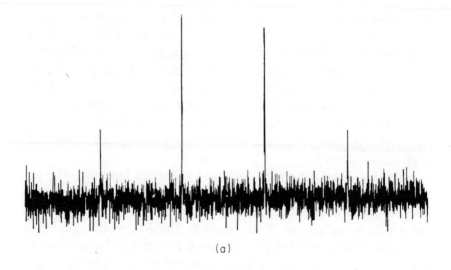

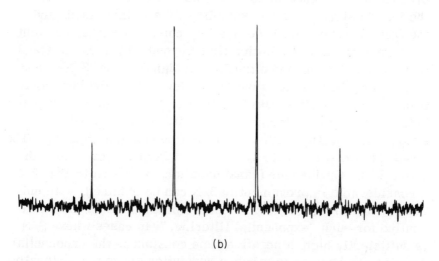

Figure 2-4.    Effect of an "exponential filter" on the $^{13}$C spectrum of $^{13}$CH$_3$I.   (a) No filter.   (b) With a digital exponential filter.   Note broadening of lines that accompanies reduction in noise.

tween pulses.  Generally when $T_1$ is long, it is preferable to use pulse widths shorter than those used to produce 90° pulses. The magnitude of each FID is reduced by the smaller projection of M on the $y'$ axis, but this loss of signal is more than compensated by the more rapid restoration of M to its equilibrium position along the $z'$ axis.

Most FT NMR experiments are conducted with a small on-line computer that acquires the data, carries out the Fourier transformation, and plots the resultant spectrum.  The computer often carries out a number of other functions as well. For example, the timing of pulse widths and repetition rates, and in more complex experiments the choice of various pulse sequences, may be under computer control.  We mentioned previously one kind of data processing—application of an exponential filter function to the data before Fourier transformation.  After transformation the frequency domain data consist of a real spectrum and an "imaginary" spectrum, 90° out of phase with the former.  In each of these, however, the phase is a function of frequency, so that a correction must be applied before the spectrum can be plotted.  The phase errors arise from missetting the phase sensitive detector, which produces an error independent of frequency (just as in a cw spectrum), and from unavoidable delays in beginning data acquisition after application of the pulse.  In addition, as we indicated previously, if $B_1$ is not large enough, a further frequency-dependent phase error results.  These errors usually vary only linearly with frequency, so that their correction in the computer by co-addition of proper proportions of the real and imaginary spectra is not difficult.  Other, nonlinear errors resulting from phase distortions near filter cutoff frequencies are often corrected by hardware devices in the spectrometer.

The size of the computer memory devoted to data storage is, of course, the product of the number of data points taken per second and the time during which data are acquired for each FID.  For many nuclei other than hydrogen the large range of chemical shifts requires a high data rate to avoid foldback. For example, $^{13}C$ spectra often cover more than 220 ppm—a frequency range of 5500 Hz at an observation frequency of 25

MHz. A data rate of 11,000 points/sec is thus required. If a
resolution of 1 Hz is desired, the data must be acquired for 1
sec, and an 11,000 word computer memory is required. Since
the Fourier transform program works more efficiently with $2^n$
points, a 16K memory (16,384 words) is indicated. Most NMR
systems now in use are limited in memory (many to 8K for
data), but now that the cost of memory is dropping drastically,
larger data tables should become more commonplace.

We now turn briefly to a consideration of another application of pulse FT NMR—the measurement of $T_1$ and $T_2$ for nuclei with a multiline spectrum. The most common method of
measuring $T_1$ consists of the application of the pulse sequence
$180°$, $\tau$, $90°$, where $\tau$ is a waiting period between the two
pulses. As shown in Fig. 2-5, the $180°$ pulse inverts the magnetization along the $-z'$ axis. After a time $\tau$, spin-lattice relaxation has caused M to decrease through zero if $\tau$ is long
enough to increase toward its positive equilibrium value. The
$90°$ pulse applied at time $\tau$ samples the value of M at that in-
stant by turning M to the $y'$ axis where it generates a FID, the
initial amplitude of which is proportional to M. The FID can
be Fourier transformed to obtain the spectrum in this nonequi-
librium state, the system allowed to relax back to equilibrium,
and the pulse sequence repeated with another value of $\tau$. A
plot of $\tau$ versus the log of the difference between the intensity
of each line and its equilibrium value permits the measurement
of $T_1$ for each line.

$T_2$ is usually measured by some variant of the spin-echo
method, in which the pulse sequence $90°$, $\tau$, $180°$ is used. As
shown in Fig. 2-6, the decay of the free induction signal because
of magnetic field inhomogeneity is reversed by application of
the $180°$ pulse at time $\tau$, so that an echo occurs at $2\tau$. The
height of the echo is, however, reduced below the initial amplitude of the free induction decay because of spin-spin relaxation. Application of a large series of $180°$ pulses at times $\tau$,
$3\tau$, $5\tau$, etc., causes repetition of steps $(c)$ through $(f)$ in
Fig. 2-6 and formation of echoes at times $2\tau$, $4\tau$, $6\tau$, etc.
For a nuclear spin system that has only a single spectral line
these methods can be used to determine $T_2$ quite readily. For

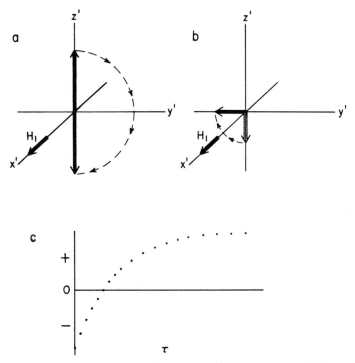

Figure 2-5.    Measurement of $T_1$ by the 180°, $\tau$, 90° method.    (*a*) 180° pulse inverts magnetization.    (*b*) 90° pulse at time $\tau$ permits measurement of magnetization remaining at that time.    (*c*) For a single line spectrum a plot of the initial amplitude of the FID following the 90° pulse versus $\tau$ permits the extraction of the value for $T_1$.    (For a multiline spectrum each point in the plot would represent a spectrum obtained by Fourier transformation of the FID.)

systems with more complex spectra FT methods can be used in principle, but there are limitations.    For example, the presence of homonuclear spin-spin coupling among the nuclei being studied causes distortions in the echoes.    In many nuclei other than hydrogen low natural abundance precludes such coupling even in complex molecules.    Little FT work has been done thus far in the determination of $T_2$ for individual lines in multiline spectra.

Other pulse sequences can be used in conjunction with FT methods for specific purposes.    For example the driven equi-

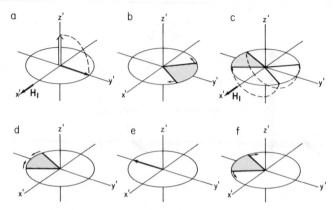

Figure 2-6.   The Hahn spin-echo experiment.   (a) A 90° pulse applied
along $x'$ at time 0 causes **M** to tip to the positive $y'$ axis.   (b) The
macroscopic magnetizations, $m_i$, of nuclei in different parts of the
sample dephase as a result of the inhomogeneity in $H_0$.   Those nuclei
precessing faster than the average (the rotation rate of the frame) appear in the rotating frame to move toward the observer.   Looking down
from the positive $z'$ axis they appear to move clockwise, whereas
those slower than average move counterclockwise.   (c) A 180° pulse
along $x'$ at time $\tau$ causes all $m_i$ to rotate 180° about the $x'$ axis.   (d)
The faster nuclei, still moving clockwise in the rotating frame, now
go away from the observer, and the slower nuclei move counterclockwise toward the observer.   (e) At time $2\tau$ the $m_i$ rephase along the
$-y'$ axis.   (f) At time $> 2\tau$ the $m_i$ again dephase.

librium Fourier transform (DEFT) method uses the sequence
90°, $\tau$, 180°, $\tau$, 90° to restore **M** to its equilibrium position
more quickly than by spin-lattice relaxation.   This sequence
has application in certain specialized situations.   The Waugh
pulse cycles, which we do not describe here, provide a clever
method for obtaining narrow-line, high-resolution type spectra
in solids, which have inherently very broad lines.   Application
of such methods has been quite limited because of the stringent
instrumental requirements.   These more complex pulse experiments, as well as other types of techniques for measuring
relaxation times, are described elsewhere.[1]   Other good discussions of FT methods in general are also available.[2]

## References

1.  See, for example, T. C. Farrar and E. D. Becker, *Pulse and Fourier Transform NMR*, Academic Press, New York, 1971.

2.  See, for example, (*a*) H. D. W. Hill and R. Freeman, *Introduction to Fourier Transform NMR*, Varian, Palo Alto, Cal., 1970; (*b*) D. A. Netzel, *Appl. Spectros.*, **26**, 430 (1972); (*c*) N. Boden, "Pulsed NMR Methods, "in *Determination of Organic Structures by Physical Methods*, F. J. Nachod and J. J. Zuckerman, Eds., Vol. 4, Academic Press, New York, 1971, p. 51.

# CHAPTER 3

# MAGNETIC MULTIPLE RESONANCE

W. McFarlane

*Chemistry Department*
*Sir John Cass School of Science and Technology*
*City of London Polytechnic*

## INTRODUCTION

Although the introduction of Fourier transform techniques has improved enormously the performance of single resonance spectrometers, these may still suffer from certain limitations owing to poor sensitivity, low natural abundance, and unfavorable relaxation times when applied to the study of nuclei other than $^1H$ and $^{19}F$; many of these limitations can be mitigated by using multiple resonance techniques. This chapter deals mainly with the theory and application of double resonance experiments (especially as applied to systems containing heteronuclei) and some triple resonance work is also mentioned. Typical uses of double resonance experiments include the removal of undesirable coupling in $^{19}F$ and $^1H$ spectra prior to spectral analysis; the elimination of broadening due to a nucleus with a quadrupole moment (e.g., $^{14}N$ or $^2D$) in $^1H$ and $^{19}F$ spectra; the determination of chemical shifts of "hidden" proton resonances; the determination of chemical shifts and coupling constants involving other nuclei from observations of the proton (or $^{19}F$) spectrum; sensitivity enhancement in the direct observation of

31

heteronuclei; the study of a wider range of time scales for exchange processes than is possible by single resonance; and the
determination of molecular geometry from the nuclear Overhauser effect. Most of these techniques can now be used in a
routine way with spectrometers that are commercially available.

The references are intended to provide a guide to the literature rather than a comprehensive survey. [1-7]

## THEORETICAL

The commonly adopted nomenclature is as follows: Polar-
izing field, $B_0$; observing, stimulating, or detecting rf field,
$B_1$ rotating at an angular frequency $\omega_1 = 2\pi\nu_1$; decoupling, irradiating, perturbing, or tickling rf field, $B_2$ rotating at angular frequency $\omega_2 = 2\pi\nu_2$. A double resonance experiment in
which nuclei of species A are observed while nuclei of species
X are simultaneously irradiated is designated[1] by A-{X}.

The most straightforward double resonance experiments
are decoupling ones in which irradiation at the resonant frequency of one nucleus removes the spin-spin splitting which it
produces in the resonances of other nuclei.[8] This occurs because the oscillatory decoupling field changes the direction of
quantization of the spins of the irradiated nuclei. In a reference frame rotating about the $z$ axis at an angular frequency
$\omega_2$ the effect of $B_0$ (applied along the *minus-z* direction) upon
nuclei of resonant frequency $\omega_2$ is reduced to zero, and the resultant field acting upon these nuclei is effectively $B_2$ which is
perpendicular to the $z$ axis. However, the $z$ axis is still the
direction of quantization for the observed spins, and the observed splitting which depends upon the scalar product $(\vec{I} \cdot \vec{I})$
vanishes, provided that $\gamma(X)B_2/2\pi$ is large compared with
$J(A-X)$. If the chemical shift difference between A and X is
not large (i.e., if both types of nuclei are of the same species)
$B_2$ may be large enough to alter the direction of quantization of
the *observed* spins, and the coupling may return. This has
been demonstrated experimentally with the AMX spin system

given by the olefinic protons of vinyl acetate, but is not important for heteronuclear work owing to the large difference between the observing and irradiating frequencies. [9]

For amplitudes of $B_2$ which are too small to give complete decoupling the spectra can be computed; in this way the results of spectral analyses can be checked. However, for very weak amplitudes of $B_2$ the behavior can be discussed in simple terms. Such experiments are called spin tickling, and the results are described by a set of rules given by Freeman and Anderson.[10, 11] The most important of these is the first, which states that irradiation at the frequency of a nondegenerate transition will split into a doublet any transitions of the other nuclei that have an energy level in common with the irradiated one. For heteronuclear experiments the width of the splitting is usually $\gamma(X)B_2/2\pi$, and small frequency offsets from exact resonance lead to asymmetric doublets. A series of such tickling experiments in which $\gamma(X)B_2/2\pi$ is substantially less than the spacing between adjacent lines in the spectrum of the irradiated nucleus can thus be used to map out its fine structure. For the relatively common situation in which a single nucleus X is coupled to several equivalent protons only a fraction of the observed proton line experiences the doublet splitting in any particular tickling experiment, and so a triplet is produced. Typically, a 3 : 2 : 3 triplet is obtained when either of the inner lines of the X quartet in an $A_3X$ system (e.g., $^{13}CH_3I$) is irradiated; irradiation of either outer line gives two 1 : 6 : 1 triplets in the A spectrum. This is illustrated in Fig. 3-1. Similar patterns with different intensity ratios are obtained with an $A_6X$ system as exemplified by the $PF_6^-$ anion (Fig. 3-2). The off-resonance behavior of these systems can be quite complex, however.

These line-splitting effects may be accompanied by intensity changes which can persist even when $B_2$ is too weak to give resolvable splitting. [12] Thus the saturation of the irradiated transition leads to equalization of the populations of a pair of energy levels and so the intensities of connected transitions change: in general one increases and the other decreases. The magnitude of the effects depends upon $\gamma(X)/\gamma(A)$ and so is relatively unimportant for experiments in which $^1H$ is observed and

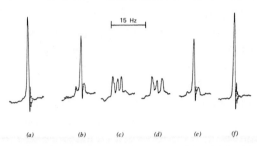

Figure 3-1. Spin-tickling experiments upon the $A_3X$ spin system given by $^{13}CH_3I$. (a) High-field $^{13}C$ satellite in proton spectrum of sample containing $^{13}C$ in natural abundance (1.1%). (b) Effect of irradiating at resonant frequency of the lowest-field component of the $^{13}C$ quartet: a 1:6:1 pattern is produced. (c) Next $^{13}C$ line irradiated: 3:2:3 intensity pattern. (d) Next $^{13}C$ line irradiated: 3:2:3 intensity pattern. (e) $^{13}C$ line at highest field irradiated: 1:6:1 intensity pattern. (f) Unperturbed spectrum.

a heteronucleus is irradiated. The phenomenon is illustrated by homonuclear experiments upon an AB spin system, as shown in Fig. 3-3.

For systems in which the relaxation paths available to the different nuclei are interrelated there may also be intensity changes due to the nuclear Overhauser effect.[13] This arises when the relaxation of the A spins is dominated by dipole-dipole interaction with the X spins, and both (or all) X transitions are saturated. The intensity of the A resonance is then enhanced by a factor of $1 + \frac{1}{2}(\gamma_X/\gamma_A)$ so the effect is greatest for A-$\{^1H\}$ experiments, and factors close to the theoretical maximum of 2.998 have been attained in direct $^{13}C$ NMR spectroscopy.[14] For homonuclear experiments the theoretical factor is 1.5; in proton work it seldom exceeds 1.25 in practice because mechanisms other than the direct dipolar interaction contribute to the relaxation. A heteronuclear application of this effect is illustrated in Fig. 3-4, which shows $^{19}F$-$\{^1H\}$ experiments on $CF_3C(O)N(CH_3)_2$. Owing to restricted rotation about the N-C(O) bond there are two methyl resonances, and irradiation of one of these increases the integrated intensity of the $CF_3$ resonance whereas irradiation of the other has little effect. The former,

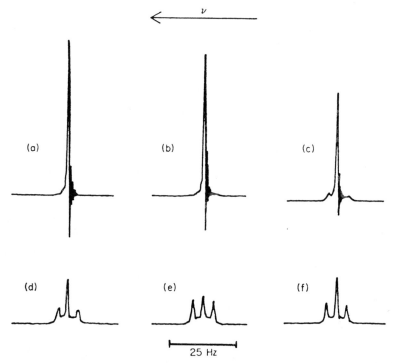

Figure 3-2.  Spin-tickling experiments upon the $A_6X$ spin system given by an aqueous solution of the $PF_6^-$ anion.  (a) High-field component of $^{19}F$ doublet.  (b) Effect of irradiating at resonant frequency of highest-field component of the $^{31}P$ septet: a $1:62:1$ pattern is produced.  (c) Next $^{31}P$ line irradiated: $3:26:3$ intensity pattern.  (d) Next $^{31}P$ line irradiated: $15:34:15$ intensity pattern.  (e) Central $^{31}P$ line irradiated: $5:6:5$ intensity pattern.  (f) Next $^{31}P$ line irradiated: $15:34:15$ intensity pattern.

therefore,  arises from the methyl group *cis* to $CF_3$.  It is important to use peak areas and not heights in experiments of this type.

The problem of producing sufficiently large rf fields to collapse the large couplings encountered in much heteronuclear work can be circumvented by modulation of $\nu_2$.  If a single coherent audio frequency given by $\gamma(X)B_2/2\pi$ is used for the modulation, quite large couplings can be collapsed with the use of only modest rf power.[15]  This is illustrated by diethyl phosphite

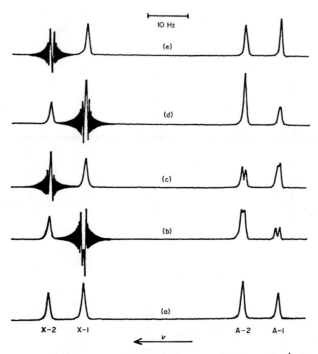

Figure 3-3. Spin-tickling experiments on a homonuclear two spin system. The point of irradiation is indicated by the beat pattern produced by direct interference between the observing and irradiating rf fields. (a) Normal single resonance spectrum. (b) and (c) $\gamma B_2/2\pi$ =1 Hz. (d) and (e) $\gamma B_2/2\pi = 0.3$ Hz. No line splitting is observed but population transfer effects alter the intensities.

in which $^1J(^{31}P\text{-H}) = 690$ Hz (see Fig. 3-5). It is important to realize that the modulation frequency used does not equal the coupling constant. When the overall spread of the X spectrum is not due to a single large coupling constant but to many small ones and/or chemical shift differences, this approach is not as successful and it is better to use random (or white) noise modulation to distribute the available power over a band which may be up to 5 kHz in width.[16] This is the technique used to produce proton-decoupled $^{13}C$ spectra, and is in part responsible for the growth of direct $^{13}C$ spectroscopy which has taken place recently.

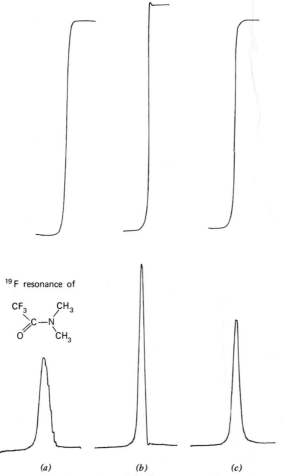

19 F resonance of

CF₃      CH₃
   \      /
    C — N
   /      \
  O        CH₃

(a)                    (b)                    (c)

Figure 3-4. Nuclear Overhauser effect experiments upon $N$, $N$-di-methyltrifluoracetamide. ($a$) Normal $^{19}$F spectrum showing unresolved fine structure due to coupling to the protons of the methyl groups, to-gether with integral. ($b$) Methyl group $cis$ to $CF_3$ irradiated. Loss of spin coupling gives an increase in peak $height$ but there is also an in-tensity change as shown by the integrated trace. ($c$) Methyl group $trans$ to $CF_3$ irradiated: loss of spin coupling again increases peak height, but the integral shows that the intensity remains essentially unaltered.

W. McFarlane

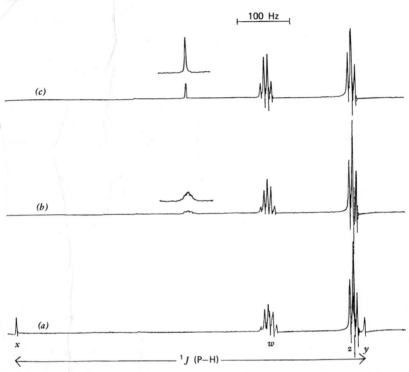

Figure 3-5. The use of modulation to improve the efficiency of decoupling. (a) Proton spectrum of $(EtO)_2PHO$ in which the proton attached to phosphorus gives peaks at $x$ and $y$ because $^1J(P-H) = 690$ Hz. (b) Attempted $^{31}P$ decoupling using maximum available rf amplitude. The proton attached to phosphorus gives a complicated pattern near the middle of the spectrum, and decoupling of the methylene protons is incomplete. (c) As for (b), but with af modulation of the $^{31}P$ irradiating radio frequency. Decoupling is essentially complete and the proton attached to phosphorus gives a sharp singlet.

The fine structure of the spectrum of a nucleus which is not observed directly may be deduced from a large number of tickling experiments, or by plotting an INDOR (Inter Nuclear DOuble Resonance) spectrum.[17] The height of an appropriate line in the observed spectrum is monitored continuously and the irradiating frequency is swept automatically through a suitable range. Provided that the amplitude $B_2$ is chosen correctly and $\gamma(A) \gg \gamma(X)$, the resulting excursions of the pen closely resemble the "normal" X spectrum. However, if $\gamma(A)$ is comparable with or less than $\gamma(X)$, the intensities are seriously perturbed and inverted lines may appear. The production of high-quality INDOR spectra places high demands on instrumental stability, but in certain cases the appearance of the X spectrum can be obtained in another way. In $A_x M_y X_z$ spin systems in which $J(A-X) \gg J(M-X)$ and $J(A-M) = 0$, the fine structure of the X resonance that arises from coupling to M can be "transferred" to the A spectrum by an A-{X} experiment[18] provided that $J(A-X) > \gamma(X) B_2 / 2\pi > J(M-X)$. This is illustrated for $(Pr^i O)_2 PHO$ in Fig. 3-6. Broadening due to the presence of a quadrupolar nucleus can be transferred similarly.[19]

## EXPERIMENTAL ASPECTS

Most commercial instruments provide facilities for homonuclear work, which is not discussed in detail here. For heteronuclear experiments a convenient source of the second radio frequency is a frequency synthesizer, which generates any desired frequency with a precision of $\pm 0.1$ Hz or better. For tickling experiments an output of considerably less than 1 W is adequate, but for decoupling an amplifier (which may be wide-band or tuned) giving up to 10 W is needed, and for certain nuclei such as $^{14}N$ even more power may be useful. The probe may be equipped with an extra coil, or the transmitter coil may be double-tuned to accept the second radio frequency. The former arrangement is probably better if high-power experiments are contemplated. In precise work it is advantageous to have the spectrometer rf oscillator and the output of the synthesizer controlled by a common frequency. If only a single nucleus is of interest in addition to the observed one, and somewhat lower precision can be tolerated, a simple crystal-controlled oscillator can be used to provide the second radio frequency. Most double resonance experiments are easiest to interpret if the recorded spectrum is a frequency sweep one;

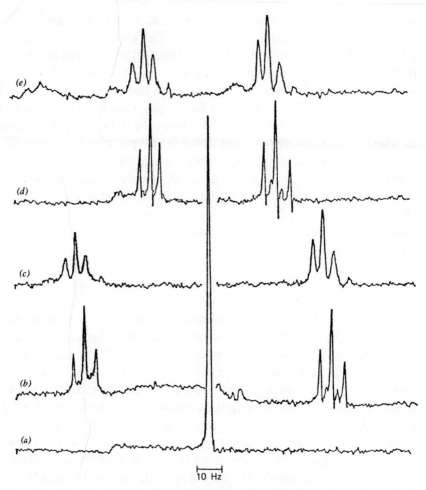

Figure 3-6. Transfer of fine structure by double resonance in $(Pr^iO)_2PHO$ in which there is no coupling between the proton attached to phosphorus and the protons of the isopropyl groups. (a) High-field component of the P–H doublet ($^1J(^{31}P\text{–}H) = 690$ Hz). (b) High-field half of $^{31}P$ spectrum irradiated with $\gamma(^{31}P)B_2/2\pi = 100$ Hz. The fine structure in the phosphorus spectrum resulting from coupling to the methine protons is transferred to the observed proton line. (c) Low-field half of $^{31}P$ spectrum irradiated with same amplitude as in (b). (d) As for (b) with $\gamma B_2/2\pi = 50$ Hz. (e) as for (c) with $\gamma B_2/2\pi = 50$ Hz.

however, since one still tends to think of peaks on the right as being at "high field" it is important to remember that there is an inverse relation between field and frequency *in this context.*

## APPLICATIONS

### Decoupling

Complete heteronuclear decoupling is now widely used for the simplification of proton and $^{19}$F spectra. Examples include the cyclic phosphite 1 which gives an AA'BB' spectrum when decoupled from $^{31}$P (Fig. 3-7), and the $BF_4^-$ anion in which $^{10}$B and $^{11}$B can be decoupled separately (Fig. 3-8). The latter example brings out the $^{10}$B/$^{11}$B isotope effect upon the $^{19}$F shielding very clearly. The technique is also valuable when quadrupolar broadening is present, as in pyrrole and methylpyridinium iodide. The latter is illustrated in Fig. 3-9 and provides an example of simultaneous $^{14}$N and $^1$H decoupling which permits accurate assignment of long-range $^1$H-$^1$H spin-spin splittings.[20] The $^1$H spectra of massively deuteriated molecules are often rather broad owing to the deuterium quadrupole moment and

1

poorly resolved couplings to deuterium, and $^1$H-$\{^2$D$\}$ experiments can be used to permit the extraction of accurate line positions and band shapes. The rf power needed for these experiments is quite small in spite of the small magnetogyric ratio of deuterium because $J$(H-D) seldom exceeds 4 Hz. Such experiments are very valuable in the study of the energetics of conformational processes.[21]

# W. McFarlane

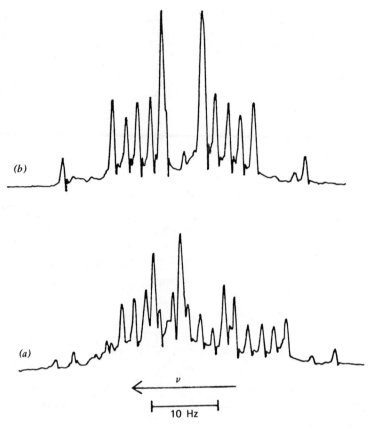

Figure 3-7. (a) Spectrum at 60 MHz given by the alicyclic protons of phenyl ethylene phosphite (1). The asymmetry arises because the two P-H coupling constants differ considerably. (b) Same spectrum with $^{31}$P decoupled: a characteristic AA'BB' pattern is produced.

When several different nuclei of the same species are responsible for observed couplings or line broadenings in the $^1$H spectrum, the decoupling experiments can give valuable diagnostic information. This has been illustrated by $^1$H-$\{^{11}B\}$ experiments upon the boron hydride derivative 2.[22] The value of complete proton decoupling in $^{13}$C spectroscopy is now well known. In this connection the technique of off-resonance de-

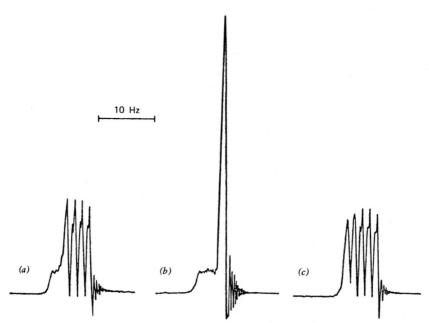

Figure 3-8.  Decoupling experiments on the $BF_4^-$ anion.  (*a*) Normal
$^{19}$F spectrum showing the four lines arising from coupling to $^{11}$B
($I = \frac{3}{2}$,  abundance = 81.2%) and some of the seven lines arising from
coupling to $^{10}$B ($I = 3$, abundance = 18.8%).  (*b*) $^{11}$B decoupled:  the multiplet from ions containing $^{10}$B is now clearly visible.  (*c*) $^{10}$B decoupled:
the $^{10}$B-$^{11}$B isotope effect upon the fluorine shielding is very apparent.

coupling can be used to identify quaternary carbon atoms since
these do not have any large couplings to protons.  An early example of this is provided by cedrol (**3**).  The nuclear Overhauser
enhancement that is obtained simultaneously is also important
and again can be used for assignment since quaternary carbons
do not experience this to any great extent.[23]

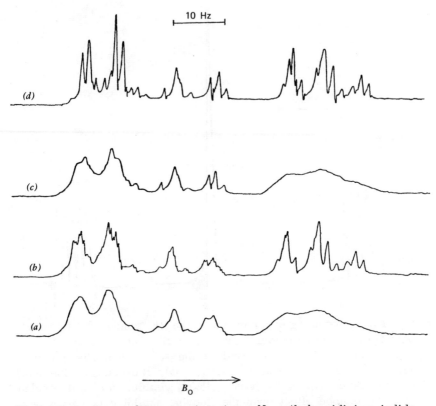

Figure 3-9. Decoupling experiments on $N$-methylpyridinium iodide. (a) Normal spectrum of aromatic protons at 60 MHz. The broad doublet at lowest field is given by the $\alpha$ protons, the central doublet by the $\gamma$ proton, and the multiplet at highest field by the $\beta$ protons. (b) $^{14}N$ decoupled to give a marked sharpening. (c) $N$-Methyl protons decoupled, thus revealing long-range coupling to the aromatic protons. (d) $^{14}N$ and the $N$-methyl protons simultaneously decoupled. By comparing (d) with (c) the magnitudes of the $^{14}N–H$ coupling constants may be estimated, and by comparing (d) with (b) the long-range interproton coupling constants can be measured.

2                                              3

## Chemical Shift Determination

The use of $^1$H-{X} double resonance experiments for determining chemical shifts has the following advantages: (a)
convenience—one spectrometer can be used for all nuclei; (b)
the observed nucleus has high sensitivity to NMR detection;
(c) high precision; (d) a single internal reference TMS can be
used; and (e) the relationship between resonances in the proton
and X spectra is obtained unequivocally.

Disadvantages are that the experiments can be rather time-
consuming, especially if the full precision of the method is to
be realized, and of course the proton (or $^{19}$F) spectrum must
display observable coupling to X. For $^{13}$C work it is probably
better to study all but the simplest molecules by direct FT
spectroscopy with proton decoupling. Because of the wide range
of chemical shifts found for other nuclei it is often adequate to
simply measure the X frequency that gives optimum decoupling
in the proton spectrum. The error will then be less than
$J$(H-X). If greater precision is wanted, then a series of tickling experiments can be used. The data may be converted to a
common standard using the equation $\Xi = \nu_{obs}[1 - (f + g - 100\delta)/10^8]$
which applies to nominal 100 MHz proton spectrometers. $\Xi$ is
the resonant frequency corrected to a field strength in which
TMS gives a proton resonance of *exactly* 100 MHz, $g$ is the
number of Hz by which the spectrometer rf oscillator exceeds
100 MHz, $f$ is the frequency in Hz of the audio side band used

W. McFarlane

TABLE 3-1. Selected resonance frequencies of various nuclei at a field strength appropriate to a TMS proton resonance of exactly 100 MHz

| Nucleus | Compound | Frequency (Hz) |
|---------|----------|----------------|
| $^1$H | $(CH_3)_4Si$ (TMS) | 100 000 000. 0 |
| $^2$H | $(CD_3)(CHD_2)SO$ | 15 350 650 |
| $^{10}$B | $BH_4^-$ | 10 743 230 |
| $^{11}$B | $BH_4^-$ | 32 082 695 |
| $^{13}$C | $(CH_3)_4Si$ | 25 145 005 |
| $^{14}$N | $NH_4^+$ | 7 223 750 |
| $^{19}$F | $C_6F_6$ | 94 078 500 |
| $^{29}$Si | $(CH_3)_4Si$ | 19 867 185 |
| $^{31}$P | $(CH_3O)_3P$ | 40 486 455 |
| $^{77}$Se | $(CH_3)_2Se$ | 19 071 520 |
| $^{119}$Sn | $(CH_3)_4Sn$ | 37 290 665 |
| $^{125}$Te | $(CH_3)_2Te$ | 31 549 800 |
| $^{183}$W | $WF_6$ | 4 161 780 |
| $^{195}$Pt | $cis$-$[(CH_3)_2S]_2PtCl_2$ | 21 421 130 |
| $^{199}$Hg | $(CH_3)_2Hg$ | 17 910 670 |
| $^{205}$Tl | $(CH_3)_2Tl^+$ | 57 893 970 |
| $^{207}$Pb | $(CH_3)_4Pb$ | 20 920 595 |

to excite the locking resonance (or for unlocked field-sweep spectra, the line observed), and $\delta$ is the proton chemical shift in ppm of locking resonance (or of the line observed in the case of unlocked field-sweep spectra).

Similar expressions can be written down for spectrometers operating at 60 or 90 MHz and the results converted into standard $\Xi$ values by dividing by 0.6 or 0.9, respectively. A larger $\Xi$ value corresponds to lower shielding. Table 3-1 can be used

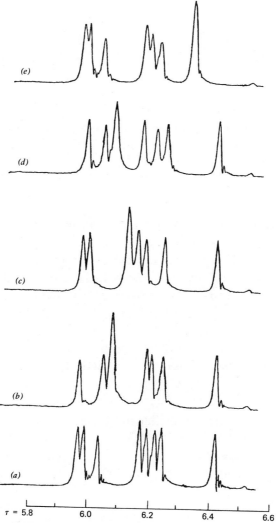

Figure 3-10.  Methoxy proton resonances of a mixture of organophos-phorus compounds.  (*a*) Normal spectrum at 60 MHz.  (*b*) With $^{31}$P irradiation at $\delta = 1$ ppm.  The doublet arising from $(MeO)_3PO$ is collapsed. (*c*) Irradiation at 31 ppm:  $(MeO)_2P(O)Me$.  (*d*) Irradiation at 71 ppm: $(MeO)_3PS$.  (*e*) Irradiation at 140 ppm:  $(MeO)_3P$.

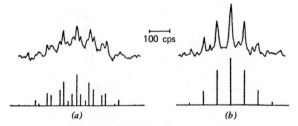

Figure 3-11.   $^{119}$Sn INDOR spectra produced by monitoring an appropriate satellite in the proton spectrum of a mixture of organotin compounds.  (*a*) Species **5**.  (*b*) Species **7**.   The stick spectra are calculated using coupling constants from the proton spectra.

to compare shifts on the Ξ scale with those obtained by conventional single resonance experiments.

    A special advantage of this technique is that it shows how
lines in the $^1$H and X spectra are related, which can be useful
in connection with preparative work.  For example, the methoxy resonance of a mixture of $(MeO)_3P$, $(MeO)_3PO$, $(MeO)_3PS$,
and $(MeO)_2P(O)Me$ is a set of eight lines which are difficult to
assign.  The $^{31}P$ decoupling experiments illustrated in Fig. 3-10
show these to be associated in pairs with phosphorus chemical
shifts of 139, 1, 71, and 31 ppm and so permit unequivocal assignment.  Similarly the rhodium complex **4** gives a complex
methyl proton resonance owing to different ring configurations,
and $^{103}$Rh decoupling experiments aid assignment.[24]  If necessary an INDOR spectrum can be plotted, as has been done to
confirm the presence[25] of **5** in a mixture of **6** and **7** (Fig. 3-11).

4

5

$$Me_2$$
$$Sn$$
$$EtN \qquad NEt$$
$$Me_2Sn \qquad SnMe_2$$
$$N$$
$$Et$$

**6**

$$Me_2$$
$$Sn$$
$$S \qquad S$$
$$Me_2Sn \qquad SnMe_2$$
$$S$$

**7**

### Selective Irradiation—the Relative Signs of Coupling Constants

A coupling constant is a measure of the interaction between a pair of nuclei and as such may be positive or negative. A positive coupling constant is defined as one for which the state of higher energy is that with the nuclear spins parallel (if the $\gamma$'s have the same sign). A knowledge of the sign is necessary if theories that claim to predict these parameters are to be assessed, and may also have diagnostic value. The absolute signs of coupling constants may be determined only with difficulty (e.g., from measurements on molecules dissolved in liquid crystals or subject to an electric field) but relative signs may be available from second-order (homonuclear) spectra, and can be obtained by double resonance experiments on all systems containing three or more different or chemically shifted nuclei. The signs of sufficient "key" coupling constants (see Table 3-2) are now known for the results of relative sign determinations to be placed upon an absolute basis.

The easiest sign-determining experiments to appreciate are selective decoupling ones which can be interpreted by subspectral analysis.[26] Consider $^{205}Tl^+Et_2$ ($^{205}Tl$ has $I = \frac{1}{2}$, abundance 70%; the other isotope $^{203}Tl$ has very similar magnetic properties and the associated lines in the proton spectrum are not separately resolved) which gives a proton spectrum consisting of two triplets (methyl protons) separated by $^3J(Tl-H)$ and two quartets (methylene protons) separated by $^2J(Tl-H)$ (see Fig. 3-12). Irradiation of the *low-field* triplet in a homonuclear decoupling experiment designed to collapse $^3J(H-H)$ affects only the *high-field* quartet, showing that these reso-

TABLE 3-2.   Absolute signs of important
coupling constants

| Coupling Constant | Sign |
|---|---|
| $^1J(^{13}C\text{-}H)$ | Positive |
| $^1J(^{13}C\text{-}^{19}F)$ | Negative |
| $^3J(H\text{---}H)$ (ethyl groups) | Positive |
| $^3J(H\text{---}H)$ (aromatic) | Positive |
| $^3J(F\text{---}F)$ (aromatic) | Negative |
| $^1J(^{31}P\text{-}H)$ | Positive |

nances arise from ions with the same spin orientation of $^{205}$Tl.
That is, the two Tl-H coupling constants are of opposite sign.
This result can be confirmed by other homonuclear experiments. Attempts to compare either of these signs with that of
$^3J(H\text{-}H)$ by similar $^{205}$Tl decoupling experiments will fail, however, because the two Tl-H coupling constants are so large,
and it is necessary to have recourse to tickling experiments.
These are most conveniently illustrated by reference to the
AMX spin system given by the olefinic protons of vinyl acetate
(Fig. 3-13). $J$(M-X) is small and so selective decoupling experiments can be used to show that $J$(A-M) and $J$(A-X) have the
same sign (this is actually positive). A tickling experiment in
which the line of the A double doublet at highest field is irradiated shows it to be connected to low-field M and X lines so
that $J$(M-X) is of opposite sign to the other two coupling constants; that is, it is negative. This result can be obtained by
writing down the spin orientations of the other nuclei beneath
the lines of the spectrum, or by constructing an energy-level
diagram according to Freeman and Anderson's rules. More
complex systems are best dealt with by breaking down into appropriate subspectra and then constructing the partial energy-
level diagrams where necessary. In some complicated cases
it may be helpful to use a computer to sort out the various al-
ternatives.

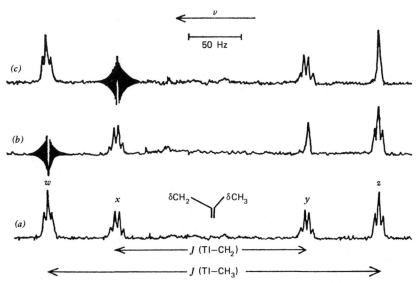

Figure 3-12. Selective decoupling experiments on $Et_2Tl^+$. (a) Normal 60 MHz proton spectrum. (b) With irradiation of *low-field* triplet (methyl resonance) at $w$ as indicated by the beat pattern. The *high-field* quartet (methylene resonance) at $y$ is collapsed, thus showing that $^2J(\mathrm{Tl}\cdots\mathrm{H})$ and $^3J(\mathrm{Tl}\cdots\mathrm{H})$ are of opposite sign. This is confirmed in (c) in which irradiation of the *low-field* quartet at $x$ decouples the *high-field* triplet at $z$.

In heteronuclear systems not only are the signs of the various coupling constants obtained, but also the magnitudes between nuclei that are not observed directly. Thus experiments shown in Fig. 3-14 on $Me_3P$ give the sign and magnitude of $^1J(^{31}P-^{13}C)$ in addition to the sign of $^2J(^{31}P-H)$ relative to $^1J(^{13}C-H)$, and many other systems have been studied similarly.

More complex spin systems in which energy levels and/or transitions may be degenerate can also be treated, but some caution must be exercised, as Freeman and Anderson's tickling rules may apply only in a modified form. The results obtained indicate that for directly bound elements the coupling constants are generally positive unless an element of high electronegativity is present. This is thought to be due to dominance by the Fermi contact interaction. When both elements are very

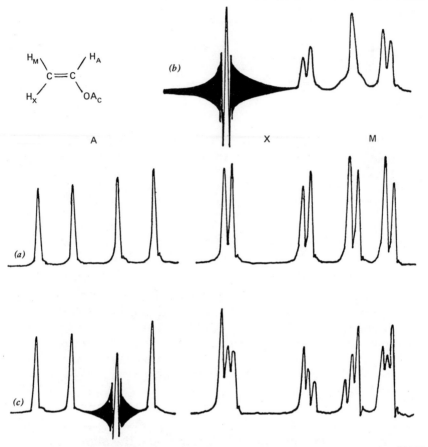

Figure 3-13. Homonuclear double resonance experiments on the AMX spin system given by the olefinic protons of vinyl acetate. (a) Normal 60 MHz proton spectrum. Note that $|J(M\text{-}X)|$ is much smaller than $|J(A\text{-}M)|$ or $|J(A\text{-}X)|$. (b) Selective collapse of $J(M\text{-}X)$ in half the molecules by irradiation in the X spectrum, showing $J(A\text{-}M)$ and $J(A\text{-}X)$ to be of like sign. (c) Tickling an A line and splitting half of the X and half of the M lines. This experiment shows that $J(A\text{-}M)$ and $J(X\text{-}M)$ are of opposite sign, and that $J(A\text{-}X)$ and $J(M\text{-}X)$ are also of opposite sign.

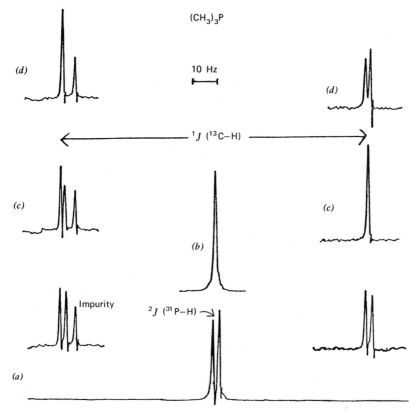

Figure 3-14. Heteronuclear double resonance experiments on tri-methyl phosphine, Me$_3$P. (a) Normal 60 MHz proton spectrum with $^{13}$C satellites shown at higher gain. (b) $^{31}$P decoupled in molecules containing no $^{13}$C by irradiation at the center of the phosphorus spectrum. (c) $^{31}$P splitting in *high-field* $^{13}$C satellite collapsed by irradiating at a point 7 Hz to *low-field* of center of phosphorus spectrum. (d) $^{31}$P splitting in *low-field* $^{13}$C satellite collapsed by irradiating at a point 7 Hz to *high field* of center of phosphorus spectrum. Thus $^1J(^{31}P-^{13}C)$ has a magnitude of 14 Hz and is of opposite sign to $^1J(^{13}C-H)$.

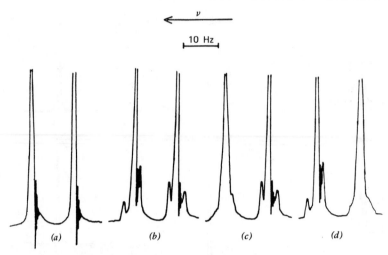

Figure 3-15.   Heteronuclear triple resonance experiments on
$(MeO)_2P(O)D$.   (a) Normal 60 MHz proton spectrum showing methoxy
doublet, the appearance of which is unaffected by irradiation at the
deuterium resonant frequency because $^4J(^2D \cdots H) = 0$.   (b) With weak
irradiation of one of the septets in the phosphorus spectrum.   (c) As
for (b) but with simultaneous irradiation of the *low-field* line in the
deuterium spectrum.   *The low-field* component of the methoxy doublet
is affected so $^1J(^{31}P-^2D)$ and $^3J(^{31}P \cdots H)$ are of like sign.   This is confirmed in (d), which shows the effect of irradiating at the frequency of
the *high-field* deuterium line.

electronegative the situation is very complex.   The case of
phosphorus is important because of the way in which the coupling constants change sign as the valence and coordination
number of phosphorus alter.

Triple Resonance

An example of a triple resonance experiment involving
simultaneous irradiation at two proton and one $^{14}N$ resonant frequencies has already been discussed, and it is also easy to perform
spin tickling experiments upon a decoupled spectrum.   A more
sophisticated application is the study of AMX and related spin

systems in which $J$(A-X) is zero. This frustrates the normal comparison of the signs of $J$(A-M) and $J$(M-X) by double resonance (unless $\nu_A$-$\nu_X$ is very small) because certain A and X transitions are degenerate in related pairs. However, this degeneracy can be lifted by irradiation with a weak field at the M resonant frequency and a normal A-{X} double resonance experiment can then succeed.[27] The technique has been applied to a number of homo- and heteronuclear systems and is illustrated here by $(CH_3O)_2P(O)D$ in which $^4J$(H---D) = 0 (Fig. 3-15). A $^1$H---{$^2$D} double resonance experiment has no effect upon the appearance of the methoxy proton resonance, but an effect is obtained when there is simultaneous irradiation at the $^{31}$P resonant frequency. True heteronuclear triple resonance experiments of this type also give the resonant frequencies of nuclei that are not coupled to protons and so may be used for chemical shift determinations. $^{77}$Se shielding in a range of organophosphorus selenides has been studied in this way.[28]

## References

1.   J. D. Baldeschwieler and E. W. Randall, *Chem. Rev.*, **63**, 81 (1963).
2.   D. F. Evans, *Lab. Pract.*, 57 (1965).
3.   R. A. Hoffman and S. Forsen, *Progress in NMR Spectroscopy*, J. W. Emsley, J. Feeney, and L. H. Sutcliffe, Eds., Vol. 1, Pergamon Press, Oxford, 1966, p. 15.
4.   W. McFarlane, *Annual Reviews of NMR Spectroscopy*, E. Mooney, Ed., Academic Press, London, Vol. 1, p. 135 (1968); Vol. 5, p. 353 (1972).
5.   W. McFarlane, *Chemistry in Britain*, **5**, 142 (1969).
6.   W. von Phillipsborn, *Angew. Chem.*, **10**, 472 (1971).
7.   W. McFarlane, *Determination of Organic Structures by Physical Methods*, J. Zuckerman, Ed., Vol. 4, Academic Press, New York, 1971, p. 139.
8.   V. Royden, *Phys. Rev.*, **93**, 944 (1954).
9.   T. M. Connor, D. H. Whiffen, and K. A. McLauchlan, *Mol. Phys.*, **13**, 221 (1967).

10. R. Freeman and W. A. Anderson, *J. Chem. Phys.*, **37**, 2053 (1962).
11. W. A. Anderson and R. Freeman, *J. Chem. Phys.*, **39**, 806 (1963).
12. R. Kaiser, *J. Chem. Phys.*, **39**, 2435 (1963).
13. A. W. Overhauser, *Phys. Rev.*, **92**, 411 (1953).
14. K. F. Kuhlmann and D. M. Grant, *J. Am. Chem. Soc.*, **90**, 7355 (1968).
15. W. A. Anderson and F. A. Nelson, *J. Chem. Phys.*, **39**, 183 (1963).
16. R. Ernst, *J. Chem. Phys.*, **45**, 3845 (1966).
17. E. B. Baker, *J. Chem. Phys.*, **37**, 911 (1962).
18. R. Freeman and B. Gestblom, *J. Chem. Phys.*, **47**, 1472 (1967).
19. R. F. Freeman, R. R. Ernst, and W. A. Anderson, *J. Chem. Phys.*, **46**, 1125 (1967).
20. H. C. E. McFarlane and W. McFarlane, *Org. Magn. Resonance*, **4**, 161 (1972).
21. R. E. Klinck, D. H. Mair, and J. B. Stothers, *Chem. Commun.*, 409 (1967).
22. P. M. Tucker and T. Onak, *J. Am. Chem. Soc.*, **91**, 9869 (1969).
23. E. Wenkert, A. O. Clouse, D. W. Cochran, and D. Doddrell, *J. Am. Chem. Soc.*, **91**, 6879 (1969).
24. W. McFarlane, *Chem. Commun.*, 700 (1969).
25. A. G. Davies, P. G. Harrison, J. D. Kennedy, T. N. Mitchell, R. J. Puddephatt, and W. McFarlane, *J. Chem. Soc. A*, 1136 (1969).
26. J. P. Maher and D. F. Evans, *Proc. Chem. Soc.*, 208 (1961).
27. A. D. Cohen, R. A. Freeman, K. A. McLauchlan, and D. H. Whiffen, *Mol. Phys.*, **7**, 45 (1963).
28. W. McFarlane and D. S. Rycroft, *Chem. Commun.*, 902 (1972).

# CHAPTER 4

## SOME CHEMICAL SHIFT CALCULATIONS FOR NITROGEN NUCLEI

G. A. Webb

*Department of Chemical Physics*
*University of Surrey*
*Guildford*

## INTRODUCTION

Experimental measurement of NMR chemical shifts can be interpreted in terms of a screening constant, $\sigma$, whose average isotropic value is available from experiments on nonviscous liquids. The screening constant is usually discussed on the basis of the formulation proposed by Ramsey,[1] which relates it to the electronic structure of the molecule containing the nucleus. The relationship is given by

$$\sigma = \frac{e^2}{3mc^2} \left\langle 0 \left| \sum_K \frac{1}{r_K} \right| 0 \right\rangle \frac{-4\beta^2}{3} \sum_n \left\langle 0 \left| \sum_K L_K \right| n \right\rangle$$

$$\times \left\langle n \left| \sum_K \frac{L_K}{r_K^3} \right| 0 \right\rangle \cdot (E_n - E_0)^{-1}, \tag{1}$$

where $\beta$ is the Bohr magneton, $\langle 0 |$ refers to the molecular ground state wave function in the absence of the magnetic field used in the NMR experiment with energy $E_0$, and $\langle n |$ refers to the electronic excited state with energy $E_n$. The separation of the nucleus considered from electron K is $r_K$, and $L_K$ is the orbital angular momentum operator for this electron. The other symbols in equation 1 have their usual significance.

The first term in equation 1 is called the "diamagnetic" term. It corresponds to free rotation of the electrons in the molecule around the nucleus in question. This rotation is hindered by the presence of the other nuclei and electrons in the molecule. The hindrance is represented by the second term in equation 1, called the "paramagnetic" term.

In order to be able to evaluate $\sigma$ from equation 1 there are two major difficulties to be overcome. The first is that for molecules of a reasonable size the "diamagnetic" and "paramagnetic" terms become large and nearly cancel, thus rendering $\sigma$ as the small difference between two large terms, which is unsatisfactory. This has been largely overcome by breaking $\sigma$ down into a number of local components and dealing with these individually.[2] The local components are considered to be separated from the overall motion of the electrons in the molecule.

$$\sigma_A = \sigma_A^{dia} + \sigma_A^{para} + \sigma_{AB} \text{ (long range)} , \qquad (2)$$

where $\sigma_A^{dia}$ and $\sigma_A^{para}$ refer to "diamagnetic" and "paramagnetic" contributions to the screening of nucleus A, $\sigma_A$, from electrons associated with the atom containing A. $\sigma_{AB}$ represents contributions from electron currents on neighboring atoms in the molecule. In proton NMR spectroscopy of conjugated systems $\sigma_{AB}$ can include a ring current component. For nitrogen nuclei in a diamagnetic environment the experimental range of chemical shifts is ~ 900 ppm; since ring currents do not usually produce a shift greater than a few ppm at most, they are ignored in nitrogen NMR spectroscopy. Solvent effects, shielding anisotropies, and neighboring electric field effects are also neglected when discussing nitrogen chemical shift differences in molecules. Consequently when considering differences in the screening of nitrogen nuclei in series of related molecules we are concerned only with changes in $\sigma_N^{dia}$ and $\sigma_N^{para}$.

The second difficulty encountered in using Ramsey's expression is due to the necessity of having some knowledge of all the eigenfunctions and eigenvalues of the excited electronic states of the molecule in order to evaluate the "paramagnetic" term. These are involved in an infinite summation as shown in equation 1. This problem is usually dealt with by means of the approximations introduced by Pople.[3] He has used linear combinations of gauge invariant atomic orbitals in a molecular orbital description of the "diamagnetic" and "paramagnetic" terms.

## SEMIEMPIRICAL CALCULATIONS

By means of Pople's approximations, the expression for the "diamagnetic" contribution to the screening of nucleus A becomes

$$\sigma_A^{dia} = \frac{e^2}{3mc^2} \sum_\mu P_{\mu\mu} \langle r_\mu^{-1} \rangle, \tag{3}$$

where $P_{\mu\mu}$ is the electron occupation of atomic orbital $\mu$, and the summation is taken over all occupied atomic orbitals. Consequently as the electron density around nucleus A increases, the screening from $\sigma_A^{dia}$ increases and the NMR signal moves to higher values of the applied magnetic field at constant applied frequency. In nitrogen NMR screening contributions from $\sigma_N^{dia}$ appear to be small.[4] Consider the nitrogen chemical shift difference between pyridine and the pyridinium ion as an example. The experimental difference is $113 \pm 2$ ppm. The results of an all-valence electron calculation[5] give contributions of 10.8 ppm and 128.8 ppm from $\sigma_N^{dia}$ and $\sigma_N^{para}$, respectively. The calculated chemical shift difference of 118 ppm is in very good agreement with the experimental data. Further evidence for the rather insignificant contribution arising from changes in $\sigma_N^{dia}$ for a series of related molecules is furnished by CNDO calculations on some diazines.[6] Within the series changes in $\sigma_N^{dia}$ are calculated to be about 0.5 ppm, whereas the observed nitrogen chemical shift range is about 100 ppm.

CNDO calculations of $\sigma_N^{dia}$ have also been reported for some small inorganic ions[7] (Table 4-1), where shifts to low field are positive.

TABLE 4-1. Comparison of calculated changes in $\sigma_N^{dia}$ and experimental $^{14}N$ chemical shifts for some simple ions, taking $NO_3^-$ as an arbitrary reference

| Ion | Calculated Change in $\sigma_N^{dia}$ (ppm) | Experimental $^{14}N$ Chemical Shift (ppm) |
|---|---|---|
| $NO_3^-$ | 0 | 0 |
| $NO_2^-$ | $-55.8$ | $+274$ |
| $N_3^+$ (central) | $-66.7$ | $-128$ |
| $N_3^+$ (terminal) | $-86.3$ | $-277$ |
| $NCO^-$ | $-91.7$ | $-300$ |
| $NH_4^+$ | $-154.7$ | $-354$ |

Comparison of the experimental $^{14}N$ chemical shifts and calculated changes in $\sigma_N^{dia}$ shows that with the exception of $NO_2^-$ the two sets of data vary in opposite directions. This casts some doubt on the validity of the calculations.

It seems reasonable to conclude, however, that in the majority of cases concerning series of closely related molecules, changes in $\sigma_N^{dia}$ may be neglected in discussions of nitrogen chemical shift differences. Attempts to account for the experimental chemical shift differences are usually based on changes in $\sigma_N^{para}$ only.

Within the approximations of Pople's treatment the "paramagnetic" term becomes

$$\sigma_A^{para} = -\frac{2}{3}\frac{e^2\hbar^2}{m^2c^2}\langle r^{-3}\rangle_{2p}\sum_i^{occ.}\sum_j^{unocc.}(\Delta E_{i \to j})^{-1}(c_{i,xA}c_{j,yA} - c_{i,yA}c_{j,xA})$$

$$\times \sum_B(c_{i,xB}c_{j,yB} - c_{i,yB}c_{j,xB}) + (c_{i,yA}c_{j,zA} - c_{i,zA}c_{j,yA})$$

$$\times \sum_B(c_{i,yB}c_{j,zB} - c_{i,zB}c_{j,yB}) + (c_{i,zA}c_{j,xA} - c_{i,xA}c_{j,zA})$$

$$\times \sum_B(c_{i,zB}c_{j,xB} - c_{i,xB}c_{j,zB}), \tag{4}$$

where $i$ and $j$ refer to occupied and unoccupied molecular orbitals, and $c_{i,yA}$ is the coefficient of the $2Py$ orbital on atom A in the $i$th molecular orbital. The summation over all nuclei B includes A. The orbital expansion term, $\langle r^{-3} \rangle_{2p}$, is the mean value of the reciprocal cube of the radius of the $2p$ orbital on nucleus A.

Equation 4 incorporates the replacement of the infinite sum over excited states by a restricted summation, but the excited state wave functions are still required in order to evaluate the coefficients $c_{j,yA}$, etc. By considering the ordering of these coefficients in equation 4 it becomes apparent that the energies $\Delta E_{i \cdot j}$ do not correspond to $\pi \rightarrow \pi^*$ transitions but possibly may be estimated from transitions involving sigma or nonbonding electrons. This problem is often avoided by replacing the summation over $\Delta E_{i \cdot j}$ by a parameter $\Delta E_{av}$ which is not related to any of the actual electronic excitation energies of the molecule.[8] This is called the average excitation energy (AEE) approximation; when considered together with the closure relation

$$\overset{\text{occ.}}{\underset{i}{\sum}} C_{\mu i} C_{\nu i} + \overset{\text{unocc.}}{\underset{j}{\sum}} C_{\mu j} C_{\nu j} = \delta_{\mu\nu} \tag{5}$$

and applied to equation 4, the Karplus-Pople equation is produced:

$$\sigma_A^{\text{para}} = - \frac{e^2 \hbar^2}{2 m^2 c^2} \langle r^{-3} \rangle_{2p} (\Delta E_{av})^{-1} \underset{B}{\sum} Q_{AB} \, ,$$

where the summation over B includes A, and

$$Q_{AB} = \tfrac{4}{3} \delta_{AB} (P_{xAxB} + P_{yAyB} + P_{zAzB}) - \tfrac{2}{3} (P_{xAxB} P_{yAyB} + P_{xAxB} P_{zAzB}$$

$$+ P_{yAyB} P_{zAzB}) + \tfrac{2}{3} (P_{xAyB} P_{xByA} + P_{xAzB} P_{xBzA} + P_{yAzB} P_{yBzA}) . \tag{6}$$

The terms $P_{\mu\nu}$ are the elements of the bond-order matrix for the atomic orbitals comprising the molecular orbitals:

$$P_{\mu\nu} = 2 \overset{\text{occ.}}{\underset{i}{\sum}} C_{\mu i} C_{\nu i} \, . \tag{7}$$

The presence of the Kronecker delta, $\delta_{AB}$, ensures that the first term in the $Q_{AB}$ summation of equation 6 is concerned with electron charge densities on the atom containing the nucleus in question.

Unfortunately the Karplus-Pople equation is not gauge invariant in the general case; consequently the value of $\sigma_A^{para}$ obtained from it depends upon the origin chosen for the coordinate system. Alternative averaging procedures that maintain gauge invariance have been reported for some diatomic molecules. Calculations based on these procedures give agreement with the proton NMR spectra of the diatomic molecules. However, this approach has so far not been applied to larger molecules.

The Karplus-Pople equation has been used widely in order to interpret the chemical shifts of molecules larger than diatomics. Its usefulness largely depends upon whether or not $\Delta E_{av}$ may be assumed constant for a series of molecules. Although $\Delta E_{av}$ is not related to any of the actual electronic excitation energies of a molecule it is somewhat surprising that in some cases negative values of $\Delta E_{av}$ have been reported. [9, 10]

We now consider the relative importance of changes in $\langle r^{-3} \rangle_{2p}$, $Q_{AB}$, and $\Delta E_{av}$ in deciding changes in $\sigma_A^{para}$ as given by equation 6.

For nitrogen the orbital expansion term is often evaluated from Slater's rules by

$$\langle r^{-3} \rangle_{2p} = \frac{1}{3} \left( \frac{Z_{2p}}{2a_0} \right)^3 , \tag{8}$$

where $a_0$ is the Bohr radius and $Z_{2p}$ is the effective nuclear charge for the $2p$ electron. Equation 8 shows that $\sigma_A^{para}$ is proportional to $Z^3$; hence $\sigma_A^{para}$ is expected to be much more important than $\sigma_A^{dia}$ in deciding the chemical shifts of heavy nuclei such as nitrogen.

An increase in electron density on the atom containing the nucleus in question leads to a decrease in $Z$, for a given electron, and hence from equation 8 to a decrease in $\langle r^{-3} \rangle_{2p}$. Consequently $\sigma_A^{para}$ tends to decrease in magnitude as the electron density on the atom increases. Therefore changes in the orbital expansion term act in the same sense as changes in $\sigma_A^{dia}$.

The major contribution to $\langle r^{-3} \rangle_{2p}$ arises from atomic orbitals close to the nucleus where Slater's rules are thought to be unreliable. Hence it is advisable to use more satisfactory wave functions in estimating $\langle r^{-3} \rangle_{2p}$. By using SCF wave functions Valenik and Lynden-Bell[11] have reported for nitrogen:

$$\langle r^{-3} \rangle_{2p} = 3.099 - 0.732 Q_N , \tag{9}$$

where $Q_N$ is the total electronic charge on the nitrogen atom. By means of equation 9 it has been shown that changes in the orbital expansion term are negligible when compared with those in the other components of $\sigma_N^{para}$ for a series of small molecules and ions.[11]

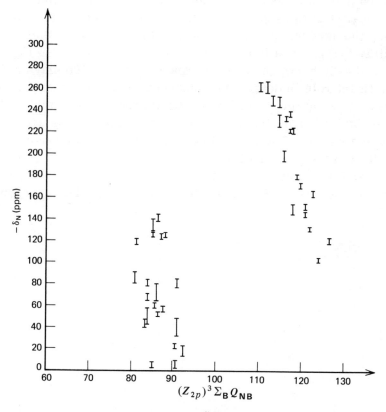

Figure 4-1. Plot of $-\delta_N$ against $(Z_{2p})^3 \sum_B Q_{NB}$ for some five-membered ring conjugated heterocycles.[12]

The major contribution to changes in $\sigma_A^{para}$, for a series of closely related molecules, is usually considered to arise from the summation of $Q_{AB}$ expressions in equation 6. The bond orders and charge densities comprising $Q_{AB}$ are readily available from molecular orbital computer programs of various degrees of sophistication. For large molecules most calculations have been performed on the $\pi$ electron systems only; however, with the widespread use of CNDO and INDO programs it is anticipated that $\sigma$ electrons will be directly included in many future calculations. Figures 4-1 and 4-2 show the reasonable correlations found between [14]N chemical shift and $(Z_{2p})^3 \Sigma_B Q_{NB}$ for series of five-membered[12] and six-membered[13] ring conjugated heterocycles. This shows the close relationship of chemical shift to $\Sigma_B Q_{NB}$ for these series of molecules. However, the agreement found from the plot of chemical shift against $\Sigma_B Q_{NB}$ alone is not as good as those reported in Figs. 4-1 and 4-2; hence the orbital expansion term also plays a significant role in determining changes in $\sigma_N^{para}$ for these molecules. The charge densities and bond orders used in this work were obtained from Pariser-Parr-Pople calculations and hence they are $\pi$ only. The distribution of $\pi$ electrons in a molecule can lead to the polarization of the $\sigma$ electrons which should also be taken into account. For an $sp^2$-hybridized nitrogen atom bonded to atoms A, B, and C the $\sigma$ orbitals may be represented as follows.

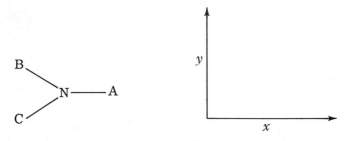

$$\psi_{NA} = \frac{1}{\sqrt{2}} \left[ \sqrt{1+a}\, \frac{1}{\sqrt{3}}(s_N - \sqrt{2}p_{xN}) + \sqrt{1-a}\, \frac{1}{\sqrt{3}}(s_A + \sqrt{2}p_{xA}) \right],$$

$$\psi_{NB} = \frac{1}{\sqrt{2}} \left[ \sqrt{1+b}\, \frac{1}{\sqrt{3}}(s_N + \frac{1}{\sqrt{2}}p_{xN} + \frac{\sqrt{6}}{2}p_{yN}) + \sqrt{1-b}\, \frac{1}{\sqrt{3}} \right.$$

$$\left. \times (s_B - \frac{1}{\sqrt{2}}p_{xB} - \frac{\sqrt{6}}{2}p_{yB}) \right], \tag{10}$$

$$\psi_{NC} = \frac{1}{\sqrt{2}} \left[ \sqrt{1+c}\, \frac{1}{\sqrt{3}}(s_N + \frac{1}{\sqrt{2}}p_{xN} - \frac{\sqrt{6}}{2}p_{yN}) + \sqrt{1-c}\, \frac{1}{\sqrt{3}} \right.$$

$$\left. \times (s_C - \frac{1}{\sqrt{2}}p_{xC} + \frac{\sqrt{6}}{2}p_{yC}) \right],$$

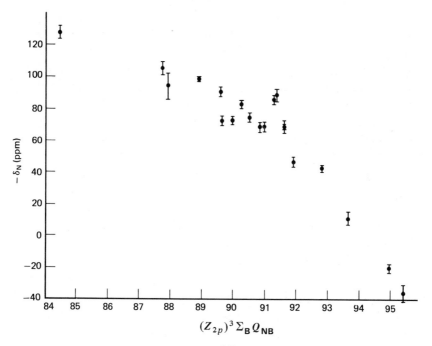

Figure 4-2. Plot of $-\delta_N$ against $(Z_{2p})^3 \sum_B Q_{NB}$ for some six-membered ring conjugated heterocycles.[13]

where $a$, $b$, and $c$ are the polarization parameters for the $\sigma$ orbitals whose wave functions are $\psi_{NA}$, $\psi_{NB}$, and $\psi_{NC}$. By including the effects of $\sigma$ bond polarization $\sum_B Q_{NB}$ becomes

$$\sum_{N, A, B, C} Q_{NB} = 2 - \tfrac{4}{9}(a + b + c)(q_N{}^\pi - 1) - \tfrac{2}{9}(ab + bc + ac)$$
$$+ \tfrac{4}{9}(\sqrt{1-a}\,{}^2 p_{NA}{}^\pi + \sqrt{1-b}\,{}^2 p_{NB}{}^\pi + \sqrt{1-c}\,{}^2 p_{NC}{}^\pi). \quad (11)$$

If A is a lone electron pair, $n$, then $a = 1$, and

$$\sum_{N, n, B, C} Q_{NB} = \tfrac{22}{9} + \tfrac{2}{9}(b + c - bc) - \tfrac{4}{9}(1 + b + c)q_N{}^\pi$$
$$+ \tfrac{4}{9}(\sqrt{1-b}\,{}^2 p_{NB}{}^\pi + \sqrt{1-c}\,{}^2 p_{NC}{}^\pi),$$

where $q_N{}^\pi$ refers to the $\pi$ charge density on the nitrogen atom and $p_{NB}{}^\pi$ to the $\pi$ bond order of the bond between atom B and the nitrogen atom. By means of equation 11 reasonable accounts of the chemical shift differences have been obtained for several series of molecules including nitroalkanes, nitramines, alkyl nitrates, alkyl isocyanates, isothiocyanates and azides, [14] nitriles and isonitriles, [15] linear triatomic molecules and ions, [16] nitrocarbanions, [17] and some substituted pyridines. [18]

Equation 11 shows that an increase in $q_N{}^\pi$ leads to a high-field nitrogen chemical shift and an increase in $p_{NB}{}^\pi$ to a low-field shift. These predictions have been verified from a consideration of the electron-donating and electron-withdrawing properties of the substituents in the above series of molecules.

In the case of the substituted pyridines the correlation between nitrogen chemical shift and $\langle r^{-3} \rangle_{2p} \sum_B Q_{NB}$ is improved by including the effects of $\sigma$ bond polarization. Based upon equation 11, $\sum_B Q_{NB}$ includes a value of $b + c \approx 30\%$ which has been estimated from NQR data[18] (Fig. 4-3).

Compared with Figs. 4-1 and 4-2, the finer scale divisions in Fig. 4-3 show the relatively small shift differences between the various substituted pyridines. Hence the subtle effects of $\sigma$ bond polarization become more noticeable in these molecules than in the five-membered and six-membered ring conjugated heterocycles.

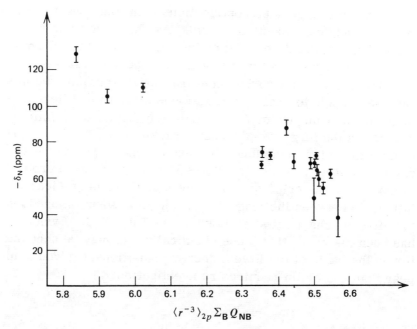

Figure 4-3.   Plot of $-\delta_N$ against $\langle r^{-3} \rangle_{2p} \sum_B Q_{NB}$ for some substituted pyridines. [18]

Although the AEE approximation allows for reasonable agreement between the calculated and experimental values of nitrogen chemical shifts in a number of cases, intuitively the approximation is more reasonable for $^{13}C$ than for nitrogen due to the presence of the lone-pair electrons in the latter.  On account of possible low-lying $n \rightarrow \pi^*$ transitions the lone-pair electrons are likely to dominate $\Delta E_{av}$, rendering it a variable with slight changes in electronic structure at the nitrogen atom. Consequently the different excitation energy (DEE) approximation has been introduced. [6] This incorporates the lowest-known excited states according to equation 4 and uses an average value for the remaining excited states.  The DEE approximation has been employed with some success for some six-membered ring conjugated heterocycles. [6] However, by considering a more extended series of conjugated heterocycles it is less satisfactory in accounting for their nitrogen chemical shifts.  This probably arises from exaggerated contributions from the lowest excited states to the "paramagnetic" term. As shown in Table 4-2 there is no apparent correlation between the nitrogen chemical shift and the lowest $n \rightarrow \pi^*$ transition for these molecules.

Within the framework of the AEE approximation attempts have been made to explain nitrogen chemical shift differences in terms of changes in $\Delta E_{av}$ which are assumed to follow changes in the longest-wavelength transitions.[19-23] These are usually those involving lone-pair excitations. Only in cases of extremely low-lying excited states is this explanation probable. In the case of a series of azo compounds there is no clear relationship between the longest-wavelength absorption band and the nitrogen chemical shift as shown in Table 4-3. Even so, it has been claimed[23] that these chemical shifts may be accounted for on the basis of the lowest energy transitions. It would appear that this claim requires reinvestigation.

TABLE 4-2.

| Molecule | Nitrogen Chemical Shift (ppm)[a] | Lowest $n \rightarrow \pi^*$ Transition (nm) |
|---|---|---|
| s-Triazine | − 98 | 2720 |
| Quinazoline | − 90 (mean) | 3300 |
| Pyrimidine | − 82 | 2980 |
| Pyridine | − 68 | 2700 |
| Quinoxaline | − 46 | 3390 |
| Pyrazine | − 42 | 3280 |
| Phthalazine | − 11 | 3570 |
| Pyridazine | + 20 | 3400 |
| Cinnoline | + 36 (mean) | 3890 |

[a]Low field chemical shifts are positive

TABLE 4-3.  Comparison of nitrogen chemical shifts, relative to pyridazine as an arbitrary reference, and the longest-wavelength absorption band for a series of azo compounds

| Compound | Nitrogen Chemical Shift (ppm)[a] | Longest-Wavelength Absorption Band (Å) |
|---|---|---|
| Pyridazine | 0 | 3400 |
| trans-FN=NF | −38 | <2000 |
| ⁻ON=NO⁻ in NaOH | −52 | 3000 |
| HON=NOH in $HClO_4$ | −55 | 2430 |
| trans-Azobenzene | −52 to −102 | 4200−4400 |
| $CF_3N=NCF_3$ | −109 | 3550 |
| $CH_3CH_2N=NCH_2CH_3$ | −124 | 3550 |

[a]Low field chemical shifts are positive

Valenik and Lynden-Bell have reported[11] screening constant calculations which do not incorporate the AEE approximation and have found a reasonable correlation between theory and experiment for some nitrogen chemical shifts.  Their calculations are based upon the Ramsey expression without making the separation given by equation 2.  The molecular orbitals are evaluated by an extended Hückel technique using a constant value for the coulomb integrals, no variation being allowed for changes in the state of hybridization or with the nature of neighboring atoms.

Figure 4-4 shows that reasonable agreement between theory and experiment is found but that different correlation lines are obtained for pyramidal, planar three coordinate, planar two coordinate, and for linear nitrogen environments.  It is anticipated that these difficulties would be removed in an SCF calculation, but it appears that CNDO/2 calculations do not lead to an improvement,[24] which is rather surprising.

Calculations of a similar order of approximation have been based upon coupled Hartree-Fock perturbation theory.[5,25,26]

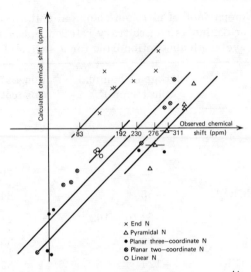

Figure 4-4. Correlation of calculated and observed [14]N chemical shifts for some small molecules and ions.[11]

The expressions used are gauge invariant and the molecular orbitals have been evaluated by the INDO approximations. These calculations show the relative importance of changes in the excitation energy and angular momentum parameters to changes in $\sigma_N^{para}$.

For example, $NF_3$ has its [14]N NMR signal at 123 ppm to low field of the signal given by $ONF_3$, and $NO_2^-$ has its signal at 250 ppm to low field of the signal from $NO_3^-$. The structural difference in both cases consists of replacing a lone pair of electrons by an oxygen to nitrogen bond. The low-field signal from $NF_3$ compared with $ONF_3$ is caused largely by an increase in the angular momentum terms in equation 1. In the case of the $NO_2^-$ and $NO_3^-$ ions the low-field [14]N shift shown by $NO_2^-$ is due mainly to a very low energy transition involving the lone-pair electrons in this ion. This appears to be an extreme case of the effect of lone-pair electrons on nitrogen chemical shifts. Since these calculations allow a clear insight into the electronic changes causing chemical shift differences it is to be hoped that further results will be reported soon.

TABLE 4-4. Comparison of some experimental nitrogen chemical shifts with values calculated by ab initio procedures

| Molecule | Experimental Nitrogen Chemical Shift (ppm)[a] | Calculated Nitrogen Chemical Shifts (Diamagnetic + Paramagnetic Components) = Total Shift (ppm) | | | | |
|---|---|---|---|---|---|---|
| | | STO-5G St. | STO-5G Opt. | LEMAO-5G St. | LEMAO-5G Opt. | 4-31G |
| $NH_3$ | 0 | 0 | 0 | 0 | 0 | 0 |
| $CH_3NH_2$ | +5 | $(-26+25)$ $=-1$ | $(-26+32)$ $=+6$ | $(-26+17)$ $=-9$ | $(-26+27)$ $=+1$ | $(-25+22)$ $=-3$ |
| $(CH_3)_2NH$ | +12 | $(-70+42)$ $=-28$ | $(-69+51)$ $=-18$ | $(-71+55)$ $=-16$ | $(-70+79)$ $=+9$ | $(-68+47)$ $=-21$ |
| $HCONH_2$ | +115 | $(-28+58)$ $=+30$ | $(-29+62)$ $=+33$ | $(-29+99)$ $=+70$ | $(-30+91)$ $=+61$ | $(-29+80)$ $=+51$ |
| $CH_3CN$ | +246 | $(+7+261)$ $=+268$ | $(+7+266)$ $=+273$ | $(+8+416)$ $=+424$ | $(+7+427)$ $=+434$ | $(+11+320)$ $=+331$ |

[a]Low field chemical shifts are positive

71

## AB INITIO CALCULATIONS

From a theoretical viewpoint ab initio calculations are more satisfying than semiempirical ones. Pople and his co-workers[27,28] have recently reported the results of some ab initio calculations of the screening of nitrogen nuclei in small molecules using the method of finite perturbation theory.[29]

The STO-5G calculations in Table 4-4 are obtained from a set consisting of a single function for hydrogen and five Gaussian functions for the other atoms, using either standard (St.) or optimized (Opt.) values of the $\zeta$ parameters. The LEMAO-5G sets are composed of least-energy minimal atomic orbitals, each found from five Gaussian functions, except for the hydrogen $1s$ function which is a single function. In the 4-31G set the $1s$ shell of the heavy atom, for example, nitrogen, is represented by a single function composed of four Gaussians. However the valence orbitals are described by inner and outer parts which are the sums of three and four Gaussian functions, respectively.

Table 4-4 shows that the calculated nitrogen chemical shifts are generally not in good agreement with the observed ones, an exception being the large downfield shift of the $CH_3CN$ resonance. Apparently the lack of agreement rests with the calculated value of the "diamagnetic" component of the screening, which is often large, and whose trend is opposite in direction to that of the observed chemical shifts. The results obtained with the LEMAO-5G Opt. and 4-31G sets of functions most closely follow the experimental shifts. The major contribution to changes in the chemical shift arise from differences in the "paramagnetic" component of the screening constant. Consequently it is not unreasonable to account for chemical shift differences solely in terms of changes in this component in the semiempirical methods of calculation.[30]

In conclusion it seems that there is an optimum level of agreement between experimental and theoretical values of chemical shift at a particular level of approximation depending upon the molecules being considered. For small molecules the more exact calculations appear to be the more suitable ones although ab initio attempts are not too encouraging so far. For larger molecules satisfactory agreement can often be obtained by a judicious use of the AEE approximation, especially for series of closely related molecules.

## References

1. N. F. Ramsey, *Phys. Rev.*, **78**, 699 (1950).
2. A. Saika and C. P. Slichter, *J. Chem. Phys.*, **22**, 26 (1954).
3. J. A. Pople, *J. Chem. Phys.*, **37**, 53, 60 (1962).
4. M. Witanowski and G. A. Webb, in *Annual Reports on NMR Spectroscopy*, E. Mooney, Ed., Vol. 5a, Academic Press, London, New York, 1972, p. 395.
5. H. Kato, H. Kato, and T. Yonezawa, *Bull. Chem. Soc. Japan*, **43**, 371 (1970).
6. T. Tokuhiro and G. Fraenkel, *J. Am. Chem. Soc.*, **91**, 5005 (1969).
7. A. Sadlej, *Org. Magn. Resonance*, **2**, 63 (1970).
8. L. C. Snyder and R. G. Parr, *J. Chem. Phys.*, **34**, 837 (1961).
9. A. D. McLachlan and M. R. Baker, *Mol. Phys.*, **4**, 255 (1961).
10. D. W. Davies, *Mol. Phys.*, **13**, 465 (1967).
11. A. Valenik and R. M. Lynden-Bell, *Mol. Phys.*, **19**, 371 (1970).
12. M. Witanowski, L. Stefaniak, H. Januszewski, Z. Grabowski, and G. A. Webb, *Tetrahedron*, **28**, 637 (1972).
13. M. Witanowski, L. Stefaniak, H. Januszewski, and G. A. Webb, *Tetrahedron*, **27**, 3129 (1971).
14. M. Witanowski, *J. Am. Chem. Soc.*, **90**, 5683 (1968).
15. M. Witanowski, *Tetrahedron*, **23**, 4299 (1967).
16. J. E. Kent and E. L. Wagner, *J. Chem. Phys.*, **44**, 3530 (1966).
17. M. Witanowski and S. A. Shevelev, *J. Mol. Spectrosc.*, **33**, 19 (1970).
18. M. Witanowski, T. Saluvere, L. Stefaniak, H. Januszewski, and G. A. Webb, *Mol. Phys.*, **23**, 1071 (1972).

19.    T. K. Wu, *J. Chem. Phys.*, **49**, 1139 (1968).
20.    T. K. Wu, *J. Chem. Phys.*, **51**, 3622 (1969).
21.    J. Mason and W. Van Bronswyk, *J. Chem. Soc. A*, **1763** (1970).
22.    L. O. Anderson, J. Mason, and W. Van Bronswyk, *J. Chem. Soc. A*, 296 (1970).
23.    J. Mason and W. Van Bronswyk, *J. Chem. Soc. A*, 791 (1971).
24.    J. N. Murrell and A. J. Harget, *Semiempirical SCF-MO Theory of Molecules*, Wiley-Interscience, London, New York, 1972.
25.    F. G. Herring, *Can. J. Chem.*, **48**, 3498 (1970).
26.    F. Aubke, F. G. Herring and A. M. Qureschi, *Can. J. Chem.*, **48**, 3504 (1970).
27.    R. Ditchfield, D. P. Miller, and J. A. Pople, *J. Chem. Phys.*, **53**, 613 (1970).
28.    R. Ditchfield, D. P. Miller, and J. A. Pople, *J. Chem. Phys.*, **54**, 4186 (1971).
29.    J. A. Pople, J. W. McIver, and N. S. Ostlund, *J. Chem. Phys.*, **49**, 2960 (1960).
30.    G. A. Webb and M. Witanowski, in *Nitrogen NMR*, M. Witanowski and G. A. Webb, Eds., Plenum Press, London, 1973.

# CHAPTER 5

# $^{14}$N NMR OF SOME ORGANOMETALLIC AZIDES

Joachim Müller

*Department of Chemistry*
*University of Marburg*

Very little experimental work has been done on the $^{14}$N NMR spectroscopy of azides.[1] In the 1960s, Kanda[2] and Forman[3] recorded the spectrum of the azide ion $N_3^-$ in aqueous solution, and Witanowski[4] published the spectra of the covalent compounds methyl- and ethylazide. In contrast to the azide ion, where two resonances with an intensity ratio of $1:2$ are observed, methylazide shows three resonance peaks (Fig. 5-1), corresponding to the three different nitrogen atoms.

The correct assignment was made by Witanowski, who considered the distribution of the electric charges within the molecule. In this way the central nitrogen atom $N_\beta$ is the most positive one and its resonance is found at lowest field. The nitrogen atom $N_\alpha$, which is the neighbor of the heteroelement, is the most negative due to inductive effects—the electrons are withdrawn from the azido group by addition of the substituent—and so its signal is found at highest field.

Recently, Beck and co-workers[5] reported the $^{14}$N NMR data of a large number of azido complexes and of organoarsenic azides. In nearly all cases three resonances are observed as expected.

To assign the resonance lines in the correct manner one

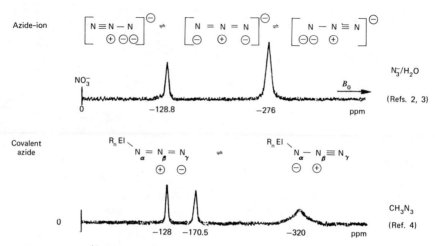

Figure 5-1.  $^{14}$N NMR spectra of an ionic and a covalent azide.

may also look at the line widths.  In the case of $^{14}$N NMR spectroscopy the line widths are usually not governed by the inhomogeneity of the field, the transmitter, or external conditions.
The very large line widths (several hundred Hz) are often
caused by a faster relaxation process.  The nitrogen-14 nucleus has an electric quadrupole moment and may also have—by
the steric arrangement of its neighbors—an electric field gradient.  On consideration of the symmetry of the three nitrogen
atoms in azides one finds that the central nitrogen atom is *sp*
hybridized.  It has the highest symmetry and should have the
narrowest resonance line.  The $N_\alpha$ atom, which has three different neighbors—one nitrogen atom, another element, and a
lone pair—must have the broadest line and this is the line that
occurs furthest upfield (Fig. 5-1).

Figure 5-2 and Table 5-1 show the $^{14}$N NMR spectra, the
chemical shifts, and the line widths of the azides discussed in
this chapter.  At first sight it is noticed that not all these compounds show three lines.  Secondly, the line-shape relations
between the center peaks and the side bands are not always the
same—in a few cases so much rf power had to be applied to get
a spectrum that the center bands were saturated.  Furthermore, if one looks at the chemical shifts, it is seen that all

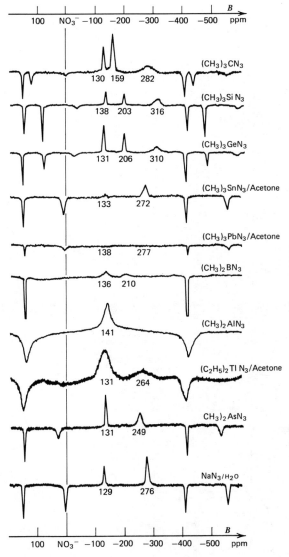

Figure 5-2. $^{14}$N NMR spectra of azides. The inverted lines are the 180° phase shifted sidebands ($\nu = 2000$ Hz).

77

TABLE 5-1. Chemical shifts and line widths of $^{14}N$ NMR spectra of azides[a,b]

| Compound | $N_\alpha$ | | $N_\beta$ | | $N_\gamma$ | | Solvent |
| --- | --- | --- | --- | --- | --- | --- | --- |
| | Chemical Shift, $\delta$ | Line Width, $\Delta\nu_{1/2}$ | Chemical Shift, $\delta$ | Line Width, $\Delta\nu_{1/2}$ | Chemical Shift, $\delta$ | Line Width, $\Delta\nu_{1/2}$ | |
| $N_3^-$ | −276 | 52.5 | −128.8 | 24 | −276 | 52.5 | $H_2O$ |
| $CH_3N_3$ | −320 | 100 | −128 | 17 | −171 | 19 | — |
| $C_2H_5N_3$ | −305 | 122 | −129 | 22 | −167 | 28 | — |
| $(CH_3)_3CN_3$ | −282 | 180 | −130.5 | 30 | −159 | 40 | — |
| $(CH_3)_3SiN_3$ | −316.5 | 123 | −137.8 | 15.7 | −203.5 | 12.3 | — |
| $(CH_3)_3GeN_3$ | −310.5 | 220 | −131.3 | 18 | −205.8 | 63 | — |
| $(CH_3)_3SnN_3$ | −272 | 130 | −133.3 | 24 | −272 | 130 | Acetone |
| $(CH_3)_3PbN_3$ | −277 | 100 | −138 | 20 | −277 | 100 | Acetone |
| $(CH_3)_2BN_3$ | | | −136 | 25 | −210 | 500 | Toluene |
| $(CH_3)_2AlN_3$ | | | −141 | 100 | | | (Molten) |
| $(C_2H_5)_2TlN_3$ | −264 | 700 | −130.8 | 109 | −264 | 700 | Acetone |
| $(CH_3)_2AsN_3$ | −249 | 180 | −131.5 | 20 | −249 | 180 | — |
| $NO_3^-$ | 0 | 12 | | | | | $H_2O$ |

[a]All chemical shift data are reported with respect to $NO_3^-$; high-field shifts have negative values.
[b]The total error of the chemical shift is about ± 1 ppm for resonance lines with half line widths below 100 Hz.

shifts of the central nitrogen atom $N_\beta$ are about 130 ppm to high field relative to aqueous concentrated sodium nitrate solution. There is a high electronic mobility in the azide group, especially in the $\pi$ system; consequently almost all of the charge withdrawn by inductive effects on one side of $N_\beta$ is replaced by charge from the other side. Nevertheless there are small differences between the azides reported, for example, a slight increase in shift within the fourth main group with increasing electronegativity of the elements. Only silicon fails to follow this rule. In this case the high-field shift of the signal must be explained by back-donation, which becomes very probable for silicon on the assumption of a $p_\pi$-$d_\pi$ bond. In some cases one observes only one or two resonance lines. For the ionic sodium and diethylthallium azides this is to be expected. For the trimethyltin and trimethyllead azides, however, the structure of similar compounds must be considered. Thus planar trimethyltin groups with bridging anions forming long chains are often found. The vibrational spectra of the azide of trimethyltin suggest a similar structure. From the $^{14}$N NMR data for this compound structure 1 is proposed, where the two outer nitrogen atoms are magnetically equivalent and a spectrum similar to that of an ion is recorded.

$$
\begin{array}{c}
\diagdown \phantom{xx} \diagup \\
\text{Sn} \\
\diagup \phantom{x} \diagdown \\
\diagdown \phantom{xxx} \diagup \\
\phantom{xx} \text{N} = \text{N} = \text{N} \\
\diagup \phantom{xxx} \diagdown \\
\text{Sn} \\
\diagup \phantom{x} \diagdown \\
\diagdown
\end{array}
$$

1

The nonappearance of the third line or the second and third lines in the spectra of the boron and aluminum compounds, respectively, can be explained only by the presence of these two elements. They also possess an electric quadrupole moment giving rise to a further broadening of the lines to such an extent that they can no longer be found. Only the $^{14}$N NMR spectrum of dimethylarsenic azide is inexplicable. This compound is a monomeric volatile liquid and there is no evidence of permanent association. One can only suggest that a rapid exchange takes place similar to that of the halides of trivalent phosphorus—rapid on the NMR, but not on the infrared time scale.

## References

1.   J. Müller, *J. Organometal. Chem.*, **51**, 119 (1973).
2.   T. Kanda, Y. Saito, and K. Kawamura, *Bull. Chem. Soc. Japan*, **35**, 172 (1962).
3.   R. A. Forman, *J. Chem. Phys.*, **39**, 2393 (1963).
4.   M. Witanowski, *J. Am. Chem. Soc.*, **90**, 5683 (1968).
5.   W. Beck, W. Becker, K. F. Chew, W. Derbyshire, N. Logan, D. M. Revitt, and D. B. Sowerby, *J. Chem. Soc. Dalton*, 245 (1972).

# CHAPTER 6

# STRUCTURAL EFFECTS ON THE ONE-BOND $^{15}$N-H COUPLING CONSTANT

Theodore Axenrod

*Department of Chemistry*
*The City College of the City University of New York*

## INTRODUCTION

Steric, electronic, and geometric effects have a profound influence on spin-coupling properties and an understanding of the relationship between these parameters is an important consideration in the elucidation of molecular structures. Interproton couplings[1] and, to a lesser extent, one-bond $^{13}$C-H couplings[2] have been exploited with greatest advantage in structure determinations, but the factors that influence the one-bond $^{15}$N-H coupling constant have received considerably less attention. There are several straightforward reasons for this.

Of the two naturally occurring isotopes of nitrogen, $^{14}$N and $^{15}$N, only the less abundant (0.36%) $^{15}$N nucleus is generally amenable to NMR study. This is particularly true where the determination of spin-coupling constants is of interest; except for a few instances where the field gradient at $^{14}$N is low,[3] the quadrupole-induced relaxation of the latter nucleus leads to broadened resonance lines and even complete collapse of spin-spin multiplets. Similar considerations apply to carbon; the $^{12}$C nucleus has no nuclear spin and only the less abundant $^{13}$C

nucleus (1.1%) can be observed in an NMR experiment. Low
sensitivity is a characteristic shared by both $^{13}$C and $^{15}$N nuclei, and the measurement of their respective couplings to hydrogen is most conveniently done from the proton spectra.
However, in the case of $^1J(^{13}C\text{-H})$ the $^{13}$C satellites are relatively easy to observe, whereas the threefold lower natural
abundance of $^{15}$N necessitates the use of enriched samples in
order to determine $^1J(^{15}N\text{-H})$.[4]

Another difficulty that is frequently encountered in the de-
termination of $^1J(^{15}N\text{-H})$, but not in the case of $^1J(^{13}C\text{-H})$, arises
from the exchange of protons attached to nitrogen.[5] Elimina-
tion of this exchange can usually be achieved by rigorous drying
of the sample,[6] resorting to reduced temperatures for spectral
measurement, and removal of traces of catalytic agents present
in the solvents employed. However, even these measures may
fail with the more basic nitrogen compounds where the exchange may be largely autocatalytic.[7]

This paper surveys the variety of different compounds in
which structural factors have been found to influence the mag-
nitude of $^1J(^{15}N\text{-H})$. Correlations between $^1J(^{15}N\text{-H})$ and certain
structural parameters have been established which in many in-
stances can readily be accounted for in terms of present theory.[4] Examples drawn from the recent literature are presented
to illustrate the application of $^{15}$N-H coupling information to
some problems of molecular structure determination.


## $^{15}$N-H COUPLING IN DIFFERENT FUNCTIONAL GROUPS

It is now fairly well accepted that the principal contribution
to the nuclear spin-spin interaction between elements of the
first row of the periodic table arises from the Fermi contact
term.[8] For coupling between directly bonded $^{15}$N-H nuclei,
this view is reinforced by the excellent correspondence that
has been found between the amount of $s$ character in the nitrogen orbitals and the coupling constant.[9] Table 6-1 lists some
representative $^1J(^{15}N\text{-H})$ values for typical compounds in which

TABLE 6-1.   Some typical $^{15}$N-H coupling constants

| Compound | Solvent | Hybridization | $^1J(^{15}$N-H),(Hz) |
|---|---|---|---|
| —$^{15}\overset{+}{N}H_3$ | HFSO$_3$ | $sp^3$ | 76.9 |
| | HFSO$_3$ | $sp^2$ | 96.0 |
| —C≡$^{15}$N$^+$-H | HFSO$_3$-SbF$_5$-SO$_2$ | $sp$ | 136.0 |

the nitrogen hybridization is varied.  A plot of these data is found to be linear as illustrated in Fig. 6-1.  Indeed, the usefulness of $^1J(^{15}$N-H) values as indices of the nitrogen hybridization has been pointed out by at least two independent groups of workers who advanced similar empirical equations.  Binsch and co-workers[9] adduced the relationship

$$\%s = 0.43\,^1J(^{15}\text{N-H}) - 6 \,,$$

whereas Bourn and Randall[10] using slightly different model compounds proposed an equation

$$\%s = 0.34\,^1J(^{15}\text{N-H}) \,,$$

which has no intercept.

Both solvent and substituents do have an effect on $^1J(^{15}$N-H), which is not obvious from these empirical equations.  Table 6-2 illustrates the effect of N-alkyl substitution in amines.[11] Although the values of $^1J(^{15}$N-H) for a series of methylamines all fall in the range expected for nominal $sp^3$ hybridization of the nitrogen, careful examination does, however, reveal that $^1J(^{15}$N-H) increases slightly with increasing methyl substitu-

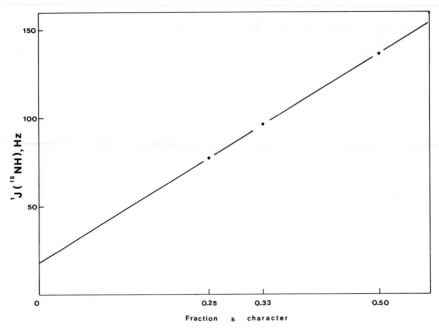

Figure 6-1.  Plot of $^1J(^{15}N\text{-}H)$ vs nitrogen hybridization.

TABLE 6-2.  Effect of $N$-alkyl substitution
on $^1J(^{15}N\text{-}H)$ in amines

| Compound | $^1J(^{15}N\text{-}H)$, (Hz) |
| --- | --- |
| $^{15}NH_3$ | 61.2 |
| $CH_3$-$^{15}NH_2$ | 64.5 |
| $CH_3$-$^{15}NH$-$CH_3$ | 67.0 |
| ⬡—$^{15}NH_2$ | 82.6 |
| ⬡—$^{15}NH$-$CH_3$ | 87.0 |

84

TABLE 6-3. Some representative $^{15}$N-H
coupling constants in ring-
substituted aniline derivatives

| $G$ —$^{15}$NH$_2$ | $^1J(^{15}$N-H), (Hz) (in DMSO) |
|---|---|
| 2,4-(NO$_2$)$_2$ | 92.6 |
| 4-NO$_2$ | 89.6 |
| 3-NO$_2$ | 86.2 |
| 3-Cl | 85.1 |
| 4-Br | 84.0 |
| 3-CH$_3$O | 83.0 |
| H | 82.6 |
| 4-CH$_3$ | 81.4 |
| 4-CH$_3$O | 79.4 |
| 4-N(CH$_3$)$_2$ | 78.8 |

tion. The same also appears to be true for aniline derivatives.[12] Several possible explanations for this effect have been
considered. Although the trend can be accounted for satisfactorily by postulating a small increase in the $s$ character of the
$^{15}$N-H bond in going from $^{15}$NH$_3$ to (CH$_3$)$_2$$^{15}$NH, precise bond-
angle data are not available to support this view. Changes in
the effective nuclear charge[13] or variations in the electronic
excitation energy are other possibilities.

The one-bond $^{15}$N-H coupling constants in aniline and its
ring-substituted derivatives have received considerable attention[14, 15]; a representative compilation appears in Table 6-3.
These $^{15}$N-H coupling constants are found to be intermediate
between those expected for tetrahedral ($sp^3$) and trigonal ($sp^2$)
nitrogen. Ring substituents that diminish $\pi$-electron density

on the nitrogen atom cause an increase in $^1J(^{15}N\text{-}H)$ which is consistent with a change in hybridization of the nitrogen atom and an alteration in the geometry of the amino group. The chemical shifts of the amino protons and of $^{15}N$ are interpreted in a similar manner. [15] It is interesting to note that a methyl substituent attached directly to nitrogen has an effect on $^1J(^{15}N\text{-}H)$ opposite in direction to that found for a *para*-methyl substituent.

The effects of solvents on $^1J(^{15}N\text{-}H)$ are intuitively reasonable, but are not readily subject to quantitative explanation. Paolillo and Becker[7] studied the effect of varying the solvent medium from one of low to high dielectric; their data are summarized in Table 6-4. Although $^1J(^{15}N\text{-}H)$ increases in solvents of increasing hydrogen-bonding ability and dielectric constant, there is no simple correlation between these parameters and it is clear that other less well defined solvent interactions are important.

Solvent effects on $^1J(^{15}N\text{-}H)$ in *ortho*-substituted anilines have been examined[16] and these show an interesting trend which contrasts with the results of similar studies of *meta*- and *para*-substituted anilines. [15] In *para*-substituted anilines, $^1J(^{15}N\text{-}H)$ is found to be consistently (about 4 Hz) larger in dimethyl sulfoxide solution than in chloroform. For the *ortho*-substituted anilines listed in Table 6-5, the difference in the coupling constants in the two solvents, $\Delta J(^{15}N\text{-}H)$, is found to decrease regularly as the hydrogen-bonding ability of the *ortho* substituent increases, finally reaching a limiting value of 0.7 Hz in *ortho*-nitroanilines. These findings have been interpreted in terms of the amino group being intermolecularly hydrogen bonded to the solvent in dimethyl sulfoxide, whereas in chloroform a strong intramolecular hydrogen bond between the *ortho* substituent and the amino group persists. Accordingly, the $\Delta J(^{15}N\text{-}H)$ value is taken as a measure of the hydrogen-bonding ability of the *ortho* substituent.

Delocalization of the lone pair of electrons on nitrogen in amides produces substantial double bond character in the central C-N bond and the associated barrier to internal rotation

TABLE 6-4. Solvent effects on $^1J(^{15}\text{N-H})$ in aniline

| Solvent | $\epsilon$ | $^1J(^{15}\text{N-H})$, (Hz) |
|---------|-----------|------------------------------|
| $C_6D_{12}$ | 2.02 | 78.0 |
| $CCl_4$ | 2.22 | 78.0 |
| $CDCl_3$ | 5.05 | 78.0 |
| Dioxane-$d_8$ | 2.20 | 80.6 |
| Pyridine-$d_5$ | 12.3 | 81.4 |
| Acetone | 20.7 | 82.1 |
| DMF-$d_7$ | 36.7 | 82.3 |
| DMSO-$d_6$ | 48.9 | 82.3 |

TABLE 6-5. Solvent effects on $^{15}\text{N-H}$ coupling constants in *ortho*-substituted anilines

$^{15}\text{NH}_2$

| G | $^1J(^{15}\text{N-H})$, (Hz) | | $\Delta J$ |
|---|------------------------------|---|------------|
| | in $CDCl_3$ | in DMSO | |
| 2-$NO_2$, 4-Cl | 91.1 | 91.8 | 0.7 |
| 2-$NO_2$ | 90.3 | 91.0 | 0.7 |
| 2-COPh | 88.1 | 89.3 | 1.2 |
| 2-Cl, 4-$NO_2$ | 89.2 | 90.5 | 1.3 |
| 2-$CF_3$ | 83.6 | 86.5 | 2.9 |
| 2-Br | 81.4 | 84.3 | 2.9 |
| 2-F | 80.1 | 82.5 | 3.4 |
| H | 78.6 | 82.6 | 4.0 |

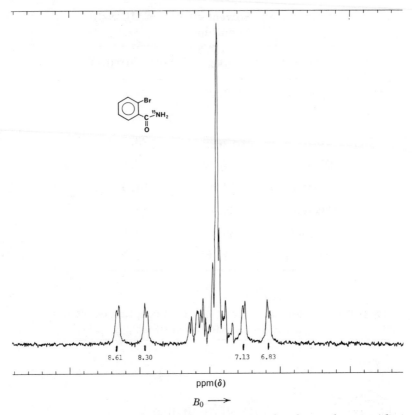

Figure 6-2.   60 MHz proton NMR spectrum of *o*-bromobenzamide-$^{15}$N in DMSO at +10 °C.

leads to nonequivalence of the substituents on nitrogen. The phenomenon is fairly general and well recognized,[17] as shown in the spectrum of *ortho*-bromobenzamide in Fig. 6-2. Two $^{15}$N-H couplings corresponding to the hydrogen atoms oriented *cis* and *trans* to the carbonyl oxygen atom are present, as is evidence of geminal $H_a$-$^{15}$N-$H_b$ coupling. Data for several benz-amide derivatives are summarized in Table 6-6.[12] Although the *trans* coupling is invariably slightly larger, there appears to be no trend in either the *cis* or *trans* $^1J(^{15}$N-H) values that can be related to an electronic effect arising from the ring substitu-ents. This is surprising in view of the fact that the $^{15}$N reso-

TABLE 6-6.  Substituent effects on $^{15}$N-H coupling
constants in benzamide derivatives

| G | $^1J(^{15}$N-H), (Hz) | |
|---|---|---|
| | cis | trans |
| 2-NO$_2$ | 88.4 | 89.6 |
| 2-Br | 88.2 | 90.0 |
| 3-Br | 88.2 | 88.8 |
| 4-I | 88.1 | 89.2 |
| 4-F | 88.2 | 88.6 |
| 2-CF$_3$ | 88.1 | 89.2 |
| 3-CF$_3$ | 88.7 | 89.1 |
| 3,5-(CH$_3$)$_2$ | 88.0 | 88.8 |

nances in these compounds do show a small but regular downfield shift induced by electronegative substituents.[18]

Another interesting demonstration of restricted rotation is
provided by the low-temperature spectrum of 2-nitro-3,4-di-
methylaniline shown in Fig. 6-3.  In this case $\pi$ bonding between
amino nitrogen and the phenyl group leads to restricted rotation
about the aryl-nitrogen bond and the nonequivalence of the two
amino protons.[19]  Although intramolecular hydrogen bonding
might be expected to favor the nonequivalence of the amino protons, the small difference in chemical shift (< 1 ppm) suggests
that hydrogen bonding does not play a major role.

In strong acid media nitriles undergo N protonation to give
nitrilium ions in which the hybridization of the nitrogen is $sp$.

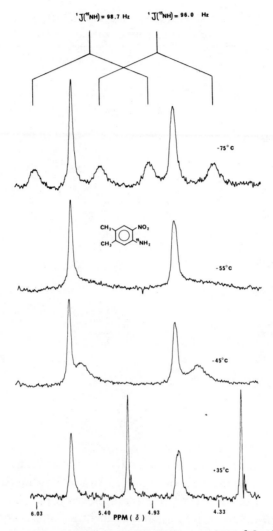

Figure 6-3. Temperature-dependent NMR spectra of 2-nitro-4,5-dimethylaniline in acetone-glyme solvent. Chemical shift values are reported in ppm downfield from acetone.

TABLE 6-7.  Substituent effects on $^{15}$N-H
coupling constants in protonated
benzonitrile derivatives

| G | $^1J(^{15}\text{N-H})$, (Hz) |
|---|---|
| 4-CH$_3$ | 136.2 |
| H | 136.2 |
| 2-Br | 135.1 |
| 4-CH$_3$O | 134.7 |
| 4-NO$_2$ | 132.6 |
| 4-NO$_2$, 2-Cl | 132.5 |
| 3,5-(CH$_3$O)$_2$ | 131.8 |

A series of benzonitrile derivatives has been studied[20] in HFSO$_3$-SbF$_5$-SO$_2$ solution at low temperatures and the $^1J(^{15}\text{N-H})$ data for these compounds are shown in Table 6-7.  In contrast to the benzamides, $^1J(^{15}\text{N-H})$ in these benzonitrilium ions is found to vary with the nature of the ring substituent.  Electron-withdrawing substituents bring about a regular, although small, decrease in the magnitude of $^1J(^{15}\text{N-H})$.  The explanation for this is not clear, especially in view of the accompanying observations that both the $^{15}$N and $^{15}$N-bound proton resonances experience deshielding as a function of electron-withdrawing substituents. Moreover, it is intriguing that the substituent effect on the corresponding $^1J(^{13}\text{C-H})$ in the isoelectronic phenylacetylenes is opposite in direction to that on $^1J(^{15}\text{N-H})$ in the benzonitrilium ions.[20]

## APPLICATIONS

The composition of the tautomeric equilibrium in Schiff bases has been successfully studied by means of $^{15}N$ substitution.[21] Since the keto (1a) and enol (1b) forms arise from an intramolecular proton exchange, the observed $^{15}N$-H coupling constant may be taken as the weighted average ($P_N$ and $P_O$ are the respective mole fractions) of the coupling constant in each tautomer:

1a                                    1b

$$J(\text{obs}) = J(^{15}NH)P_N + J(OH)P_O \ .$$

The method affords a direct measure of the composition of the tautomeric equilibrium.

In this manner $N$-methyl-2-hydroxy-1-naphthaldehydeimine-$^{15}N$ has been shown to exist predominantly in the ketoamine tautomer (2), whereas substitution of methyl by phenyl makes the enolimine tautomer (3) the dominant species at equilibrium.

2                                    3

Finally, it is frequently of interest to determine the basic site in a molecule, and where nitrogen is involved the problem is amenable to isotopic substitution. For example, in strong acid double-labeled urea has been shown to form a dication corresponding to **4**, where both oxygen and nitrogen sites are protonated.[22] Two different $^1J(^{15}$N-H) values are found (96.8 and 76.6 Hz) which are in agreement with the presence of $sp^2$ and $sp^3$ nitrogen atoms.

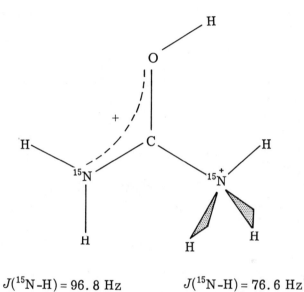

$J(^{15}$N-H) = 96.8 Hz          $J(^{15}$N-H) = 76.6 Hz

## References

1.  S. Sternhell, *Quart. Rev. London,* **23,** 236 (1969).
2.  G. C. Levy and G. L. Nelson, *Carbon-13 Nuclear Magnetic Resonance for Organic Chemists,* Wiley-Interscience, New York, 1972, p. 56.
3.  I. D. Kuntz, Jr., P. von R. Schleyer, and A. Allerhand, *J. Chem. Phys.,* **35,** 1533 (1961).
4.  T. Axenrod, "Correlations of $^{15}$N Coupling Constants with Molecular Structure," in *Nitrogen NMR,* G. Webb and M. Witanowski, Eds., Plenum Press, London, 1973.

5. C. S. Springer and D. W. Meek, *J. Phys. Chem.*, **70**, 481 (1966).
6. K. L. Henold, *Chem. Commun.*, 1340 (1970).
7. L. Paolillo and E. D. Becker, *J. Magn. Resonance*, **2**, 168 (1970).
8. W. McFarlane, *Quart. Rev. London*, **23**, 187 (1969).
9. G. Binsch, J. B. Lambert, B. W. Roberts, and J. D. Roberts, *J. Am. Chem. Soc.*, **86**, 5564 (1964).
10. A. J. R. Bourn and E. W. Randall, *Mol. Phys.*, **8**, 567 (1964).
11. M. Alei, Jr., A. E. Florin, W. Litchman, and J. F. O'Brien, *J. Phys. Chem.*, **75**, 932 (1971).
12. T. Axenrod, P. S. Pregosin, and M. J. Wieder, unpublished results.
13. D. M. Grant and W. M. Litchman, *J. Am. Chem. Soc.*, **87**, 3994 (1965).
14. M. R. Bramwell and E. W. Randall, *Chem. Commun.*, 250 (1969).
15. T. Axenrod, P. S. Pregosin, M. J. Wieder, E. D. Becker, R. B. Bradley, and G. W. A. Milne, *J. Am. Chem. Soc.*, **93**, 6536 (1971).
16. T. Axenrod and M. J. Wieder, *J. Am. Chem. Soc.*, **93**, 3541 (1971).
17. T. H. Siddall and W. E. Stewart, in *Progress in Nuclear Magnetic Resonance Spectroscopy*, J. W. Emsley, J. Feeney, and L. H. Sutcliffe, Eds., Vol. 5, Pergamon Press, Oxford, 1969, p. 33.
18. P. S. Pregosin, E. W. Randall, and A. I. White, *J. Chem. Soc. B*, 513 (1972).
19. T. Axenrod and M. T. Wieder, unpublished results.
20. T. Axenrod and F. Macchia, unpublished results.
21. G. O. Dudek and E. P. Dudek, *J. Am. Chem. Soc.*, **86**, 4283 (1964); **88**, 2407 (1966).

CHAPTER 7

# SPIN COUPLING BETWEEN DIRECTLY BONDED $^{15}$N NUCLEI

Suryanarayana Bulusu and Joseph R. Autera

*Explosives Division*
*Feltman Research Laboratory*
*Picatinny Arsenal*
*Dover, New Jersey*

and

Theodore Axenrod

*Department of Chemistry*
*The City College of The City University of New York*

## INTRODUCTION

The study of the NMR parameters of the $^{15}$N nucleus holds considerable potential as a means of elucidating structural details of organic compounds. However, such investigations have until recently been limited by the low natural abundance and unfavorable intrinsic sensitivity of this nucleus. The advent of pulse Fourier transform NMR spectroscopy now renders the $^{15}$N nucleus reasonably amenable to study.

Substantial interest in $^{15}$N NMR has been evident in recent years, and the field has been the subject of several comprehensive reviews. [1,2] In addition to $^{15}$N chemical shifts in a variety of organic compounds, the available data consist mainly of coupling constants determined for nitrogen and hydrogen nu-

clei separated by one to three bonds.[3] Only very limited excursions have been made into the area of coupling constant determinations between nitrogen and nuclei other than hydrogen.[3]
The latter have largely been confined to $^{13}C$.

With the exception of trans-azoxybenzene[4] and the recently
reported p-hydroxyazobenzene,[5] coupling constants between directly bonded $^{15}N$ nuclei are unknown and the structural features
that influence this coupling have not been systematically investigated. The established relationships of $^{1}J(^{15}N\text{-}H)$ to nitrogen
bond hybridizations[3] and the emerging patterns relating electronic, geometric, and steric interaction effects to $^{13}C\text{-}^{13}C$
couplings[6] encourage the view that the study of $^{15}N\text{-}^{15}N$ couplings
would be an attractive approach to the elucidation of bonding
details in the N-N bond.

Accordingly, a study was undertaken in these laboratories
aimed initially at gathering coupling constant data for a variety
of N-N bonds in different functional groups and diverse molecular environments. With this information available, it is hoped
that greater insight into the fundamental nature of the N-N bond
can be gained. Moreover, as such data are accumulated they
should be useful in making spectral assignments in other $^{15}N$
NMR studies.

To facilitate the observation and accurate measurement of
the $^{15}N\text{-}^{15}N$ spin-spin interactions, it is necessary to employ
compounds having each of the nitrogen centers enriched in the
$^{15}N$ isotope. These multiple-labeled materials are prepared by
the conventional synthetic procedures and the $^{15}N$ isotope enrichment is a minimum of 95%; in many cases 99% enrichment
is achieved. The results of these investigations together with
a preliminary discussion of the indicated trends are presented
here.

## Results and Discussion

At the present time a total of 10 directly bonded $^{15}N\text{-}^{15}N$
couplings are available and eight of these are derived from the
present investigation. Of these latter compounds, five are

nitramines which were either prepared or were readily available from related studies in these laboratories concerned with the structural chemistry of the nitramino group and its associated explosive properties. The measured $^{15}$N-$^{15}$N coupling constants for these five nitramines are summarized in Table 7-1. In Table 7-2 three additional $^{15}$N-$^{15}$N couplings studied in the present investigation along with two obtained from the literature are presented.

Several observations can be made from the $^{15}$N-$^{15}$N coupling constant data presented in these tables. First, despite the fact that these are coupling constants between directly bonded $^{15}$N nuclei the magnitude of the coupling constants is considerably smaller than that found for the one-bond $^{15}$N-H coupling (60–140 Hz). This is not unexpected in view of the smaller gyromagnetic ratio of the $^{15}$N nucleus compared with that of hydrogen.

TABLE 7-1.   $^{15}$N-$^{15}$N spin coupling constants in nitramines

| Nitrogen-15 Labeled Nitramine[a] | $J(^{15}$N-$^{15}$N) (Hz) |
|---|---|
| Dimethylnitramine | 4.9[c] |
| 1,3,5-Trinitro-1,3,5-triaza-cyclohexane (RDX)[b] | 8.9[d] |
| 1,3,5,7-Tetranitro-1,3,5,7-tetrazacyclooctane (HMX) | 4.5 |
| 3,7-Dinitro-1,3,5,7-tetraza-bicyclo (3,3,1)nonane (DPT) | 8.5 |
| Methyl (2,4,6-trinitrophenyl)-nitramine (Tetryl) | 4.9 |

[a]All nitrogens are $^{15}$N labeled.
[b]Names in parentheses are those used in the explosives field.
[c]Solvent is deuterochloroform.
[d]For this and the following three compounds the solvent is dimethyl sufoxide-$d_6$.

**TABLE 7-2.** $^{15}$N–$^{15}$N NMR coupling constants in different structures

| Nitrogen-15 Labeled Compound[a] | | $J(^{15}$N–$^{15}$N) (Hz)[b] | Approximate Hybridization of N–N Bond |
|---|---|---|---|
| Phenylhydrazine | $\phi - NH - NH_2$ | 6.7 | $sp^3 - sp^3$ |
| p-Nitrobenzaldehyde-phenylhydrazone | $\phi - NH - N{=}CH - \phi - (p-)NO_2$ | 10.7 | $sp^3 - sp^2$ |
| trans-Azoxybenzene[5] | $\phi - N{=}N - \phi$ with O | 13.7 (in ether) | $sp^2 - sp^2$ |
| p-Hydroxyazobenzene[6] | $\phi - N{=}N - \phi - (p-)OH$ | $15.0 \pm 1$ (in acetone) | $sp^2 - sp^2$ |
| Dibenzylnitrosamine | $(\phi CH_2)_2 N^+{=}N - O^-$ | 19.0 | $sp^2 - sp^2$ |

[a]All N–N bonds are fully $^{15}$N labeled.
[b]Solvent is dimethyl sulfoxide-$d_6$ except where noted.

The range of $^{15}$N-$^{15}$N coupling constants is about 5–20 Hz and it is comparable to that observed for similar couplings between directly bonded $^{15}$N and $^{13}$C nuclei.[3]

If the principal contribution to the nuclear spin-spin interaction between elements of the first row of the periodic table is considered to arise from the Fermi contact term as predicted from theory and MO calculations,[7] the coupling would be related to the $s$-electron densities at each center. A consequence of such a dependence on the valence electron densities at the nuclei is that the measured coupling constants can in principle be correlated with the hybridization and/or effective nuclear charge at the nuclei.[7] Although the data are sparse, a trend of this nature is indicated in Table 7-2.

For example, a comparison of the observed $^{15}$N-$^{15}$N coupling in phenylhydrazine (6.7 Hz) with that found in $p$-nitrobenzaldehyde phenylhydrazone (10.7 Hz), shows a significant increase (4 Hz) in the coupling. This presumably is a result of the incorporation of the terminal amino nitrogen of phenylhydrazine in an imine bond. Thus the change toward increased $sp^2$ hybridization at the latter nitrogen atom parallels the increased coupling. In *trans*-azoxybenzene or $p$-hydroxyazobenzene where both nitrogen atoms may be taken as $sp^2$ hybridized the magnitude of the coupling is enhanced even further (15 Hz) and finally in the case of dibenzylnitrosamine for which there is much evidence for substantial $p$-$\pi$ bonding between the nitrogens,[8] the largest coupling (19 Hz) to date has been found. If this interpretation is correct, one may predict even greater $^{15}$N-$^{15}$N coupling in the phenyldiazonium ion where both nitrogens are presumably $sp$ hybridized. This coupling, however, is not known at present.

It is noteworthy from the available data that the $^{15}$N-$^{15}$N coupling found in dibenzylnitrosamine is strikingly larger than in the nitramines (5–9 Hz). A possible explanation is that in dimethylnitramine, for example, the requirement of adjacent positive charges renders the contributing polar structure 1 much less important than the analogous structure 2 in the case of dibenzylnitrosamine. Such a difference is also consistent

**1**                  **2**

with the results of electron diffraction studies, which show that the N-N bond distance is considerably shorter in dimethylnitrosamine (1.41 Å) than in dimethylnitramine (1.57 Å).[9] Also, in spite of the apparently similar N-N bonding in *trans*-azoxybenzene and dibenzylnitrosamine as shown in these structures, the former shows a significantly different coupling constant, presumably due to the effect of phenyl conjugation on the hybridization.

The data presented in Table 7-1 on the five nitramines may be divided into groups, one in which the $^{15}N$-$^{15}N$ couplings are found to be about 5 Hz and the other about 9 Hz. The former includes the eight-membered ring compound 1, 3, 5, 7-tetranitro-1, 3, 5, 7-tetrazacyclooctane (**3**), and the latter includes the structurally similar six-membered ring compound 1, 3, 5-trinitro-1, 3, 5-triazacyclohexane (**4**), as well as the bridged tetrazabicyclononane (**5**). It is of particular interest that (**3**) and (**4**) show a significant difference in $^{15}N$-$^{15}N$ coupling (4.5 and 8.9 Hz, respectively), although they are structurally very similar. These differences appear to arise from the different average conformations that these ring systems can assume.

**3**               **4**               **5**

Various physical measurements[10] and MO calculations[11] indicate that the molecular geometry of (4) in solution has a higher degree of symmetry than in the crystalline state. In solution the molecule has approximately $C_{3v}$ symmetry with some distortion associated with the nitro groups. Over a wide range of temperature and in different solvents the $^1$H NMR spectrum shows the methylene hydrogens[10] and the ring nitrogens[12] to be chemically equivalent. Thus in solution the molecule is viewed as corresponding to an idealized chair conformation with three planar $_C^C{>}\text{N-N}{<}_O^O$ groups and the ring undergoes rapid interconversion with the cyclic nitrogens oscillating about the planar $(sp^2)$ position.[10]

Although the crystal structure of (3) has been determined,[13, 14] much less is known about its molecular geometry in solution. The presence of only a single resonance in the proton spectrum[15] indicates that the ring system is undergoing rapid interconversion. However, molecular models reveal that the ring system cannot accommodate four planar $_C^C{>}\text{N-N}{<}_O^O$ groups without introducing severe transannular nonbonded interactions. In contrast to the situation in (4), this may result in greater average distortion or rotation about the N-N bond, diminishing the $\pi$ bonding between the nitrogens. This explanation is consistent with the smaller $^{15}$N-$^{15}$N coupling found in (3). Moreover, if one considers the tetrazabicyclononane (5), the $_C^C{>}\text{N-N}{<}_O^O$ groups are constrained into a rigid chair conformation and molecular models indicate little or no impediment to planarity within the latter groups. On this basis, one would expect $^{15}$N-$^{15}$N coupling comparable with that found in (4), and the experimental result supports this view.

It should be pointed out that a planar framework is also expected in both dimethylnitramine[16, 17] and in methyl (2, 4, 6-trinitrophenyl) nitramine,[18] but the observed $^{15}$N-$^{15}$N coupling constants in these compounds appears to be inconsistent with the values and the interpretations suggested for the cyclic nitramines. There is some question about whether a direct comparison with the cyclic nitramines is warranted in view of the structural dissimilarities that exist. Further work is in progress to test the generality of these observations.

## References

1.  E. W. Randall and D. G. Gillies, in *Progress in Nuclear Magnetic Resonance Spectroscopy*, J. W. Emsley, J. Feeney, and L. H. Sutcliffe, Eds., Vol. 6, Pergamon Press, New York, 1971, p. 119.

2.  R. Lichter, in *Determination of Organic Structures by Physical Methods*, J. J. Zuckerman and F. C. Nachod, Eds., Vol. 4, Academic Press, New York, 1972.

3.  T. Axenrod, in *Nitrogen NMR*, G. Webb and M. Witanowski, Eds., Plenum Press, London, 1973.

4.  G. Binsch, J. B. Lambert, B. W. Roberts, and J. D. Roberts, *J. Am. Chem. Soc.*, **86**, 5564 (1964).

5.  N. N. Bubnov, K. A. Bilevitch, L. A. Poljakova, and O. Yu Okhlobystin, *Chem. Commun.*, 1058 (1972).

6.  J. Bartuska and G. Maciel, *J. Magn. Resonance*, **5**, 2 (1971).

7.  W. McFarlane, *Quart. Rev. London*, **23**, 187 (1969).

8.  T. Axenrod, P. S. Pregosin, and G. W. A. Milne, *Chem. Commun.*, 702 (1968).

9.  R. Stolevik and P. Rademacher, *Acta Chem. Scand.*, **23**, 660, 672 (1969).

10. A. Filhol, C. Clement, M-T. Forel, J. Paviot, M. Rey-Lafon, G. R. Choux, C. Trinquecoste, and J. Cherville, *J. Phys. Chem.*, **75**, 2056 (1971).

11. M. K. Orloff, P. A. Mullen, and F. C. Rauch, *J. Phys. Chem.*, **74**, 2189 (1970).

12. M. Witanowski, T. Urbanski, and L. Stefaniak, *J. Am. Chem. Soc.*, **86**, 2569 (1964).

13. H. H. Cady, A. C. Larson, and D. T. Cromer, *Acta Cryst.*, **16**, 617 (1963).

14. C. S. Choi and H. P. Boutin, *Acta Cryst.*, **B26**, 1235 (1970).

15.  A. Lamberton, I. O. Sutherland, J. E. Thorpe, and H. M. Yusuf, *J. Chem. Soc.*, **1968**, 6.
16.  (*a*) J. Stals, C. G. Barraclough, and A. C. Buchanan, *Trans. Faraday Soc.*, **65**, 904 (1969); (*b*) J. Stals, *Aust. J. Chem.*, **22**, 2505 (1969).
17.  W. Costain and E. G. Cox, *Nature*, **160**, 826 (1947).
18.  H. H. Cady, *Acta Cryst.*, **23**, 601 (1967).

# CHAPTER 8

# INFLUENCE OF LONE-PAIR ORIENTATION ON SOME NUCLEAR SPIN-SPIN COUPLING CONSTANTS. INDO MOLECULAR ORBITAL CALCULATIONS OF NITROGEN-PROTON COUPLING CONSTANTS OVER TWO AND THREE BONDS

Rod Wasylishen*

*Department of Chemistry*
*University of Manitoba*
*Winnipeg*

## INTRODUCTION

A considerable body of experimental data is now available which indicates that nuclear spin-spin coupling constants involving nitrogen or phosphorus nuclei are often markedly influenced by the spatial orientation of their lone-pair elec-

*Holder of an NRCC postdoctoral fellowship, 1972–1973. Present address: National Institute of Arthritis, Metabolism and Digestive Diseases, National Institutes of Health, Bethesda, Maryland.

trons.[1-6]* For example, in the case of $^{15}$N-formaldoxime (1) Lehn and co-workers[7] have reported $^2J(^{15}$N-H$_A) = -13.88$ Hz while $^2J(^{15}$N-H$_B) = 2.68$ Hz. For benzyl nitrite, Axenrod, Wieder, and Milne[8] have observed $^3J(^{15}$N-H$) \simeq 0.0$ Hz in (2a) and $\pm 2.4$ Hz in (2b).

2a                         2b

Several other examples of stereospecific $^{15}$N-H coupling constants are presented later.

One of the first examples of stereospecific $^2J(^{31}$P-H) values was reported by Gagnaire et al.[9] For 3, 4-dimethyl-1-phenyl-phosphacyclopentene (3) they observed $^2J(^{31}$P-H$_A) = \pm 25$ Hz and

3

$^2J(^{31}$P-H$_B) = \mp 6$ Hz. Bushweller and co-workers[10] have recently reported stereospecific $^3J(^{31}$P-H) values in *tert*-butyldichloro-phosphine (4). At $-160\,°$C, rotation of the *tert*-butyl group is

4

*In this chapter the lone pair is treated as having directional character, in line with a simple hybridization model.

slow and $^3J(\underline{P}\text{-}*C\underline{H}_3)_{trans} = \pm 4.5$ Hz and $^3J(\underline{P}\text{-}C\underline{H}_3)_{gauche} = \pm 20.9$ Hz.

Recently Pople's group[11] has demonstrated that an SCF finite perturbation method using INDO wave functions is capable of correctly predicting the relative magnitudes of the stereo-specific $^1J(^{13}C\text{-}H)$ values in 5a and 5b.

|                          | 5a         | 5b         |
|--------------------------|------------|------------|
| $^1J(^{13}C\text{-}H)_{obs}$  | 163.0 Hz   | 177.0 Hz   |
| $^1J(^{13}C\text{-}H)_{calc}$ | 160.1 Hz   | 182.4 Hz   |

In view of the success of Pople's SCF MO theory in this and other cases[12] we have employed this same method for the calculation of $^{15}N\text{-}H$ couplings in a wide variety of compounds that are usually considered as having a localized lone pair. $J(N\text{-}H)$ values have been computed as a function of various angles in order to study the detailed angular dependence of these couplings. The calculated results are compared with available experimental data and an attempt is also made to compare some of the trends calculated for N-H coupling constants with those observed for $^{31}P\text{-}H$ couplings. Results of a more detailed study of calculated nitrogen-proton coupling constants have been given elsewhere.[4]

## CALCULATION OF COUPLING CONSTANTS

In fluids nuclear spins interact with one another via intervening electrons.[13] Ramsey[13] has shown that three types of interactions between the electron and nuclear spins are possible: (1) a magnetic dipole-dipole interaction, (2) an orbital-dipole interaction, and (3) a Fermi contact interaction. When protons are involved the Fermi contact mechanism predominates[14-17]; hence most attempts at calculating X-H coupling constants are based on this mechanism alone. In the calculations discussed here only this interaction is considered.

In the finite perturbation formalism[17-19] the calculation of the Fermi contact contribution to a nuclear spin coupling, $J(A-B)$, involves the calculation of an open-shell molecular orbital wave function under the influence of the perturbation

$$h_B = \left(\frac{8\pi}{3}\right) \beta \mu_B s_B^2(0)$$

due to the presence of a nuclear moment, $\mu_B$. $\beta$ is the Bohr magneton and $s_B^2(0)$ is the $s$-orbital electron density of atom B at the nucleus. The perturbation, $h_B$, at atom B has the effect of introducing a very small spin density, $\rho^{spin}$, throughout the molecular electron system. The nuclear spin-spin coupling constant, $J(A-B)$, is proportional to the value of the spin density $\rho_{sAsA}^{spin}$ transmitted to the second nucleus, A.

$$J(A-B) = \frac{\hbar}{2\pi}\left[\left(\frac{8\pi}{3}\right)\beta\right]^2 \gamma_A \gamma_B s_A^2(0) s_B^2(0)\left[\left(\frac{\partial}{\partial h_B}\right)\rho_{sAsA}^{spin}(h_B)\right]_{h_B=0} . \quad (1)$$

$\rho_{sAsA}^{spin}(h_B)$ is the diagonal element of the spin-density matrix (in the presence of the contact perturbation $h_B$) corresponding to the valence $s$ orbital of atom A, $\gamma_A$ and $\gamma_B$ are the magnetogyric ratios of nuclei A and B, and $s_A^2(0)$ is the valence $s$-orbital electron density of atom A at the nucleus. Using the method of finite differences, Pople et al.[19] have shown that the derivative in equation 1 can be approximated by $\rho_{sAsA}^{spin}(h_B)/h_B$. Thus the coupling constant (in Hz) is approximated by the expression:

$$J(A-B) = \hbar/2\pi\left[\left(\frac{8\pi}{3}\right)\beta\right]^2 \gamma_A \gamma_B s_A^2(0) s_B^2(0)\left[\frac{\rho_{sAsA}^{spin}(h_B)}{h_B}\right] . \quad (2)$$

The error in the approximation has a minimum when $h_B$ is approximately $10^{-3}$ hartrees. Values of $\rho_{sAsA}^{spin}(h_B)$ have been calculated using open-shell semiempirical molecular orbitals at the INDO level of approximation. The coupling constants were calculated using equation 2 with the same parameterization used previously by Pople et al.[12,17] The convergence criteria and geometries used for all calculations have been given elsewhere.[4]


## TWO-BOND NITROGEN-PROTON COUPLING CONSTANTS

In Table 8-1 observed and calculated stereospecific $^2J(^{15}N\text{-}H)$ values are presented for several compounds. Notice that if the lone-pair lobe is directed *cis* to the C-H bond then $^2J(^{15}N\text{-}H)$ is observed and calculated to be large and negative; if the nitrogen lone pair and C-H bond lie *trans* to one another then the absolute value of $^2J(^{15}N\text{-}H)$ is small. To test the generality of this rule the detailed angular dependence of $^2J(^{15}N\text{-}H)$ was calculated for methylamine, methylisocyanate, aziridine, methylenimine, and oxaziridine.[4] Only the results calculated for methylamine and aziridine are presented here.


### Methylamine

In Fig. 8-1, $^2J(^{15}N\text{-}H)$ is plotted as a function of the angle $\theta$, defined in 6a and 6b. This angle can also be described as the dihedral angle defined by the lone-pair N-C fragment and the N-C-H fragment.

6a                                      6b

TABLE 8-1. Observed and calculated stereospecific two-bond [15]N–H coupling constants[a]

| | Experimental Results | | | Calculated Results | |
|---|---|---|---|---|---|
| | Compound | $^2J(^{15}N-H)_{obs}$ | Reference | Compound | $^2J(^{15}N-H)_{calc}$ |
| 1. | (structure) | $-4.8$ to $-5.4$ | 20 | (structure) $\theta = 64°$ | $-5.78$ |
| 2. | (structure) | $\sim 0.0$ | 20 | (structure) $\theta = 64°$ | $0.95$ |
| 3. | (structure) | $H_A\ (-)\ 4.9$ $H_B\ (-)\ 0.3$ | 21 | (structure) $\theta = 68°$ | $H_A\ -9.14$ $H_B\ 0.62$ |
| 4. | (structure) | $H_A\ 2.68$ $H_B\ -13.88$ | 7 | (structure) | $H_A\ -0.77$ $H_B\ -10.90$ $H_C\ -5.50$ |

110

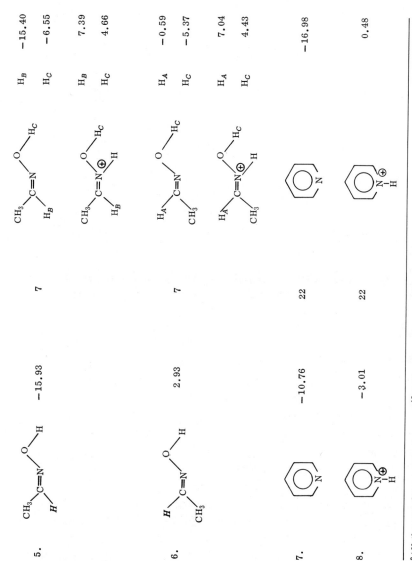

5. 

$CH_3$—C=N—O—H  $H$  −15.93  7

$H_B$  −15.40
$H_C$  −6.55

6. 

$H$—C=N—O—H  $CH_3$  2.93  7

$H_B$  7.39
$H_C$  4.66

$H_A$  −0.59
$H_C$  −5.37

$H_A$  7.04
$H_C$  4.43

7.  −10.76  22  −16.98

8.  −3.01  22  0.48

[a]All the coupling constants are $J(^{15}N–H)$ values in Hz; $J(^{14}N–H)$ values have been converted to $J(^{15}N–H)$ values by multiplying by −1.402757.[23]

111

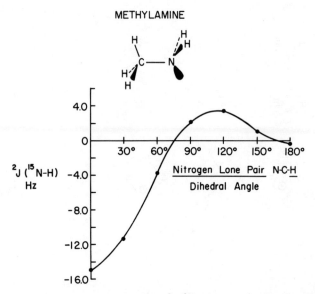

Figure 8-1.  A plot of the calculated $^2J(^{15}N\text{-H})$ versus the dihedral angle defined by the lone-pair N–C–H fragment in methylamine.

For $\theta = 0°$ one of the C–H bonds eclipses the lone pair, and $^2J(^{15}N\text{-H})$ for the proton in that bond is calculated as large and negative, $-14.90$ Hz.  For $\theta = 180°$, one of the C–H bonds sits *trans* to the lone pair and $^2J(^{15}N\text{-H})$ is $-0.40$ Hz.  Notice that $^2J(^{15}N\text{-H})_{cis}$ and $^2J(^{15}N\text{-H})_{trans}$ predicted for methylamine are similar to those observed for unsaturated compounds such as formaldoxime (compound 4 in Table 8-1).

To compare experimental and calculated couplings in methylamine, we need only consider the staggered conformation because the barrier to rotation about the C–N bond is 1.98 kcal/mole.[23]  For the staggered conformation one calculates an average of $-2.70$ Hz while the observed value is $-1.0$ Hz.[24]

Few observations are available for two-bond $^{15}N\text{-H}$ coupling constants in unstrained saturated systems of fixed geometry.  However, it is of some interest to compare calculations for methylamine with those observed for the saturated 1, 3-oxazine (7).  Riddel and Lehn[25] have measured $^2J(^{15}N\text{-H}_{2a})$ and $^2J(^{15}N\text{-H}_{4a})$ as approximately zero while $^2J(^{15}N\text{-H}_{4e})$ is about

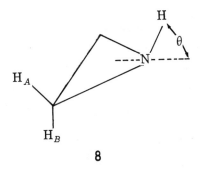

**7**

1.5 Hz. The sign of the latter coupling is almost certainly negative since on protonation of the nitrogen atom these couplings
are no longer resolved.* Notice that the relevant angle is approximately 180° for $H_{2a}$ and $H_{4a}$ but for $H_{4e}$ it is approximately
60°. For $\theta = 180 \pm 5°$ and $\theta = 60 \pm 5°$ INDO data gives $^2J(^{15}N\text{-}H)$
as $- 0.4 \pm 0.1$ and $- 3.8 \pm 1.3$ Hz, respectively, for methyl-
amine.

## Aziridine

In Fig. 8-2 the calculated values of $^2J(^{15}N\text{-}H)$ in aziridine
(**8**) are plotted as a function of the angle $\theta$ between the bisector

**8**

---

*The choice of sign is based on the observation that $^2J(^{15}N\text{-}H)$
usually increases on protonation.[7] This observation is supported by MO calculations. For example, in protonated methylamine INDO calculations give $^2J(^{15}N\text{-}H) = 2.43 \pm 0.05$ Hz, independent of the H-N-C-H dihedral angle.

AZIRIDINE

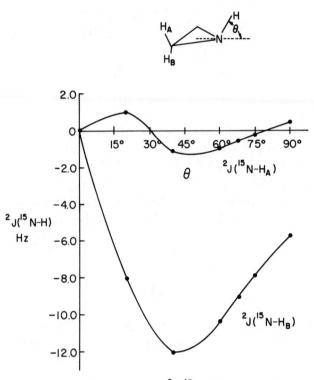

Figure 8-2.  A plot of the calculated $^2J(^{15}N-H)$ in aziridine versus the angle $\theta$ between the N-H bond and the extension of the bisector of the C-N-C angle.

of the C-N-C angle and the N-H bond.  The INDO results give a minimum energy when $\theta \simeq 68°$, and then $^2J(^{15}N-H_A) = -0.62$ and $^2J(^{15}N-H_B) = -9.14$ Hz.

For oxaziridine (compounds 1 and 2 in Table 8-1), $^2J(^{15}N-H)$ was also calculated as a function of $\theta$.  Its angular dependence is very similar to that obtained for aziridine although $^2J(^{15}N-H)$ in oxaziridine is more positive than in aziridine.  When $\theta = 64°$, $^2J(^{15}N-H_A) = 0.95$ and $^2J(^{15}N-H_B) = -5.78$ Hz, in excellent agreement with the measurements of Jennings et al.[20] on some substituted oxaziridines.

Notice that in the case of aziridine the hybridization of the nitrogen atom changes with variations in $\theta$. This is not so for methylamine in which the hybridization of the nitrogen atom remains essentially unchanged for all values of $\theta$. Because of this difference one does not expect the same angular depen-dence of $^2J(^{15}N\text{-}H)$ in methylamine as in aziridine.

It is of interest to compare the influence of the nitrogen and phosphorus lone pair on $^2J(^{15}N\text{-}H)$ and $^2J(^{31}P\text{-}H)$, respec-tively. For a series of cyclic phosphines, Albrand et al.[26] have plotted the observed $^2J(^{31}P\text{-}H)$ value against the dihedral angle involving the phosphorus lone pair and the P-C-H frag-ment. These data are replotted in Fig. 8-3 along with a plot of $^2J(^{15}N\text{-}H)$ as calculated for methylamine in Fig. 8-1. The simi-lar shape of these two curves suggests a similar mechanism by

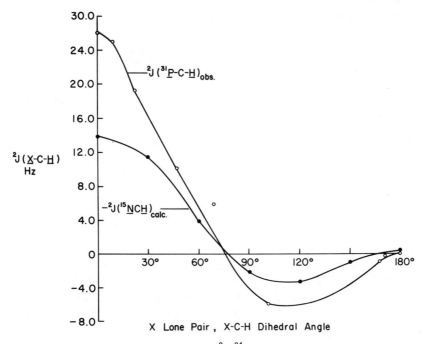

Figure 8-3. A plot of the observed $^2J(^{31}P\text{-}H)$ in some tervalent phos-phorus compounds[26] and of the calculated $-^2J(^{15}N\text{-}H)$ in methylamine versus the dihedral angle defined by the lone-pair X-C-H fragment.

which spin information is transferred between the nuclei. Other examples of two-bond $^{31}$P-H coupling constants that depend on lone-pair orientation have appeared in the literature.[9,27]


## THREE-BOND NITROGEN-PROTON COUPLING CONSTANTS

In Fig. 8-4 the N-C-C-H dihedral angle, $\theta$, in ethylamine is plotted against calculated $^{14}$N-H coupling constants.* Two conformations of the amino group are considered; in the $E$ conformation (9) the N-H bonds and the C-H bonds of the adjacent methylene group are eclipsed, whereas in the $S$ conformation (10) they are staggered.$^\dagger$ We note that in the $E$ conformation the nitrogen lone pair lies $cis$ to the methyl group while in the $S$ conformation it lies $trans$ to the methyl group. The $E$ conformation is calculated to be more stable than the $S$ conformation for all values of the angle $\theta$.

**9**

$E$ conformation
$\theta = 180°$ (H$_3$); $60°$ (H$_1$ and H$_2$)


*Since the experimental plots[28] present $^3J(^{14}$N-H) versus the N-C-C-H dihedral angle, the same convention is followed here. This facilitates a comparison between $^3J(^{14}$N-H) and the analogous couplings $^3J$(H-H), $^3J$(F-H), and $^3J$(P-H); all these nuclei have positive gyromagnetic ratios.
$^\dagger$In this discussion, $E$ and $S$ denote eclipsed and staggered conformations of the amino group, respectively.

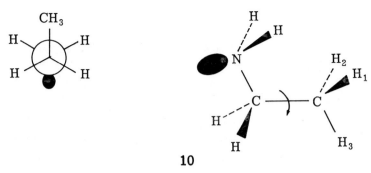

**10**

S conformation

$\theta = 180°$ ($H_3$); $60°$ ($H_1$ and $H_2$)

ETHYLAMINE

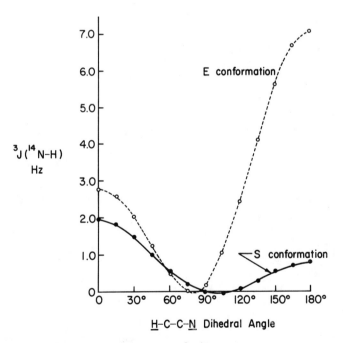

H–C–C–N Dihedral Angle

Figure 8-4. A plot of the calculated $^3J(^{14}$N–H) versus N–C–C–H dihedral angle for the *E* and *S* conformations of ethylamine. The *E* and *S* conformations are depicted in **9** and **10**.

From Fig. 8-4 it is apparent that $^3J(^{14}N\text{-}H)$ is dependent both on the orientation of the nitrogen lone pair and on the N-C-C-H dihedral angle.  Both the $E$ and $S$ conformations of ethylamine are most stable when the C-H bonds of the methyl and methylene groups are staggered, that is, for $\theta = 60°$ ($H_1$ and $H_2$) and $180°$ ($H_3$).  Then $^3J(^{14}N\text{-}H)_{av}$ is 2.65 Hz for the $E$ conformation, and 0.61 Hz for $S$.  In rigid compounds a similar dependence of $^3J(N\text{-}H)$ on lone-pair orientation is observed. [8,29-31]  For example, in $N$, $N$-dimethylnitrosamine (11), $^3J(^{15}N\text{-}H) = \pm 2.2$ Hz when the methyl group lies $cis$ to the lone pair while $^3J(^{15}N\text{-}H) = \pm 0.8$ Hz when the methyl group lies $trans$ to the lone pair. [31]

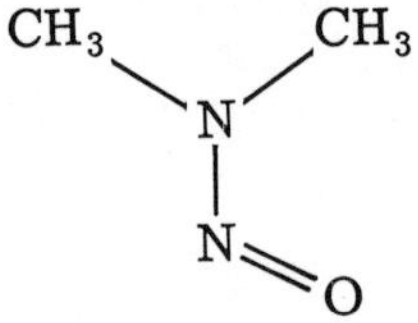

CH$_3$       CH$_3$

N

N=O

**11**

For protonated ethylamine the calculated angular dependence of $^3J(^{14}N\text{-}H)$ for the $E$ and $S$ conformations is very similar; this is in contrast to the behavior calculated for the unprotonated molecules (Fig. 8-4).

Calculations on $cis$-acetaldoxime (12a) and $trans$-acetaldoxime (12b) give the following results:

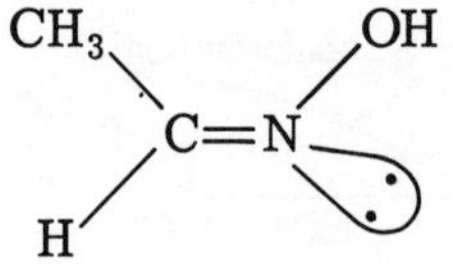

CH$_3$   OH        H   OH

C=N         C=N

H            CH$_3$

**12a** 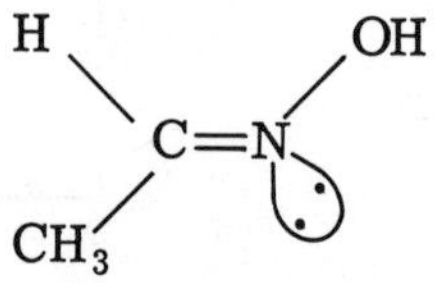             **12b**

|  | 12a | 12b |
|---|---|---|
| $^3J(^{14}N\text{-}H)_{obs}$ | 1.57 Hz | 2.86 Hz |
| $^3J(^{14}N\text{-}H)_{calc}$ | 2.65 Hz | 3.57 Hz |

Although the calculated values of $^3J(^{14}N\text{-}H)$ are too large, the relative magnitudes calculated for the *cis* and *trans* isomers are in agreement with experiment.[7]

The value of $^3J(^{31}P\text{-}H)$ in aminophosphines depends on the orientation of the phosphorus lone pair.[32-34] Below $-120\,^\circ C$ the proton NMR spectrum of methylaminobis(trifluoromethyl)-phosphine, $CH_3NH \cdot P(CF_3)_2$, indicates the presence of two rotational isomers 13a and 13b. The most abundant isomer is assigned to 13a where $^3J(^{31}P\text{-}H) = 13$ Hz while in 13b, $^3J(^{31}P\text{-}H) \simeq 4.0$ Hz.[32] In phosphines and in nitrogen compounds containing a lone pair, $^3J(^{31}P\text{-}H)$ and $^3J(^{14}N\text{-}H)$ apparently display a similar dependence on the orientation of the lone pair.

13a                    13b

## SOME RECENT EXAMPLES OF LONE-PAIR EFFECTS
## ON $J(N\text{-}C)$ AND $J(P\text{-}C)$

The application of noise decoupling and time averaging or pulsed Fourier transform techniques[5, 6, 35, 36] has recently allowed observation of stereospecific N-C and P-C coupling constants.

### $^2J(P\text{-}C)$

For 2, 2, 3, 4, 4-pentamethylphosphetaxes, Gray and Cramer[5] have observed that $^2J(P\text{-}C\text{-}C)$ is stereospecific.* The coupling

*See *J. Org. Chem.*, **37**, 3470 (1972) for a complete account of this work.

is large (27–37 Hz) when the methyl group is *cis* to the phosphorus lone pair and small (0–5 Hz) when *trans* to the lone pair. For example, for **14** $^2J(\underline{P}\text{-}\underline{C}H_3(A)) = \pm 30.5$ Hz while $^2J(\underline{P}\text{-}\underline{C}H_3(B)) = \pm 2.1$ Hz.

    Similarly Breen et al.[35] report $^2J(P\text{-}C\text{-}C) = \pm 32$ Hz in **15a** and $^2J(P\text{-}C\text{-}C) \simeq 0$ Hz in **15b**.

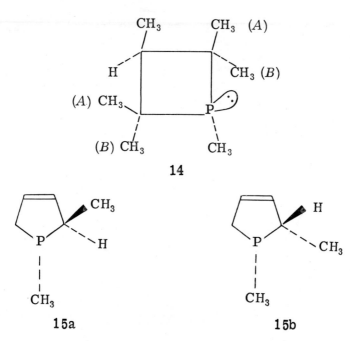

**14**

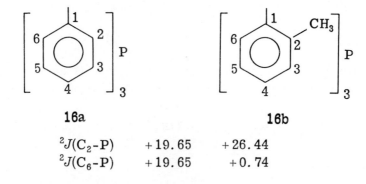

**15a**             **15b**

    Sørensen, Hansen, and Jakobsen[6] have observed stereospecific $^2J(^{31}P\text{-}^{13}C)$ values in several aromatic phosphines. The results they obtained for **16a** and **16b** are depicted below. These

**16a**                  **16b**

| | 16a | 16b |
|---|---|---|
| $^2J(C_2\text{-}P)$ | +19.65 | +26.44 |
| $^2J(C_6\text{-}P)$ | +19.65 | +0.74 |

results, along with other experimental data, indicate a pre-ferred conformation for the *ortho*-substituted arylphosphine in which the ring planes are twisted so as to align the substituent towards the phosphorus lone pair of electrons.[6]

$$^2J(N-C)$$

Pregosin, Randall, and White[36] have reported recently that $^2J(^{15}N-^{13}C_3) = \pm 2.7$ Hz while $^2J(^{15}N-^{13}C_8) = \pm 9.3$ Hz for quinoline (17a). For protonated quinoline (17b) they have observed that $^2J(^{15}N-^{13}C_3) \simeq ^2J(^{15}N-^{13}C_8) \simeq 1.0$ Hz.

17a                    17b

The two-bond $^{15}N-^{13}C$ coupling constants in *N*, *N*-dimethyl-nitrosamine (11) also depend on lone-pair orientation. $^2J(^{15}N-^{13}C)$. $= \pm 7.5$ Hz when the methyl group lies *cis* to the lone pair while $^2J(^{15}N-^{13}C) = \pm 1.4$ Hz when the methyl group lies *trans* to the lone pair.[1b]

## CONCLUSIONS

In all the examples considered here the magnitudes of the stereospecific coupling constants $^nJ(N-X)$ or $^nJ(P-X)$ are con-siderably larger when the nitrogen or phosphorus lone pair lies *cis* to X as opposed to lying *trans* to X. Furthermore, it ap-pears that a proximate lone pair makes a positive contribution to the reduced coupling constants discussed here. This empir-ical rule should be very useful in conformational and structural work.

Although the INDO molecular orbital calculations are not in quantitative agreement with experiment, they reproduce the overall trends observed and offer some insight into the nature of these stereospecific coupling constants. For meaningful calculations of $J(N-C)$ it will probably be necessary to include the dipolar and spin-orbital terms.[15,37-39]

## ADDENDA

Since this manuscript was written, several other relevant papers have been published.

Bendrude and Tan[40] have shown that $^3J(\underline{H}\ CO\underline{P})$ depends on both the HCOP dihedral angle and lone pair orientation.

Large differences of $^1J(P, P)$ in pentamethylcyclopentaphosphine, $(CH_3P)_5$, have been attributed to the stereochemical orientation of the phosphorus lone-pair.[41] Indeed calculated $^1J(N, N)$ values in hydrazine are found to be very dependent on lone pair orientation.[42]

Lichter $et\ al.$[43] have examined the influence of lone-pair orientation on $^nJ(^{15}N, ^{13}C)$ $(n = 1, 2$ and $3)$ in a wide variety of oximes. Many of the experimental trends are reproduced by the INDO calculations.

## ACKNOWLEDGMENTS

The author would like to thank Dr. T. Schaefer for several enlightening discussions pertaining to this work, and Dr. E. D. Becker for critically reading and commenting on this manuscript. The author also wishes to thank NRC of Canada for the fellowship.

Figures 8-1 to 8-4 were reproduced by permission of the National Research Council of Canada from the $Canadian\ Journal\ of\ Chemistry$, **50**, 2989–3008 (1972).

## References

1.  (a) E. W. Randall and D. G. Gillies, in *Progress in Nuclear Magnetic Resonance Spectroscopy*, J. W. Emsley, J. Feeney, and L. H. Sutcliffe, Eds., Vol. 6, Pergamon Press, Oxford, 1971, p. 119. (b) R. L. Lichter, in *Determination of Organic Structures by Physical Methods*, F. C. Nachod and J. J. Zuckerman, Eds., Vol. 4, Academic Press, New York, 1971, p. 195. (c) T. Axenrod, "Correlations of $^{15}N$ Coupling Constants with Molecular Structure, " in *Nitrogen NMR*, G. Webb and W. Witanowski, Eds., Plenum Press, London, 1973.

2.  T. Yonezawa, I. Moroshima, K. Fukuta, and Y. Ohmori, *J. Mol. Spectrosc.*, **31**, 341 (1969).

3.  V. M. S. Gil and S. J. S. Formosinho-Simões, *Mol. Phys.*, **15**, 639 (1968).

4.  R. Wasylishen and T. Schaefer, *Can. J. Chem.*, **50**, 2989 (1972).

5.  G. A. Gray and S. E. Cramer, *Chem. Commun.*, 367 (1972).

6.  S. Sørensen, R. S. Hansen, and H. J. Jakobsen, *J. Am. Chem. Soc.*, **94**, 5900 (1972).

7.  D. Crépaux, J. M. Lehn, and R. R. Dean, *Mol. Phys.*, **16**, 225 (1969).

8.  T. Axenrod, M. J. Wieder, and G. W. A. Milne, *Tetrahedron Lett.*, 1397 (1969).

9.  D. Gagnaire, J. B. Robert, and J. Verrier, *Chem. Commun.*, 819 (1967).

10.  C. H. Bushweller, J. A. Brunelle, W. G. Anderson, and H. S. Bilofsky, *Tetrahedron Lett.*, 3261 (1972).

11.  G. E. Maciel, J. W. McIver, Jr., N. S. Ostlund, and J. A. Pople, *J. Am. Chem. Soc.*, **92**, 1 (1970).

12.  G. E. Maciel, J. W. McIver, Jr., N. S. Ostlund, and J. A. Pople, *J. Am. Chem. Soc.*, **92**, 11, 4151, 4497, 4506 (1970).

13.  N. F. Ramsey, *Phys. Rev.*, **91**, 303 (1953).

14.  W. McFarlane, *Quart. Rev.*, **23**, 187 (1969).

15.  A. D. C. Towl and K. Schaumburg, *Mol. Phys.*, **22**, 49 (1971).

16. C. J. Jameson and H. Gutowsky, *J. Chem. Phys.*, **51**, 2790 (1969).

17. J. A. Pople and D. L. Beveridge, *Approximate Molecular Orbital Theory*, McGraw-Hill, New York, 1970.

18. J. A. Pople, J. W. McIver, Jr., and N. S. Ostlund, *Chem. Phys. Lett.*, **1**, 465 (1967).

19. J. A. Pople, J. W. McIver, Jr., and N. S. Ostlund, *J. Chem. Phys.*, **49**, 2960, 2965 (1968).

20. W. B. Jennings, D. R. Boyd, C. G. Watson, E. D. Becker, R. B. Bradley, and D. M. Jerina, *J. Am. Chem. Soc.*, **94**, 8501 (1972).

21. M. Ohtsuru and K. Tori, *Tetrahedron Lett.*, 4043 (1970).

22. R. L. Lichter and J. D. Roberts, *J. Am. Chem. Soc.*, **93**, 5218 (1971).

23. T. Nishikawa, T. Itoh, and K. Shimoda, *J. Chem. Phys.*, **23**, 1735 (1955).

24. L. Paolillo and E. D. Becker, *J. Magn. Resonance*, **3**, 200 (1970).

25. F. B. Riddell and J. M. Lehn, *J. Chem. Soc. B*, **1968**, 1224.

26. J. P. Albrand, D. Gagnaire, J. Martin, and J. B. Robert, *Bull. Soc. Chim. France*, **1969**, 40.

27. H. Goldwhite and D. G. Rowsell, *Chem. Commun.*, 1665 (1968).

28. A. A. Bothner-By and R. H. Cox, *J. Phys. Chem.*, **73**, 1830 (1969).

29. T. Axenrod, P. S. Pregosin, and G. W. A. Milne, *Chem. Commun.*, 702 (1968).

30. T. Axenrod, M. J. Wieder, and G. W. A. Milne, *Tetrahedron Lett.*, 401 (1969).

31. T. Axenrod, *Spectrosc. Lett.*, **3**, 263 (1970).

32. A. H. Cowley, M. J. S. Dewar, W. R. Jackson, and W. B. Jennings, *J. Am. Chem. Soc.*, **92**, 1085 (1970).

33. A. H. Cowley, M. J. S. Dewar, W. R. Jackson, and W. B. Jennings, *J. Am. Chem. Soc.*, **92**, 5206 (1970).

34. J. Nelson, R. Spratt, and B. J. Walker, *Chem. Commun.*, 1509 (1970).

35. J. J. Breen, S. I. Featherman, L. D. Quin, and R. C.

Stocks, *Chem. Commun.*, 657 (1972).

36. P. S. Pregosin, E. W. Randall, and A. I. White, *J. Chem. Soc.*, *Perkin Trans.*, II, 1 (1972).

37. A. C. Blizzard and D. P. Santry, *J. Chem. Phys.*, **55**, 950 (1971).

38. I. Morishima, A. Mizuno, T. Yonezawa, and K. Goto, *Chem. Commun.*, 1321 (1970).

39. K. Hirao, H. Nakatsuji, H. Kato, and T. Yonezawa, *J. Am. Chem. Soc.* **94**, 4078 (1972).

40. W. G. Bentrude and H. W. Tan, *J. Am. Chem. Soc.*, **95**, 4666 (1973).

41. J. P. Albrand, D. Gagnair and J. B. Robert, *J. Am. Chem. Soc.*, **95**, 6498 (1973).

42. R. Wasylishen, unpublished results.

43. R. L. Lichter, D. E. Dorman, and R. Wasylishen, submitted for publication.

CHAPTER 9

# THE EFFECTS OF PARAMAGNETISM ON NMR SPECTRA
# OF NUCLEI OTHER THAN PROTONS

G. A. Webb

*Department of Chemical Physics*
*University of Surrey*
*Guildford*

## INTRODUCTION

The presence of a paramagnetic center usually alters both
the position and width of an NMR signal. Although most of the
published work deals with proton NMR studies we are concerned primarily with the smaller range of reported data on
other nuclei in paramagnetic environments.[1,2]

The line width of the NMR signal of a nucleus, $x$, in a
paramagnetic environment is proportional to $\gamma_x^2$. Now the ratio
of $\gamma_H^2$ to $\gamma_D^2$ is 42.4; hence by replacing a proton by a deuteron
in a paramagnetic sample and examining the deuterium NMR
line width, it should be narrower than the width of the line from
the corresponding proton by a factor of about 42.4. This has
been demonstrated for some transition metal complexes.[3] The
observed narrowing is most appreciable for the broadest proton lines in the spectrum. The quadrupolar broadening expected
in the spectrum of the deuteron becomes significant when the
deuteron replaces a proton with a relatively narrow line. In
this case the line width decreased by less than the expected
factor of 42.4.

The majority of papers dealing with NMR of paramagnetic systems has been concerned with changes in line position rather than line width. The change in line position can arise from either or both of two mechanisms relating to the interactions between the nuclear spins and the spins of the unpaired electrons. One is the contact interaction which provides information relating to the electronic structure of the sample molecules; the other is the pseudocontact interaction which can be interpreted in terms of the geometric structure of the sample molecules.

## CONTACT INTERACTIONS

The contact interaction between an electron and a nuclear spin in a large applied magnetic field is a scalar quantity. It depends upon $a_N$, the nuclear hyperfine interaction constant of nucleus N, which is defined in gauss by

$$a_N = \frac{8\pi}{3} g_N \beta_N |\psi(0)_N|^2 , \tag{1}$$

where $\psi(0)_N$ is the wave function governing the distribution of the unpaired electron evaluated at the nucleus N. From equation 1 it is apparent that $a_N$ has finite values only for those cases in which there are electrons with nodeless wave functions at the site of the nucleus, allowing a direct "contact" between the electron and nucleus. Since only $S$ orbitals are nodeless at the nucleus, for the contact mechanism to operate it is necessary for the unpaired electrons to have some $S$ character. This is usually assumed to arise through electron correlation procedures.[4] For nuclei attached to those with the unpaired spin density the interaction is interpreted by an equation similar to that proposed by McConnell,[5]

$$a_N = Q\rho , \tag{2}$$

where $\rho$ is the unpaired spin density on the neighboring nucleus and $Q$ is a proportionality constant. This interaction produces

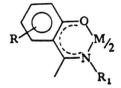

Figure 9-1.  Structure of substituted salicylaldimines.

a change, $\Delta B$, in the value of the applied field, $B_0$, at resonance, where

$$\frac{\Delta B}{B_0} = \frac{a_N g_e^2 \beta_e^2 S(S+1)}{g_N \beta_N \cdot 3KT} . \tag{3}$$

The other symbols in equation 3 have their usual significance. McGarvey[6-8] has recently shown that equation 3 is incomplete. However, the revised form of this equation has not been widely applied and the differences between electronic structure predicted by it and equation 3 are in many cases rather small. Consequently we consider only equation 3 in the interpretation of $a_N$ in terms of the electronic structure of the system being studied.

In transition metal complexes contact shifts are considered to arise from the transfer of unpaired electron spin density between the atom on the ligand whose nucleus is being considered and the central metal ion. It is usual to discuss electron spin delocalization according to whether the unpaired spin resides primarily in the $\sigma$ or $\pi$ orbital system of the ligand.

The early work in this area is based on proton NMR data.[1] Among the ligand systems studied are salicylaldimines (Fig. 9-1), aminotroponeimines (Fig. 9-2), and 1:3 diimines (Fig. 9-3). When these ligands are coordinated to cobalt(II) or nickel(II) the corresponding contact shift patterns in their NMR

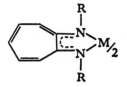

Figure 9-2.  Structure of substituted aminotroponeimines.

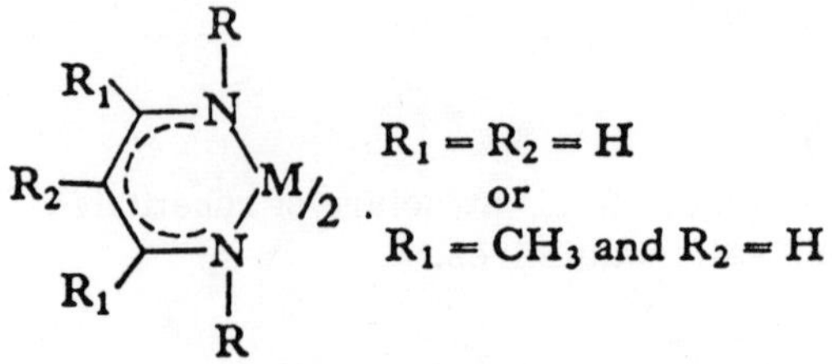

Figure 9-3.   Structure of substituted 1 : 3 diimines.

spectra have been attributed to spin delocalization in the $\pi$ orbital system of the ligands.   This conclusion has been reached for the following reasons:

1.  The observed shifts alternate in sign between adjacent positions on a conjugated fragment of the ligand.  This is consistent with an alternating change in sign of the $\pi$ spin density in equation 2, as obtained from valence bond and McLachlan type molecular orbital calculations on the $\pi$ system. [9]

2.  In general the magnitudes of the shifts do not attenuate appreciably as the protons become further removed from the metal ion.  This is in agreement with a largely nonattenuating $\pi$ spin density.

3.  By replacing a proton with a methyl group on the ligand, the proton resonance signal of the methyl group's protons is contact shifted in the opposite direction to that of the replaced proton.  This is consistent with the hyperconjugative model for the interaction between a $P_\pi$ electron and methyl protons, corresponding to $Q$ in equation 2 taking a negative value for a -CH fragment and a positive one for a $-C-CH_3$ fragment.

These three features are generally taken to be typical of a $\pi$ spin delocalization.  In contrast to this the downfield shifts experienced by pyridine protons, decreasing in magnitude in the order $\alpha > \beta > \gamma$, when pyridine is coordinated to Ni(acetylacetonate)$_2$, is generally taken to be diagnostic of a $\sigma$ spin delocalization (Fig. 9-4), which is consistent with the $\sigma$ symmetry of the highest-filled molecular orbital in pyridine.  However,

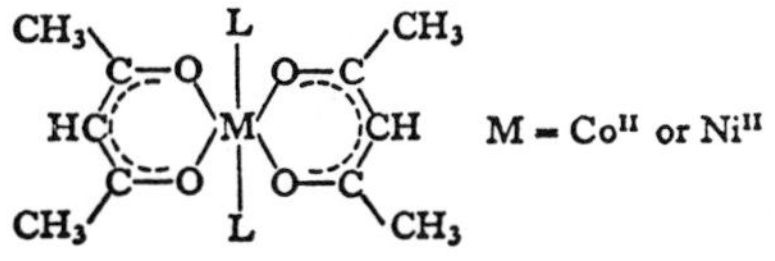

Figure 9-4. Structure of some complexes formed by the addition of a base L to metal diacetylacetonates.

these conclusions based upon proton NMR data need to be re-evaluated in the light of recent $^{13}$C NMR studies and more advanced molecular orbital calculations.

$^{13}$C NMR data have been reported for some nickel(II) aminotroponeimines[10, 11] (Fig. 9-5). If the spin density in the ligand system is entirely in the $\pi$ orbitals then the proton and $^{13}$C contact shift data should be compared as

$$\frac{\Delta B \, ^{13}C_i}{\Delta B \, ^1H_i} \approx +6.2 - 2.5 \, \frac{(\rho C_h^{\pi} + \rho C_j^{\pi})}{\rho C_i^{\pi}}$$

$$\frac{\Delta B \, ^{13}CH_3}{\Delta B C^1H_3} \approx +2.1 \,,$$

(4)

where $lC_i^{\pi}$ is the spin density in the $P_{\pi}$ orbital centered on carbon atom $i$. By comparing the ratios of the contact shifts predicted by equation 4 for C-H and C-CH$_3$ fragments, with the experimentally obtained shift ratios, it is possible to see whether or not the assumption of $\pi$ delocalization only is a

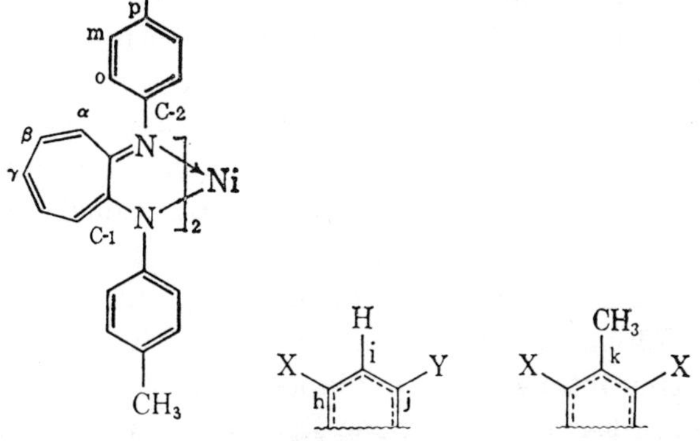

Figure 9-5. Structure and lettering scheme for some nickel(II) amino-troponeimines.[11]

TABLE 9-1.    Comparison of some $^{13}$C and $^{1}$H contact shifts
for some nickel(II) aminotropeneimines

| | $^{13}$C Shift Compared With | | | | |
| --- | --- | --- | --- | --- | --- |
| | $\alpha$-H | $\beta$-H | $o$-H | $m$-H | CH$_3$ |
| Predicted | +8.4 | +18.1 | +10.6 | +12.0 | +2.0 |
| Observed | +14.9 | +10.5 | +16.9 | +6.7 | +2.3 |

reasonable one.  In Table 9-1 an upfield shift has a negative
sign and a downfield shift a positive sign.  The predicted contact shifts are obtained from equation 4.

In all cases the observed ratios are positive as predicted.
However, the agreement of the magnitudes of the predicted and
observed ratios is not good in general.  Improvement is obtained by postulating some $\sigma$ spin density in addition to the $\pi$
spin density.  A positive $\sigma$ spin density at the $\beta$ and *meta* positions would reduce the predicted upfield shifts for the carbon
nuclei in these positions and increase the downfield proton
shifts, thereby producing better agreement with the experimental ratios.  For the $\alpha$ and *ortho* positions a positive $\sigma$ spin
density would increase the downfield shift of the carbon nuclei
and decrease the upfield shift of the protons, which also provides a better agreement with the observed ratios.  The *para*
methyl group is sufficiently far from the nickel atom for the $\sigma$
spin delocalization mechanism to be negligibly small.  Consequently the predicted and observed shift ratios are in reasonable agreement at this position.  This investigation shows that
both $\sigma$ and $\pi$ electron delocalization must be considered to account for the $^{13}$C NMR data whereas $\pi$ delocalization only has
been proposed for these nickel(II) aminotroponeimines from
proton NMR.

Complexes of pyridine and the $\alpha$, $\beta$, and $\gamma$ picolines with
Ni(acetylacetonate)$_2$ have been studied by $^{13}$C NMR (Fig. 9-4).
The hyperfine interaction constants have been calculated for
model systems related to these ligands by an open-shell INDO

molecular orbital program. [12]  In these calculations delocalization is assumed to be confined to the $\sigma$ electron system (Table 9-2).

The ratio of $\Delta B_i / a_i$ within the $^{13}C$ sets of data should have the same value for a given ligand; a single value should also be obtained from the proton data.  The observed differences in the value of this ratio of the experimental quantity $\Delta B_i$ to the calculated quantity, $a_i$, are attributed to approximations used in the calculations, in particular to the fact that the metal ion is not included in the basis sets.

By comparing the values of $\Delta B_i$ reported in Table 9-2 it is noticeable that in general attenuation of the value of this parameter with distance from the metal ion is indicated by the proton but not by the $^{13}C$ NMR data.  Consequently the observed attenuation of the shift of a nucleus as its distance from the metal ion increases need not be typical of $\sigma$ delocalization when NMR data from nuclei other than protons are considered.  Table 9-2 also shows that the sign of $a_i$ for carbon nuclei alternates between adjacent positions in the pyridine-type ligands.  Hence this alternation cannot be said to be typical of $\pi$ delocalization when obtained from $^{13}C$ NMR data.  Finally, the observed direction of the methyl proton shifts compared with those of replaced protons at the same positions is predicted to be opposite in sign on the basis of $\sigma$ delocalization only in a number of cases.  Consequently this type of alternation in sign of contact shifts is not diagnostic of $\pi$ electron delocalization.

Thus $^{13}C$ NMR data on transition metal complexes provides a more detailed description of their electronic structure than that obtained from proton results.  The use of more sophisticated molecular orbital calculations, such as the INDO procedure, is required in order to provide a satisfactory interpretation of the observed contact shifts.

The kinetics of chemical exchange processes and solvation phenomena have also been studied by means of NMR data for nuclei other than protons in paramagnetic systems. [2, 13-15]  The $^{14}N$ results for solvent exchange of $CH_3CN$ and $NH_3$ with manganese(II), iron(II), cobalt(II), and nickel(II) ions shows some disagreement with the earlier proton NMR data. [13-15]  It seems

TABLE 9–2. Normalized $^1$H and $^{13}$C isotropic shifts, $\Delta B_i$, for pyridine-type bases coordinated to Ni(acac)$_2$, INDO/2 hyperfine constants, $a_i$, calculated for the phenyl radical analogues, and ratios of these quantities.

| Nuclei | Phenyl | | | $p$–Tolyl | | | $m$–Tolyl | | | $o$–Tolyl | | |
|---|---|---|---|---|---|---|---|---|---|---|---|---|
| | $\Delta B_i$ | $a_i$ | $\dfrac{\Delta B_i}{a_i}$ | $\Delta B_i$ | $a_i$ | $\dfrac{\Delta B_i}{a_i}$ | $\Delta B_i$ | $a_i$ | $\dfrac{\Delta B_i}{a_i}$ | $\Delta B_i$ | $a_i$ | $\dfrac{\Delta B_i}{a_i}$ |
| $\alpha$–H | +943 | +18.7 | +48.4 | +394 | +19.2 | +20.5 | +930 | +17.4 | +53.4 | +293 | +17.4 | +16.8 |
| $\alpha'$–H | | | | | | | +930 | +19.3 | +48.2 | | | |
| $\alpha'$–CH$_3$ | | | | | | | | | | −48 | −1.9 | +25.3 |
| $\beta$–H | +286 | +6.1 | +46.9 | +71 | +6.5 | +10.9 | +360 | +5.7 | +63.1 | +146 | +6.6 | +22.1 |
| $\beta'$H | | | | | | | | | | +73 | +5.3 | +13.8 |
| $\beta'$–CH$_3$ | | | | | | | +72 | +1.6 | +45.0 | | | |
| $\gamma$–H | +86 | +3.9 | +17.9 | | | | +214 | +4.3 | +49.8 | +24 | +4.2 | +5.7 |
| $\gamma$–CH$_3$ | | | | −36 | −1.2 | +30.0 | | | | | | |
| $\alpha$–C | −1000 | −4.7 | +213 | +1000 | −5.1 | +196 | −1000 | −4.9 | +204 | −1000 | −4.0 | +250 |
| $\alpha'$–C | | | | | | | −1000 | −5.0 | +200 | +488 | −3.2 | −153 |
| $\alpha'$–CH$_3$ | | | | | | | | | | +585 | +6.2 | +94 |
| $\beta$–C | +3689 | +10.8 | +342 | +1930 | +10.4 | +186 | +3214 | +10.7 | +300 | +1659 | +10.7 | +155 |
| $\beta'$–C | | | | | | | +3000 | +10.4 | +288 | +1122 | +10.8 | +104 |
| $\beta'$–CH$_3$ | | | | | | | +430 | +0.82 | +524 | | | |
| $\gamma$–C | −371 | −2.6 | +143 | −180 | −2.3 | +78.2 | −430 | −2.8 | +154 | −98 | −2.6 | +37.7 |
| $\gamma$–CH$_3$ | | | | +144 | +1.4 | +103 | | | | | | |

likely that the proton data must be reconsidered.  Solvation phenomena concerning water have been studied by $^{17}O$ NMR and are discussed in Chapter 10.

## PSEUDOCONTACT INTERACTIONS

The magnetic dipole-dipole interaction between an unpaired electron and the nucleus in question can produce a pseudocontact shift in its NMR spectrum.  For a molecule with tetragonal symmetry in solution the pseudocontact shift, $\Delta B$, is given as a function of the applied magnetic field by

$$\frac{\Delta B}{B_0} = \frac{(3\cos^2\theta - 1)}{r^3} \frac{\beta_e^2 S(S+1)}{3KT} f(g_\parallel, g_\perp) , \qquad (5)$$

where $r$ is the separation vector of the nucleus and the unpaired electron, $\theta$ is the angle made by $r$ and the principal molecular symmetry axis, and $f(g_\parallel, g_\perp)$ is a function of the anisotropy of the $g$ tensor, the exact form of which depends upon the relative magnitudes of the electronic relaxation and exchange times and the correlation time for molecular tumbling.[1]  The expression $(3\cos^2\theta - 1)/r^3$ in equation 5 is often called the geometric factor.

McGarvey has shown[6-8] that equation 5 is applicable only when only one thermally populated electronic energy level is considered, when zero-field splitting can be ignored, and when the contribution arising from the electron's orbital angular momentum is included by means of the $g$ tensor components.  For molecules with electronically nondegenerate ground terms these conditions are usually met.  However, metal complexes with $T_1$ and $T_2$ ground terms may exhibit a significant orbital contribution to the magnetic moment from spin-orbit coupling.  They may also have several thermally occupied levels to be considered as well as large zero-field splittings.  Consequently the more complete equations should be used when analyzing data for metal complexes with $T_1$ and $T_2$ ground terms.

For nuclei other than protons it may also be necessary to include a contribution to the pseudocontact shift arising from atomic $P$ orbitals on the ligand.[6] These may overlap and form molecular orbitals with the metal $t_{2g}$ set. Extensive mixing of pure spin and orbital states may occur by spin-orbit coupling allowing the ligand atomic orbitals to contribute to the dipolar interaction. An example of a ligand-centered contribution, $\Delta B^L$, to the pseudocontact shifts is furnished by the $^{14}N$ NMR data on $Na_2[Fe(CN)_5NH_3]$ where

$$\frac{\Delta B^L}{B_0} = \frac{\dfrac{16\beta^2 f^2 \langle r^{-3}\rangle p}{45KT}\left[1 + \dfrac{11}{6V}\left(1 - \exp\left\{\dfrac{-3V}{2}\right\}\right)\right]}{\left(1 + 2\exp\left\{\dfrac{-3V}{2}\right\}\right)} . \qquad (6)$$

$\beta$ is the Bohr magneton, $\langle r^{-3}\rangle p$ is the average inverse cube of the radius of the $2p$ orbitals on the nitrogen atom, $f$ is the amount of admixture of the ligand and appropriate metal orbitals in the molecular orbitals, and $V$ is the ratio of the spin-orbit coupling constant to $KT$. This is taken to have a value of 2. By taking $\langle r^{-3}\rangle p = 21.1 \times 10^{24}$ $cm^{-3}$, $f$ becomes approximately 17%, which is a reasonable estimate of the covalency in this molecule.[4] The ligand-centered contribution to the pseudocontact shift is expected to be most significant for complexes with $T_1$ and $T_2$ ground terms, which is a further reason for restricting the use of equation 5 to metal complexes with nondegenerate ground terms.

Ion-pair formation in solution has been studied by means of the isotropic shifts observed in the NMR spectra of diamagnetic cations in solutions containing paramagnetic anions. The proton NMR data of $[(Butyl)_4N][(C_6H_5)_3PMI_3]$ (Fig. 9-6), where M is either nickel(II) or cobalt(II), have been interpreted on the basis of a pseudocontact shift of the butyl protons arising from ion-pair formation in $CDCl_3$ solution.[1] By means of the geometric factor in equation 5 it is possible to predict the pseudocontact shift for every nucleus in the cation, including $^{14}N$ (Table 9-3).

The difference between the observed $^{14}N$ NMR shift and that predicted, from the geometric ratio together with the bulk susceptibility correction, gives the data in the last column of Table 9-3 for a series of ion-paired diamagnetic cations.[16]

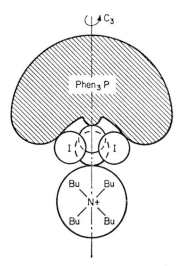

Figure 9-6.   Structure of some ion pairs, $[(Butyl)_4N][(C_6H_5)_3PMI_3]$.

This additional downfield shift shown by the nitrogen nuclei is thought to arise from the presence of unpaired spin density on the cation resulting from a weak covalent interaction between the anion and cation in the ion pair.   This probably occurs because of the overlap of the relatively diffuse orbitals of the halogens in the anions with a nitrogen $S$ orbital in the cation. Thus a small amount of spin density could be transferred between the ions.   This mechanism is supported by the fact that in general the additional downfield shift is greatest when the anion contains iodide, smaller when it contains bromide, and smallest for chloride containing anions.   Consequently we have a further case in which the NMR spectrum of another nucleus, $^{14}N$, provides a more detailed insight into the electronic structure of paramagnetic species in solution than that obtained from proton data.

In the last three years lanthanide shift reagents have been widely used to simplify proton NMR spectra by spreading them out and rendering "first order" many complicated splitting patterns.[17]   Proton data have been analyzed as arising from pseudocontact shifts according to equation 5.   However, a recent crystal structure determination of $Eu(DPM)_3(pyridine)_2$ has shown that the highest symmetry axis in this molecule is $C_2$ in

TABLE 9-3. $^{14}N$ NMR data for some ion-pair systems

| Complex | Bulk Susceptibility Shift (ppm) | Predicted $^{14}N$ Dipolar Shift (ppm) | Observed $^{14}N$ Chemical Shift (ppm) | Total Downfield Shift (ppm) |
|---|---|---|---|---|
| $[(n-C_4H_9)_4N][(C_6H_5)_3PCoI_3]$ | $-5.4$ | $-5.9$ | $+9$ | $+20$ |
| $[(n-C_4H_9)_4N][(C_6H_5)_3PNiI_3]$ | $-3.1$ | $+2.8$ | $+10$ | $+10$ |
| $[(n-C_4H_9)_4N]_2[CoCl_4]$ | $-5.8$ | $-1.7$ | $-1$ | $+6$ |
| $[(n-C_4H_9)_4N]_2[CoBr_4]$ | $-6.1$ | $-2.1$ | $+5$ | $+13$ |
| $[(n-C_4H_9)_4N]_2[CoI_4]$ | $-6.2$ | $-2.1$ | $+12$ | $+20$ |
| $[(n-C_4H_9)_4N]_2[NiCl_4]$ | $-3.8$ | $+1.3$ | $+6$ | $+8$ |
| $[(n-C_4H_9)_4N]_2[NiBr_4]$ | $-3.8$ | $+0.9$ | $+7$ | $+10$ |
| $[(n-C_4H_9)_4N]_2[NiI_4]$ | $-3.4$ | $+0.1$ | $+12$ | $+15$ |
| $[(n-C_4H_9)_4N]_2[Co(SCN)_4]$ | $-5.3$ | $0$ | $+1$ | $+6$ |
| $[(n-C_4H_9)_4N]_2[MnBr_4]$ | $-9.1$ | $+5$ | $+5$ | $+14$ |

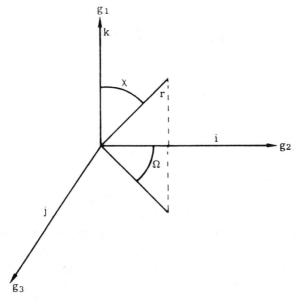

Figure 9-7.   Definition of angles and $g$ tensor components in rhombic symmetry.

the solid state.[18]  If the structure in solution is the same as this then the geometric factor appropriate to rhombic symmetry should be used in order to obtain geometric data from pseudo-contact shifts produced by shift reagents.

$$\frac{\Delta B}{B_0} = \frac{\beta^2 S(S+1)}{27KT} \frac{(g_1 + g_2 + g_3)}{r^3} [(g_1 - \tfrac{1}{2}g_2 - \tfrac{1}{2}g_3)(3\cos^2 \chi - 1)$$

$$- \tfrac{3}{2}(g_2 - g_3)(\sin^2 \chi \cos 2\Omega)] . \tag{7}$$

Equation 7 gives the pseudocontact shift for a molecule with rhombic symmetry.  The angles and $g$ tensor components are defined as in Fig. 9-7.  However, equation 5 rather than equation 7 could be appropriate to a discussion of shift reagent complexes in solution on account of the averaging processes taking place.  Irrespective of the details of the pseudocontact shift expression used, lanthanide shift reagents will continue to be useful in analyzing complicated NMR spectra.

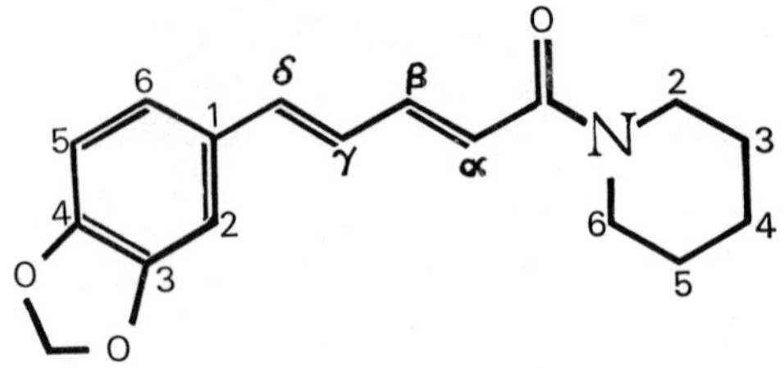

Figure 9-8.  Structure of piperine.

When nuclei other than protons are being studied then a contact contribution as well as a pseudocontact contribution may be expected from the shift reagents.[17]  This is due to the larger value of the nuclear hyperfine interaction constant per unpaired electron in an $S$ orbital for other nuclei,[19] together with the fact that the heteroatoms are the ones involved in the bonding to the metal ion in the shift reagent.

The small downfield shift observed in the $^{13}$C NMR spectrum of piperine (Fig. 9-8), for the carbonyl carbon, compared with the shifts of the other carbon resonances, is indicative of a contact contribution at this position which is the site of preferential coordination with the shift reagent.[20]  Contact shifts in $^{13}$C NMR spectra have also been shown by comparing the $^{13}$C and proton shifts induced by Eu(DPM)$_3$ in the spectra of some pyridine bases.[21]  Further evidence for the importance of contact shifts induced by lanthanide shift reagents is provided by the $^{14}$N NMR data of a series of nitrogen-containing molecules in the presence of shift reagents.[22]

We may conclude that the study of the NMR spectroscopy of nuclei other than protons in paramagnetic systems reveals a more thorough understanding of the system under investigation than is available from proton NMR data alone.


## References

1.   G. A. Webb, in *Annual Reports on NMR Spectroscopy*, E. Mooney, Ed., Vol. 3, Academic Press, London, New

York, 1970, p. 211.

2.  G. A. Webb, in *Annual Reports on NMR Spectroscopy*, E. Mooney, Ed., Vol. 7, Academic Press, London, New York, in press.

3.  A. Johnson and G. W. Everett, *J. Am. Chem. Soc.*, **94**, 1419 (1972).

4.  D. R. Davies and G. A. Webb, *Coord. Chem. Rev.*, **6**, 95 (1971).

5.  H. M. McConnell, *J. Chem. Phys.*, **28**, 1188 (1958).

6.  R. J. Kurland and B. R. McGarvey, *J. Magn. Resonance*, **2**, 286 (1970).

7.  B. R. McGarvey, *J. Chem. Phys.*, **53**, 86 (1970).

8.  B. R. McGarvey, *J. Am. Chem. Soc.*, **94**, 1103 (1972).

9.  C. L. Honeybourne and G. A. Webb, *Mol. Phys.*, **17**, 17 (1969).

10. D. Doddrell and J. D. Roberts, *J. Am. Chem. Soc.*, **92**, 4484 (1970).

11. D. Doddrell and J. D. Roberts, *J. Am. Chem. Soc.*, **92**, 5256 (1970).

12. W. D. Horrocks and D. L. Johnston, *Inorg. Chem.*, **10**, 1835 (1971).

13. W. L. Purcell and R. S. Marionelli, *Inorg. Chem.*, **9**, 1724 (1970).

14. S. F. Lincoln and R. J. West, *Aust. J. Chem.*, **24**, 1169 (1971).

15. R. J. West and S. F. Lincoln, *Inorg. Chem.*, **12**, 494 (1973).

16. D. G. Brown and R. S. Drago, *J. Am. Chem. Soc.*, **92**, 1871 (1970).

17. J. Reuben, in *Progress in NMR Spectroscopy*, J. W. Emsley, J. Feeney, and L. H. Sutcliffe, Eds., Vol. 9 Pergamon Press, Oxford, London, Part 1 (1973).

18. R. E. Cramer and K. Seff, *Chem. Commun.*, 400 (1972).

19. B. A. Goodman and J. B. Raynor, in *Advances in Inorganic Chemistry and Radiochemistry*, H. J. Emeléus and A. G. Sharpe, Eds., Vol. 13, Academic Press, London, New York, 1970, p. 135.

20. E. Wenkert, D. W. Cochran, E. W. Hagaman, R. B. Lewis, and F. M. Schell, *J. Am. Chem. Soc.*, **93**, 6271 (1971).

21. M. R. Willcott, R. E. Lenkinski, R. E. Davies, O. A. Gansow, and P. A. Loeffler, *J. Am. Chem. Soc.*, **95**, 3389 (1973).

22. M. Witanowski, L. Stefaniak, H. Januszewski, and Z. W. Wolkowski, *Tetrahedron Lett.*, 1653 (1971).

CHAPTER 10

# APPLICATIONS OF ¹⁷O MAGNETIC RESONANCE IN PARAMAGNETIC SYSTEMS

Abraham M. Achlama (Chmelnick)* and Daniel Fiat

*Department of Structural Chemistry*
*The Weizmann Institute of Science*
*Rehovot, Israel*

## INTRODUCTION

Oxygen-17 NMR is characterized by short nuclear relaxation times and large shifts. In diamagnetic liquid systems, the dominant relaxation mechanism is due to the quadrupolar interaction between the quadrupole moment of the oxygen-17 nucleus and electric field gradients in the molecule (e.g., the oxygen-17 line width in pure water at room temperature is 60 Hz[1]). In paramagnetic systems, the dominant relaxation and shift mechanisms are the scalar and dipolar interactions between the oxygen-17 nucleus and the unpaired electrons. The contribution of the scalar interaction to the transverse nuclear relaxation time, $(T_{2M})_S$, is given by[2]

$$\left(\frac{1}{T_{2M}}\right)_S = \frac{1}{3}\left(2\pi\frac{A}{h}\right)^2 S(S+1)\left[\tau_1 + \frac{\tau_2}{1 + (\omega_n - \omega_e)^2\tau_2^2}\right], \quad (1)$$

*Present address: Max-Planck-Institut für Medizinische Forschung, Abteilung Molekulare Physik, Heidelberg, Germany.

where $A$ is the hyperfine coupling constant given in ergs, $S$ is the electronic spin, and $\omega_n$ and $\omega_e$ are the nuclear and electronic Larmor frequencies, respectively. When the electronic relaxation time is shorter than the chemical exchange time, $\tau_1$ and $\tau_2$ are identified with the electronic longitudinal and transverse relaxation times ($T_{1e}$ and $T_{2e}$), respectively.

The contribution of the dipolar interaction is given by

$$\left(\frac{1}{T_{2M}}\right)_D = \frac{1}{15} S(S+1) \frac{\gamma_e^2 \gamma_n^2 \hbar^2}{r^6} \left[ 4\tau_1 + \frac{\tau_2}{1 + (\omega_n - \omega_e)^2 \tau_2^2} + \frac{3\tau_1}{1 + \omega_n^2 \tau_1^2} \right.$$

$$\left. + \frac{6\tau_2}{1 + \omega_e^2 \tau_2^2} + \frac{6\tau_2}{1 + (\omega_n + \omega_e)^2 \tau_2^2} \right] , \qquad (2)$$

where $\gamma_n$ and $\gamma_e$ are the nuclear and electronic gyromagnetic ratios, respectively, and $r$ is the average distance between the unpaired electrons and the nucleus. When the electronic relaxation time is shorter than the molecular rotation time and the exchange time, $\tau_1$ and $\tau_2$ are identified as before.[3] The Fermi contact contribution to the resonance shift is given by[4]

$$\frac{\Delta\omega}{\omega} = - S(S+1) \frac{g\beta A}{3kT\gamma_n \hbar} , \qquad (3)$$

where $T$ is the absolute temperature, and the other symbols have their usual meaning.

In this chapter the application of $^{17}O$ NMR to structure determination and the studies of electron delocalization, chemical exchange processes, and electron relaxation times are discussed. The examples are taken from studies of the hydration of the first series of transition metal ions.

## APPLICATIONS

### Structure Determination

A schematic spectrum of a system with two different magnetic sites is shown in Fig. 10-1. Under conditions of slow ex-

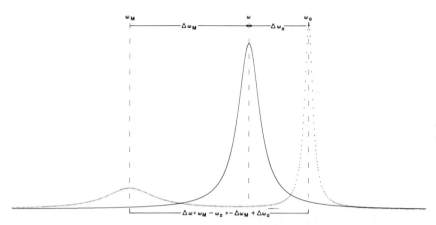

Figure 10-1. Schematic spectrum of two exchanging sites. Dashed line, slow exchange; solid line, fast exchange.

change (to be defined precisely below) two signals should in principle be observed. These signals coalesce under conditions of fast exchange. In this figure, $\omega$ is the resonance frequency of the coalesced signal. Referring specifically to the aqueous system, $\omega_M$ is the resonance frequency of water molecules coordinated to the central metal and $\omega_a$ is the resonance frequency of water molecules in pure water.

Under conditions of fast exchange, a simple relation between $\Delta\omega$ and $\Delta\omega_a$ (Fig. 10-1) holds[5]:

$$\Delta\omega = \Delta\omega_a \frac{n_t}{C.N. \times n_M} \qquad (4)$$

where $n_t$ is the total number of solvent molecules, C. N. is the coordination number of the metal, and $n_M$ is the number of metal ions.

On the other hand, $\Delta\omega$ is linearly dependent on the reciprocal of the absolute temperature, (equation 3). (We note that for structure determination no distinction is made between the contact and "pseudocontact" interactions, both being linearly dependent on $1/T$.[4])

Under conditions of slow exchange two signals are observed (Fig. 10-2). $\Delta\omega$ may then be directly measured as a function of the temperature.

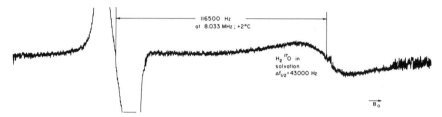

Figure 10-2.  Oxygen-17 magnetic resonance spectrum of aqueous $Ni(ClO_4)_2$.[6]

Under conditions of fast exchange,  only one signal is observed (Fig. 10-1).  Here $\Delta\omega_a$ may be measured,  relative to a suitable reference.  The results from one temperature region are then extrapolated to the other,  and the coordination number is obtained by comparing $\Delta\omega$ and $\Delta\omega_a$ for the same temperature (equation 1).  This method is illustrated in Fig. 10-3.  The system is the aqueous solution of nickelous perchlorate.  The data are converted from one set of measurements to the other by assuming a coordination number of 6.0.  The deviation from a straight line is due to exchange.

In routine NMR work,  the populations belonging to different signals are obtained by comparing areas bounded by the signals.  This method cannot usually be applied in oxygen-17 studies of paramagnetic systems because the signals are too broad and weak (Fig. 10-2).  Other methods proposed for the determination of the coordination number are the molal shift method,[7] the method of comparing line widths between systems of known and unknown coordination numbers,[8] and the method of deviation from linearity of line widths as a function of the concentration.[9]  The molal shift method is most readily applicable to diamagnetic systems,[7,10] because in paramagnetic systems the uncertainties in the results are large.[7,11]  The method of comparing line widths is controversial,[12] and some results obtained by this method are found to be incorrect.[13]  In the method of deviation from linearity,  the following relation holds:

$$c = f \cdot 10 \cdot \left[ 1 + \frac{\pi \cdot (\Delta f_{1/2})_M}{\tau_M \, \Delta\omega_M^{\,2}} \right] , \qquad (5)$$

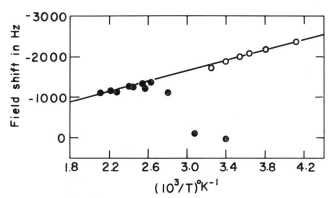

Figure 10-3. Resonance shift temperature dependence of bound ($\bigcirc$) and unbound ($\bullet$) water molecules in aqueous solution of $Ni(ClO_4)_2$.[6]

where $c$ is the molar concentration and $f$ is the relative change in line width. Again, the use of this method is practical in diamagnetic systems.

In some cases it is also possible to deduce distances within the molecule. The discussion of this point is deferred to a later section.

## Spin Delocalization

The signal of the bound water molecules in Fig. 10-2 is shifted to low field. This is also the case in the cobaltous[14] and ferrous[6] systems. In the vanadous system (Fig. 10-4), however, the bound water signal is shifted to high field, indicating negative spin density at the oxygen nuclei (the small signal appearing at high field was found to be due to asymmetric complexes existing in the $VBr_3$ solution).

The spin density, $Q_S$, is calculated from the hyperfine coupling constant by (equation 2)[16]:

$$A = -\frac{8\pi}{3} \gamma_e \gamma_n \hbar^2 \frac{Q_S}{S} \, , \tag{6}$$

$$Q_S = \tfrac{1}{2}[P_1^{\alpha}(\gamma_n) - P_1^{\beta}(\gamma_n)] \, , \tag{7}$$

Figure 10-4. Oxygen-17 magnetic resonance spectrum of aqueous VBr$_3$.[15]

where $P_1^\alpha(r_n)$ is the probability per unit volume of finding an electron with $\alpha$ spin in the vicinity $dr$ of the point $r_n$. $Q_S$ measures the excess of spin $\alpha$ over spin $\beta$ density, multiplied by the magnitude of the spin $z$ component (in $\hbar$ units).

Recently, [17] semiempirical unrestricted Hartree-Fock molecular orbital calculations of $Q_S$ have been performed on the aquo complexes of the first transition series of metal ions, in which the experimental trends (including negative spin densities) were reproduced. An example of a reciprocal effect between theory and experiment which allows steady improvement in both was given. It is concluded from the theoretical work[17] that the available experimental value for the spin density at the protons in the aqueous chromic complex is in error. Later measurements[18] have revealed a different value, in accord with theoretical expectations.

## Kinetic Parameters of Chemical Exchange

The paramagnetic contribution to the line width of the bulk molecules $T_{2p}$ is given by[19]

$$\frac{1}{T_{2p}} = \frac{1}{\tau_{H_2O}} \left[ \frac{1/T_{2M}^2 + 1/T_{2M}\tau_M + \Delta\omega_M^2}{(1/T_{2M} + 1/\tau_M)^2 + \Delta\omega_M^2} \right], \tag{8}$$

where $\tau_{H_2O}$ is the lifetime of the water ligands in the bulk solution, and $\tau_M$ is the lifetime of the ligands in the solvation sphere.

It is more convenient to study the relaxation of the bulk molecules in the limits of slow and fast exchange.

Under conditions of slow exchange, namely;

$$\Delta\omega_M{}^2 \gg \frac{1}{T_{2M}{}^2}; \ \frac{1}{\tau_M{}^2} \tag{9}$$

or

$$\frac{1}{T_{2M}{}^2} \gg \Delta\omega_M{}^2; \ \frac{1}{\tau_M{}^2} \ , \tag{10}$$

equation 8 reduces to[19]

$$\frac{1}{T_{2p}} = \frac{P_M}{\tau_M} \ , \tag{11}$$

where $P_M$ is the ratio of the number of bound molecules to the number of unbound ones.

Under conditions of fast exchange, namely, [19]

$$\frac{1}{\tau_M{}^2} \gg \Delta\omega_M{}^2; \ \frac{1}{T_{2M}{}^2} \ , \tag{12}$$

equation 8 reduces to

$$\frac{1}{T_{2p}} = P_M \tau_M \Delta\omega_M{}^2 \ . \tag{13}$$

The dependence of $\tau_M$ on the temperature is given by[20]

$$\tau_M = \left(\frac{kT}{h}\right)^{-1} \exp\left(\frac{\Delta H^{\pm}}{RT} - \frac{\Delta S^{\pm}}{R}\right) \tag{14}$$

where $\Delta H^{\pm}$ and $\Delta S^{\pm}$ are the enthalpy and entropy of activation for the exchange.

In Fig. 10-5 three regions are distinguished. The low-temperature region is described by equation 11 and the middle region by equation 13. The enthalpy of activation is obtained from the slope of the lines.

The last temperature region is discussed in the next section.

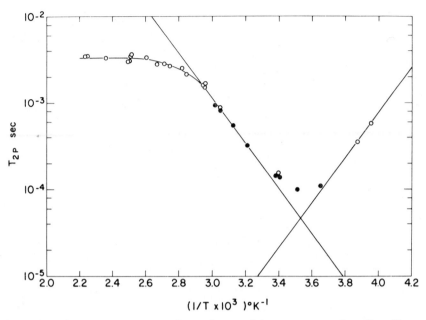

Figure 10-5.　Temperature dependence of transverse relaxation time in aqueous solution of $Co(ClO_4)_2$.[21]

## The Electronic Relaxation Time

The possibility of observing the NMR signal of the coordinated molecules depends on the electronic relaxation time. The shorter these times are, the longer is the nuclear relaxation time (equations 1 and 2). When this signal is observed (and thus $T_{2M}$ is measured directly) the electronic relaxation time can be obtained, provided that a sound assumption can be made on the distance involved. Conversely, when the electronic relaxation time is known, the distance can be obtained. At high temperatures, when the signals of the bound and unbound molecules have already coalesced, it is possible to obtain the transverse relaxation time of the bound molecules by measuring that of the unbound ones.

The horizontal line observed in Fig. 10-5 is ascribed to the shortening of $\tau_M$. When $\tau_M$ becomes short enough, extremely fast exchange conditions hold, namely,[21]

$$\frac{1}{\tau_M{}^2} \gg \frac{1}{T_{2M}\tau_M} \gg \Delta\omega_M{}^2 \ , \tag{15}$$

and equation 8 reduces to

$$\frac{1}{T_{2p}} = \frac{P_M}{T_{2M}} \ . \tag{16}$$

It is worthwhile pointing out that the electronic relaxation times obtained in this way are precisely those that cannot be obtained by the ESR method, since they are too short.


## CONCLUSION

We have seen four applications of oxygen-17 NMR in paramagnetic systems. It may be found that $^{17}$O NMR gives a better understanding of the system being studied than $^1$H NMR (e.g., the protolysis reaction is measured on top of the exchange of the whole molecule in chemical exchange studies using $^1$H NMR). However, we would like to point out that the applications discussed here are by no means peculiar to oxygen-17 and may be applied to other nuclei as well.


## References

1. H. A. Christ, P. Diehl, H. R. Schneider, and P. Dahn, *Helv. Chim. Acta*, **44**, 865 (1961).
2. A. Abragam, *The Principles of Nuclear Magnetism*, Oxford University Press, London, 1961, Chap. 8.
3. R. E. Connick and D. Fiat, *J. Chem. Phys.*, **44**, 4103 (1966).
4. H. M. McConnell and R. E. Robertson, *J. Chem. Phys.*, **29**, 1361 (1958).
5. A. M. Chmelnick and D. Fiat, *J. Chem. Phys.*, **51**, 4238 (1969).
6. A. M. Chmelnick and D. Fiat, *J. Am. Chem. Soc.*, **93**, 2875 (1971).

7.  M. Alei and J. A. Jackson, *J. Chem. Phys.*, **41**, 3402 (1964).

8.  T. J. Swift and W. G. Sayre, *J. Chem. Phys.*, **44**, 3567 (1966).

9.  J. Reuben and D. Fiat, *J. Chem. Phys.*, **51**, 4918 (1969).

10. D. Fiat and R. E. Connick, *J. Am. Chem. Soc.*, **88**, 4754 (1966).

11. D. Fiat and A. M. Chmelnick, *Proc. Colloq. Ampere*, **14**, 1157 (1967).

12. S. J. Meiboom, *J. Chem. Phys.*, **46**, 410 (1967).

13. N. A. Matwiyoff and H. Taube, *J. Am. Chem. Soc.*, **90**, 2796 (1968).

14. D. Fiat, Z. Luz, and B. L. Silver, *J. Chem. Phys.*, **49**, 1376 (1968).

15. A. M. Chmelnick and D. Fiat, *J. Magn. Resonance* **8**, 325 (1972).

16. R. McWeeny, *J. Chem. Phys.*, **42**, 1717 (1965).

17. A. M. Chmelnick and D. Fiat, *J. Magn. Resonance*, **7**, 418 (1972).

18. A. M. Achlama (Chmelnick) and D. Fiat, *J. Chem. Phys.*, in press.

19. T. J. Swift and R. E. Connick, *J. Chem. Phys.*, **37**, 307 (1962).

20. H. Eyring, *Chem. Rev.*, **17**, 65 (1935).

21. A. M. Chmelnick and D. Fiat, *J. Chem. Phys.*, **47**, 3986 (1967).

CHAPTER 11

# <sup>13</sup>C NMR SPECTROSCOPY OF ORGANO-TRANSITION METAL COMPLEXES

B. E. Mann

*The School of Chemistry** 
*The University of Leeds*

$^{13}$C NMR spectroscopy is a new tool by which we can in-
vestigate the structure of organometallic compounds; thus $^{13}$C
NMR spectra have been measured for a number of compounds
in order to try to find new information about the bonding in such
compounds.

For several years there have been attempts to explain direct coupling constants by use of the simplified Fermi contact
equation:

$$^{1}J(A-B) = \frac{\hbar}{2\pi} \frac{256\pi^2}{9} \beta^2 \gamma_A \gamma_B \frac{\langle S_A(0) \rangle^2 \langle S_B(0) \rangle^2 \alpha_A^2 \alpha_B^2}{\Delta E} , \qquad (1)$$

where $\gamma_A$ and $\gamma_B$ are the gyromagnetic ratios of nuclei A and B,
$\langle S_A(0) \rangle^2$ and $\langle S_B(0) \rangle^2$ are the valence $s$-electron densities at A
and B, $\alpha_A^2$ and $\alpha_B^2$ are the $s$ characters of the orbitals used by
the atoms containing A and B in forming the bond, and $\Delta E$ is a
mean triplet excitation energy. Although this equation has been
severely criticized and almost certainly is nonquantitative, it

*Present address Department of Chemistry, The University,
Sheffield 53 7HF.

might prove useful at the qualitative level. Consider as an example the pair of compounds $cis$-$[PtMe_2(EMe_2Ph)_2]$, E = P or As.[1] Chemically, it is usually found that arsenic bonds more weakly than phosphorus. Thus on going from the phosphorus to the arsenic compound, it would be expected that the Pt-E and E-CH$_3$ bonds might weaken, and the H$_3$C-Pt and EC-H bonds strengthen. It is found that $^1J(^{195}Pt-^{13}C)$ increases from 594 to 685 Hz and $^1J(E^{13}C-^1H)$ increases from 128.8 to 133.9 Hz, in qualitative agreement with equation 1.

We have attempted to test equation 1 more quantitatively. For metal carbonyls of the type $[M(CO)_n]$, it is assumed that all the metal valence $s$ orbital is used in bonding. In equation 1 it follows that $\alpha_A{}^2 = 1/n$, and $\langle S_A(0) \rangle^2$ has been estimated previously.[2] Thus equation 1 can be rearranged as

$$\frac{^1J(A-B)_n 2\pi}{\gamma_A \gamma_B \langle S_A(0) \rangle^2 \hbar} = \frac{256\pi^2}{9} \beta^2 \frac{\langle S_B(0) \rangle^2 \alpha_B{}^2}{\Delta E}. \tag{2}$$

TABLE 11-1.　Values of $^1J(M-^{13}C)$ and $^1J(M-^{13}C)$ $n2\pi/\gamma_A\gamma_B \langle S_A(0) \rangle^2\hbar$ for some metal carbonyls of the type $[M(CO)_n]$

| Compound | $^1J(M-^{13}C)$ | $\dfrac{^1J(M-^{13}C)2\pi_n}{\gamma_A\gamma_B \langle S_A(0) \rangle^2\hbar}$ |
|---|---|---|
| $[Mo(CO)_6]$ | 68 | 50.3–61.2 |
| $[W(CO)_6]$ | 126 | 42.7–56.6 |
| $[V(CO)_6]^{-}$ [a] | 116 | 41.7–43.9 |
| $[Fe(CO)_5]^{b}$ | 23.4 | 40.5–51.3 |
| $[Co(CO)_4]^{-}$ [c] | 287 | 49.6–55.6 |

[a]P. C. Lauterbur and R. B. King, J. *Am. Chem. Soc.*, 87, 3266 (1965).
[b]B. E. Mann, *Chem. Commun.*, 1173 (1971).
[c]E. A. C. Lucken, K. Noack, and D. F. Williams, *J. Chem. Soc. A*, 148 (1967).

Some appropriate values are given in Table 11-1.  Examination
of the data found for $^1J(\text{A-B})_n 2\pi/\gamma_A \gamma_B \langle S_A(0) \rangle^2 \hbar$ in the table shows
that this term is approximately constant, implying that it may
be valid to use equation 1 at least at the qualitative level.

Finally, we have applied equation 1 to a number of platinum complexes.  The assumption is made that in tetrahedral
or octahedral complexes platinum uses all of its valence 6s orbital in bonding, but in square-planar complexes it uses only two-
thirds of its valence 6s orbital in bonding.  Then, for
$[\text{Pt(PMe}_2\text{Ph)}_4]^{2+}$ as $^1J(^{195}\text{Pt-}^{31}\text{P}) = 2325$ Hz,

$$^1J(^{195}\text{Pt-}^{31}\text{P}) = 16000\,\alpha_{\text{Pt}}^2 , \tag{3}$$

where $\alpha_{\text{P}}^2$ is not included as it is not known.  Thus equation 3
applies only to complexes of PMe$_2$Ph, and phosphorus ligands
with similar coupling constants to platinum, for example, PEt$_3$.
A similar equation can now be derived for $^1J(^{195}\text{Pt-}^{13}\text{C})$ for
cis-$[\text{PtMe}_2(\text{PMe}_2\text{Ph)}_2]$, that is,

$$^1J(^{195}\text{Pt-}^{13}\text{C}) = 11900\,\alpha_{\text{Pt}}^2 \alpha_{\text{C}}^2 , \tag{4}$$

TABLE 11-2.  Values of $^1J(^{195}\text{Pt-}^{13}\text{C})$, $^1J(^{195}\text{Pt-}^{31}\text{P})$, $\alpha_{\text{C}}^2$, and $\alpha_{\text{Pt}}^2$ for
some complexes of platinum

| Compound | $^1J(^{195}\text{Pt-}^{13}\text{C})$ (Hz) | $^1J(^{195}\text{Pt-}^{31}\text{P})$ (Hz) | $\alpha_{\text{C}}^2$ | $\alpha_{\text{Pt}}^2$ |
|---|---|---|---|---|
| $[\text{Pt(PMe}_2\text{Ph)}_4]^{2+}$ | — | 2325 | — | 0.67 |
| $[\text{Pt(PEt}_3)_4]$ | — | 3740 | — | 0.94 |
| $[\text{Pt(PEt}_3)_3]$ | — | 4220 | — | 0.79 |
| cis-$[\text{PtMe}_2(\text{PMe}_2\text{Ph)}_2]$ | 594 | 1819 | 0.227 | 0.67 |
| cis-$[\text{PtPh}_2(\text{PEt}_3)_2]$ | 817 | 1771 | 0.333 | 0.63 |
| trans-$[\text{PtPh}_2(\text{PEt}_3)_2]$ | 594 | 2831 | 0.333 | 0.65 |
| $[\text{Pt(CN)}_4]^{2-}$ [a] | 1036 | — | 0.50 | 0.70 |

[a] H. H. Rupp, quoted by H. H. Keller, *Ber. Bunsen-ges. Phys.
Chem.*, **76**, 1080 (1972).

where in the case of methyl groups, $\alpha_C^2$ can be derived from the equation[3]

$$\alpha_C^2 = 1 - \frac{3^1J(^{13}C-^1H)}{500} \ . \tag{5}$$

Calculations on the basis of equations 3 and 4 are given in Table 11-2. Examination of the values calculated for the total valence $6s$ electron density used by the platinum, $\sum \alpha_{Pt}^2$, shows them to be consistent, with values for all the square-planar platinum(II) complexes falling between 0.63 and 0.70.

In view of the perils associated with the use of equation 1, it is dangerous to draw firm conclusions. However, the calculations given in Tables 11-1 and 11-2 imply that the qualitative and perhaps quantitative use of equation 1 may be valid for transition metal complexes.

### References

1. A. J. Cheney, B. E. Mann, and B. L. Shaw, *Chem. Commun.*, 431 (1971).
2. G. W. Smith, *Res. Publ. GMR*-444, General Motors Corporation, Warren, Mich., 1964.
3. F. J. Weigert, M. Winokur, and J. D. Roberts, *J. Am. Chem. Soc.*, **90**, 1566 (1968).

CHAPTER 12

# ASSIGNMENT TECHNIQUES IN $^{13}$C NMR SPECTROSCOPY

Felix W. Wehrli

*Varian AG*
*NMR Laboratory*
*Zug, Switzerland*

## INTRODUCTION

The structural information that proton NMR provides is based on two physical phenomena, chemical shift and scalar spin-spin coupling, where the latter is probably the most informative. In $^{13}$C NMR, however, this nuclear property can generally not be exploited because of extensive application of proton noise-decoupling techniques.[1-4] Undecoupled spectra, on the other hand, do not prove to be useful[2,3] because of their complexity due to the numerous long-range couplings. This holds at least for larger molecules. Spectra obtained with simultaneous full proton noise decoupling give rise to just one line per carbon atom whereas homonuclear coupling is not observable under normal circumstances. The low natural abundance of carbon-13, previously considered as the main obstacle to the advent of the technique, now proves fortunate, for otherwise the spectra would probably not be interpretable. Proton noise decoupling simplifies the spectrum by simultaneous removal of all $^{13}$C-H couplings. Selective elimination of some $^{13}$C-H couplings would often be desirable, but is generally hampered because the magnitude of the direct $^{13}$C-H coupling is normally large compared to the proton chemical shifts. Apart from this the proton shieldings are often unknown in complex molecules.

Roberts and co-workers[2,3] have shown the off-resonance technique[1-4] to be of great utility. Turning the noise modulation off and offsetting the decoupler frequency by a few hundred hertz away from the center of the proton resonances has the effect of removing $^{13}$C-H long-range couplings while reducing the direct couplings to a fraction of their true values. In the first step this allows a distinction to be made between $CH_3$, $CH_2$, CH, and quarternary carbons. Much more information can be obtained, however, as recently shown by Johnson.[5] This is discussed in detail in this chapter.

The chemical shift as the only assignment criterion in noise-decoupled $^{13}$C spectra is governed by effects similar to those in proton NMR. However, it covers a much wider range (about 250 ppm) and is much more sensitive to structural and conformational influences. Theoretical rationalization of $^{13}$C chemical shifts has succeeded only in the case of closely related series of molecules. On the other hand, useful empirical additivity rules[6,7] have been found, which have been successfully applied to assigning chemical shifts in steroids,[3] for example.

In general, however, it has not been possible to make a full assignment in complex organic molecules without the aid of spectral comparison of chemically related compounds. This can be achieved by introducing functional groups at various sites of a molecule or by specific labeling. $^{13}$C and deuterium labeling[8,9] has been used in biosynthetic studies. The problem there, however, is the elucidation of the biogenetic pathway rather than simply assigning the spectrum.

With the advent of the pulse Fourier technique,[10] a further important parameter in high-resolution $^{13}$C NMR has become amenable: the spin-lattice relaxation time $T_1$.[11-13] Allerhand and co-workers[14] have recently shown that from the magnitude of $T_1$ not only can information on molecular motion be obtained but also it can be utilized as an assignment aid.

The enhanced sensitivity of the Fourier spectrometer finally allows the recording of undecoupled spectra,[15,16] which, at least in simpler molecules, can provide additional information.

Assigning the lines in a $^{13}$C NMR spectrum to individual carbon atoms in a molecule obviously requires the combined application of the above-mentioned techniques. These may be summarized as follows:

1. Chemical shift
2. Off-resonance coherent decoupling
3. Selective decoupling
4. Spin-lattice relaxation
5. Spectral comparison
6. Specific labeling
7. Undecoupled spectra

## THE CHEMICAL SHIFT

In contrast to the proton chemical shift, $^{13}$C shieldings exhibit a much wider range, which is of the order of 600 ppm. Uncharged naturally occurring organic molecules usually absorb over a range of slightly more than 200 ppm. A theoretical treatment[17-19] of carbon chemical shifts has been attempted at various levels of sophistication, but only qualitative agreement has been obtained in general. By dissecting the nuclear shielding constant into a diamagnetic term, a paramagnetic term, and a term arising from electron circulations on neighboring atoms, it can be shown that the paramagnetic contribution dominates. This arises from mixing of the electronic ground state with low-level excited states.

Similar to proton chemical shifts, $^{13}$C shieldings are governed primarily by the hybridization of the carbon atom under consideration, with $sp^2$ carbons being the least shielded, followed by $sp$ and $sp^3$ type carbons. Apart from this, the electronegativity of attached substituents plays an important role. Electronegative groups can deshield carbons by as much as

30 ppm. In unsaturated systems, substituents can affect the shielding of remote centers by withdrawing or donating elec- tron density. Increasing electron density generally shifts the resonances toward higher field.

Grant and co-workers[20] have found that steric perturbation of carbons by substituents leads to increased shielding. These steric compression shifts have been shown to be a powerful as- signment tool in complex molecules.[3]

As a reference standard for $^{13}C$ chemical shifts, internal TMS is now generally accepted with positive shift values im- plying decreased shielding as in $^1H$ NMR.

How sensitive $^{13}C$ chemical shifts are with respect to the chemical environment is illustrated by Fig. 12-1, where the $^{13}C$ and $^1H$ spectra of a steroid, both in the same frequency

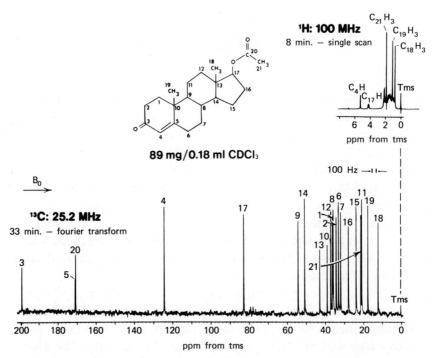

Figure 12-1. $^{13}C$ and $^1H$ spectrum of cortisone acetate shown in the same frequency scale.

scale, are shown.[21] Whereas in the proton spectrum only five assignments are possible, all 23 carbons give rise to separate signals, all of which can be designated.[3]

From the experimentally observed additivity of substituent effects, empirical rules have been derived which allow fairly accurate chemical shift predictions to be made in some classes of compounds. Grant and Paul[6] have found an empirical additivity relationship by means of which chemical shifts in linear and branched alkanes can be predicted. It is based on additive increments associated with the contribution of $\alpha$, $\beta$, $\gamma$, $\delta$, and $\epsilon$ carbons. The equation has the following form:

$$\delta C_K = B + \sum_l n_{kl} A_{kl} \, ,$$

where $n_{kl}$ stands for the number of carbon atoms in position $l$, and $B$ is a constant corresponding to the chemical shift of methane. Additional corrective terms are required in branched systems:

$$1° \ (4°)$$
$$1° \ (3°)$$
$$2° \ (4°)$$
$$2° \ (3°), \ \text{etc.,}$$

where $1° \ (4°)$, for example, is a corrective term used in order to calculate the chemical shift of a primary carbon attached to a quarternary carbon.

A similar relationship holds for the methylcyclohexanes.[7] The substituent parameters for equatorial and axial methyls differ considerably, which demonstrates the great potential of $^{13}$C NMR in conformational analysis. Figure 12-2 shows the proton noise-decoupled $^{13}$C spectrum obtained from a mixture of *cis*- and *trans*-1, 3-dimethylcyclohexane. All peaks can be assigned unambiguously and the deviations between predicted and experimental shifts are less than 1 ppm.

Roberts and co-workers[3] have applied additivity parameters to the assignment of chemical shifts in the side chain of cholestane by using 2, 6-dimethyloctane as a model compound.

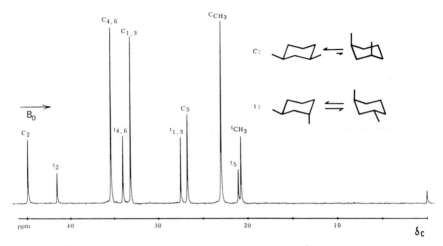

Figure 12-2.  25.2 MHz proton noise-decoupled $^{13}C$ Fourier transform spectrum of a mixture consisting of *cis*- and *trans*-1, 3–dimethylcyclo-hexane.  The lines were assigned with the aid of Grant's additivity rules.[7]

## SELECTIVE AND OFF-RESONANCE DECOUPLING

Coherent decoupling off-resonance has the effect of re-ducing the direct C-H couplings while generally removing long-range couplings, or at least reducing them to the extent that they are no longer resolved.  As is seen later, a careful ex-amination of the line shape of reduced splittings can give in-formation with respect to the degree of substitution at neighbor-ing carbon atoms.

The relationship between the observed reduced splitting $J_r$, the true C-H coupling constant $^1J(C\text{-}H)$, the offset, $\Delta f$, of the proton decoupler frequency from resonance, and the rf power level $\gamma B_2/2\pi$ has been given by Ernst[1]:

$$J_r = \frac{{}^1J(C\text{-}H) \cdot \Delta f}{\gamma B_2/2\pi} .$$

From the multiplicities of the signals obtained in this manner, the degree of substitution at each carbon can immediately be determined.

Provided there is some knowledge of the corresponding proton chemical shifts and the coupling constants $^1J$(C-H), accurate $^{13}$C chemical shift assignments may be made.[5] The magnitude of the direct coupling constant $^1J$(C-H) can be estimated fairly accurately by using Malinowski's empirical additivity rules.[22] Accordingly the magnitude of $^1J$(C-H) in molecules of the type CHXYZ can be predicted by using the following simple relationship:

$$^1J(\text{C-H}) = \zeta_X + \zeta_Y + \zeta_Z \, ,$$

where the zeta values are contributions to the coupling from the substituents X, Y, and Z.

Figure 12-3 shows the 200 ppm $^{13}$C Fourier transform spectrum of the complex molecule brucine, with noise-modulated and coherent proton decoupling. The latter was obtained by centering the decoupler frequency at $-4$ ppm (i.e., upfield) from TMS.

From the noise-decoupled spectrum only three unambiguous assignments can be made: at lowest field the carbonyl whose chemical shift of 170 ppm is typical of a lactam-type carbonyl, then the strong signal at 127 ppm which must be due to a protonated carbon and must be assigned to the olefinic carbon 21, and finally the line at 78 ppm which can only be ascribed to the ether-type methine carbon 13. A wealth of information, however, is revealed by the off-resonance spectrum. In the low-field region we find three doublets as expected. It is interesting to note the marked difference in residual splitting of the lines at 101 and 106 ppm. These have to be assigned to carbons 4 and 1, respectively. This assignment is based on the known deshielding of proton 4 as a consequence of the lactam carbonyl neighbor anisotropy effect. Since the decoupler frequency is centered to high field, a proton absorbing at lower field should be less effectively decoupled. Among the six triplets observed in the off-resonance spectrum, the one at lowest field (also having the largest residual splitting) must be carbon 22 in the $\alpha$ position to the ether oxygen. Since the protons at carbon 19 are deshielded by both the amine nitrogen and the olefinic double bond, its residual splitting should be larger than

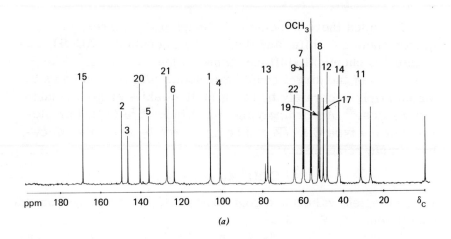

*(a)*

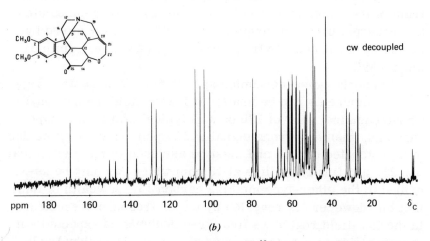

*(b)*

Figure 12-3. (*a*) Proton noise–decoupled $^{13}$C FT spectrum of brucine, 1.2 M in CDCl$_3$. The assignments are based on the off-resonance decoupled spectrum, chemical shift arguments, and spin–lattice relaxation times. (*b*) Proton single frequency off-resonance decoupled spectrum of the same sample. The irradiating frequency was centered at −4 ppm from TMS.

164

that of any other remaining triplet. This allows the line at 52.7 ppm to be assigned to carbon 19 and that at 50.2 ppm with a slightly smaller residual splitting to carbon 17. Although the protons on carbon 14 and 17 may be expected to have similar chemical shifts and the calculated $^1J$(C-H) values do not differ significantly from each other, carbon 17 must be less shielded than 14. Of the remaining three methylenes, carbon 14 can be assigned to the resonance at 42.3 ppm. No distinction can be made, however, between carbons 16 and 10. Among the methine carbons, the lines originating from carbons 7, 9, and 11 should have a larger residual splitting in the off-resonance decoupled spectrum compared to carbon 12. The doublet at 48.3 ppm, exhibiting the smallest spacing, is therefore due to carbon 12. It is shifted considerably to low field because of the large degree of $\alpha$ substitution. Among the other three doublets, carbon 11 should be more highly shielded than carbons 9 and 7, which are bonded to the electronegative nitrogen. The only quartets are those arising from the methoxyl carbons. The singlet in the high-field region must of course be assigned to the only fully substituted $sp^3$ carbon 8. The quarternary aromatic and olefinic carbons are assigned with the aid of relaxation arguments as discussed in the next section.

Selective decoupling is of limited applicability, but it is a useful tool if the proton spectrum is approximately first order. An example is provided by melampodin, whose structure was recently reported by Fischer et al.[23] The 100 MHz proton NMR spectrum, shown in Fig. 12-4, is to a sufficient extent first order with little overlap of resonances. The $^{13}$C spectrum[24] represented by Fig. 12-5 shows three lines in the high-field region between 10 and 20 ppm assignable to the three C-methyl carbons, a group of nine lines between 50 and 80 ppm which are due to the C-O carbons, a group of olefinic carbons between 115 and 135 ppm, and finally the three carbonyls between 164 and 167 ppm. The four epoxide carbons, the methoxyl, and carbon 7 can be expected to occur in the group between 50 and 60 ppm. Chemical shift arguments probably cannot contribute further to the analysis. From the off-resonance spectrum the following carbons can immediately be assigned: OCH$_3$ at 51.9

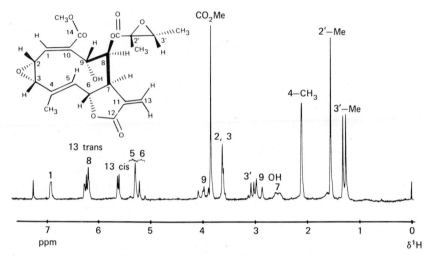

Figure 12-4. 100 MHz FT spectrum of melampodin. The assignments are those given by Fischer and co-workers.[23]

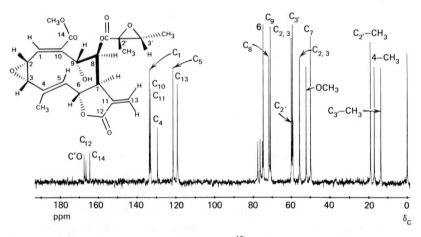

Figure 12-5. Proton noise-decoupled $^{13}$C FT spectrum of melampodin. The assignments are based on chemical shift considerations and selective decoupling experiments.

ppm, $C_2'$ at 58. 7 ppm, and $C_{13}$ at 118. 5 ppm because their multiplicities occur just once each. One of the olefinic methine carbons absorbs at 13 ppm to lower field than the two remaining methines. This is compatible only with carbon 1 which must be considerably deshielded as a consequence of mesomeric interaction with the acetyl carbon attached to carbon 10.

Although the proton chemical shift differences are as small as 0. 2 ppm, selective decoupling has proved to be possible and has permitted the assignment of all methine and methyl carbons. Figure 12-6 shows a series of selectively decoupled spectra with the decoupler centered at the proton frequencies of some methine and methyl resonances. Irradiating at 4. 0 ppm collapses the $C_9$ signal while affecting the methoxyl and epoxy carbons $C_2$ and $C_3$ less. Irradiation at 5. 2 ppm decouples protons 5 and 6. Since the corresponding carbons are separated by almost 50 ppm, $C_5$ is an olefinic, $C_6$ an aliphatic carbon, and no ambiguities arise. Irradiation at 2. 6 ppm affects $C_7$ and to a lesser extent also $C_3'$, whose proton absorbs at approximately 0. 4 ppm to lower field than that of $C_7$.

The signals due to nonproton-bearing carbons can be sorted out during the selective decoupling experiments because these carbons are subjected to long-range $^{13}$C-C-H and $^{13}$C-C-C-H couplings. By perturbing the corresponding proton resonance these quarternary carbon signals therefore experience substantial intensity enhancements. For example, when the decoupler frequency is aligned at $\delta = 2. 1$ ppm or $\delta = 2. 6$ ppm (Figs. 6a and 6b), the intensity of the carbonyl signal at 167. 0 ppm increases significantly. This allows the signal to be ascribed to $C'O$, because this is the only carbonyl exhibiting long-range couplings to high-field protons.

The application of this technique requires careful adjustment of the decoupler frequency as well as of its power level.


## SPECTRAL COMPARISON


A complete assignment in complex molecules can generally not be achieved without utilizing the tool of spectral comparison.

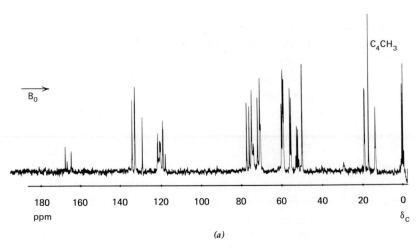

*(a)*

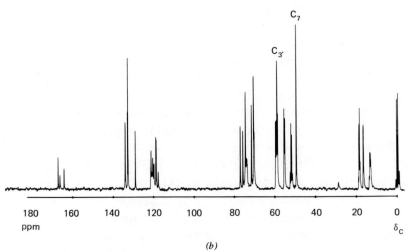

*(b)*

168

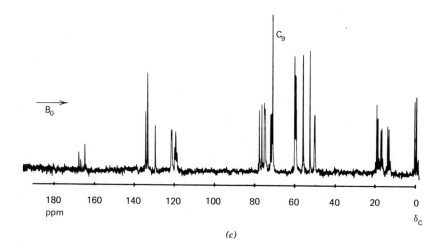

*(c)*

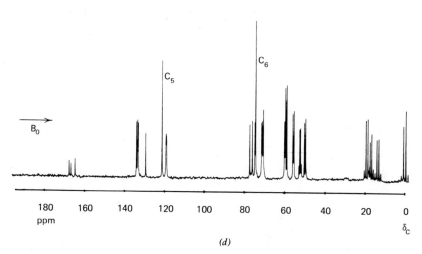

*(d)*

Figure 12-6. Selectively decoupled $^{13}$C FT spectra of melampodin.
The proton decoupler frequency was centered at *(a)* 2.1 ppm; *(b)* 2.6
ppm; *(c)* 4.0 ppm; and *(d)* 5.2 ppm. All values refer to the proton signal of TMS.

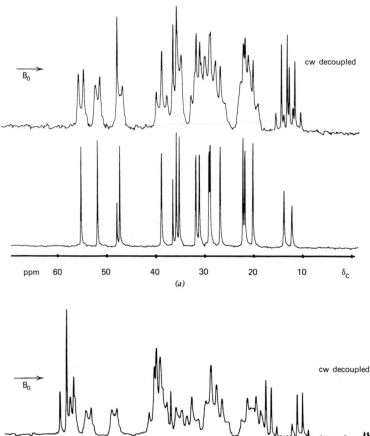

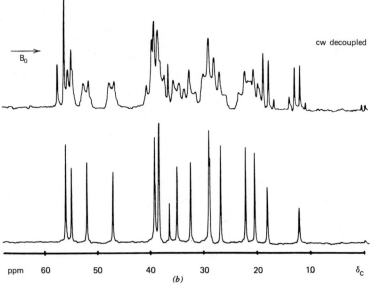

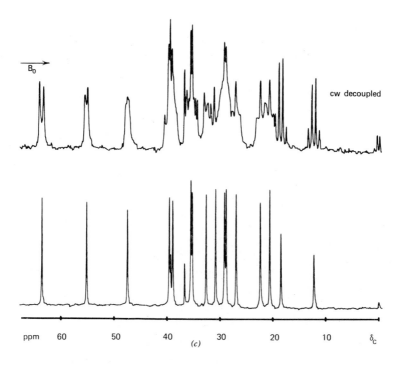

Figure 12-7. Proton noise and single frequency off–resonance de-coupled $^{13}$C FT spectra of (*a*) 17–androstanone; (*b*) 16–androstanone; and (*c*) 15–androstanone. The off–resonance decoupled spectra were obtained by centering the decoupler frequency at − 4 ppm from TMS (only high–field region shown).

This is not possible, however, unless a series of closely related compounds is available. Roberts and co-workers[3] have succeeded in fully assigning all the $^{13}C$ resonances in cholestane and a number of related steroids by introducing functional groups at various sites in the molecule and simply comparing their spectra, assuming that the perturbation of the $\alpha$ and $\beta$ carbons is greatest and no relevant effect is felt by carbons more than two bonds away.

In a carbon-13 NMR study of the complete androstanone series we have attempted to assign all 19 resonance lines with the purpose of obtaining information on how far the spectral analogy reaches in the different configurational isomers. For the ring D-substituted group of isomers there is found to be relatively little perturbation of carbons in rings A and B. To a somewhat less extent, an analogous relationship is observed within the ring A-substituted series. The greatest inconsistencies are shown to arise for the ring B isomers 6- and 7-androstanone. This is supposed to be due to the close steric proximity of the substituent to carbons in all adjacent rings and is presumably a consequence of steric interactions. In general the introduction of a carbonyl group more severely perturbs the overall geometry of the steroid scaffold than a substituent that does not alter the hybridization of the substituted carbon.

By carefully examining the single frequency off-resonance decoupled spectra it can be seen that the residual multiplets exhibit, apart from their different spacings, different line shapes. The latter is due to incompletely removed geminal C-C-H long-range couplings which can give rise to a significant broadening of lines. By comparing, for example, the lowest-field triplet in the off-resonance decoupled spectrum of 17-androstanone (Fig. 12-7) with the two quartets corresponding to the methyl carbons 18 and 19, it is evident that the latter two multiplets are much better resolved. This is because these two carbons are bonded to quarternary carbons, and thus no geminal long-range couplings are involved. By taking into consideration the fact that the methine carbons 9 and 5 bear 3 and 4 $\alpha$ hydrogens, respectively, the three low-field signals in 15-

ketoandrostane can immediately be assigned. Carbon 8, due to lower $\alpha$ substitution, is expected to be more highly shielded. Its resonance appears at 32.5 ppm, which is confirmed by the characteristic steric compression shift this carbon experiences in 15-androstanone. The spectra of the ring D-substituted series with partial assignments are shown in Fig. 12.8. The assignment of carbon 14 is verified by its downfield shift in 15-androstanone, while both carbons 5 and 9 do not significantly alter their resonance frequencies within the ring D-substituted series. In a similar manner the quaternary carbons 10 and 13 as well as the methyl carbons 18 and 19 can be assigned unambiguously. Whereas carbon 10 retains its position, carbon 13 in 17-androstanone is shifted to lower field. This is typical of carbons $\alpha$ to a carbonyl group. Carbon 15, which is known to absorb at very high field due to its strong steric interaction with carbon 19, corresponds to the line at 21.8 ppm, because this is no longer present in 15- and 16-androstanone. In the latter isomer, carbon 15 absorbs at 38.4 ppm which can be concluded from the large spacing of the residual triplet. Carbons 1–4, 6, and 11 are assigned on account of the great similarity of the corresponding chemical shifts reported for cholestane.[3] The remaining two signals are assigned to carbons 7 and 12, respectively. Carbon 12 should interact strongly with the carbonyl group in 17-ketoandrostane and can be expected to become deshielded in the 16- and 15-keto isomers, while for carbon 7 an upfield shift in the 15-keto isomer is to be anticipated. There remains, however, some ambiguity in this assignment.

Similar arguments have enabled the assignment of most of the resonance lines in the $^{13}$C spectra of 1-, 2-, 3-, and 4-androstanone as well as those of 11- and 12-androstanone. Very little spectral similarity is observed, however, between 6- and 7-androstanone.

Partial assignments obtained in the manner described above and by comparison with previously reported data are presented in Table 12-1. It is planned to extend this work with the goal of achieving complete assignment by employing rare earth shift reagents. This technique has been applied recently by several workers.[25–27]

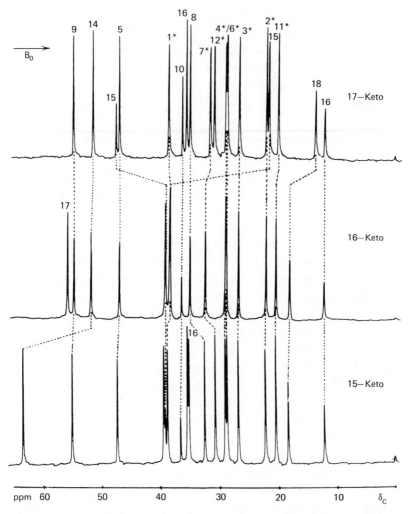

Figure 12-8. $^{13}$C spectral comparison of 15-, 16-, and 17-andros-tanone. The assignments are based on residual splittings as observed in the off-resonance decoupled spectra, and compared chemical shifts within this series of isomers. The assignments marked with an asterisk are those found in cholestane by Roberts and co-workers.[3]

TABLE 12-1. $^{13}$C chemical shifts in the androstanone series[a]

| Carbon | Androstan-$n$-one | | | | | | | | | | |
|---|---|---|---|---|---|---|---|---|---|---|---|
| | 1 | 2 | 3 | 4 | 6 | 7 | 11 | 12 | 15 | 16 | 17 |
| 1 | 214.86 | 53.97 | 38.47 | 37.65 | 38.49 | 38.39 | | 38.32 | 38.75 | 38.38 | 38.70 |
| 2 | 38.11 | 210.71 | 37.73 | 22.64 | 21.54 | 21.81 | 21.89 | 21.97 | 22.19 | 22.07 | 22.14 |
| 3 | | | 209.21 | | | 26.52 | 26.78 | 26.60 | 26.77 | 26.71 | 26.76 |
| 4 | 28.02 | 29.22 | 44.32 | 211.31 | 20.39 | 29.20 | 28.63 | 28.86 | 28.98 | 28.92 | 29.03 |
| 5 | 47.12 | 45.15 | 46.46 | 58.93 | 58.58 | 47.21 | 46.90 | 46.96 | 47.23 | 46.94 | 47.11 |
| 6 | 27.88 | 28.02 | 28.88 | 20.53 | 210.36 | | 28.51 | 28.86 | 28.68 | 28.80 | 28.80 |
| 7 | 31.42 | 32.08 | 31.97 | 30.82 | 46.89 | 211.23 | | | 30.71 | 32.38 | 31.72 |
| 8 | 36.02 | 35.23 | 35.60 | 35.38 | 38.49 | 50.44 | | 35.03 | 32.52 | 34.99 | 35.13 |
| 9 | 49.56 | 54.39 | 54.21 | 54.60 | 54.99 | 48.89 | 64.86 | 54.65 | 55.00 | 54.74 | 54.97 |
| 10 | 51.71 | 40.56 | | 42.35 | 41.46 | 36.72 | 35.97 | 36.93 | 36.51 | 36.43 | 36.44 |
| 11 | 20.39 | 20.47 | 20.39 | 20.53 | 20.58 | 20.62 | 210.42 | 37.55 | 20.46 | 20.40 | 20.11 |
| 12 | 25.40 | 25.50 | 25.40 | 25.42 | | | 56.88 | 215.21 | | | |
| 13 | 40.86 | 40.56 | 40.61 | 40.68 | 41.10 | 40.72 | 44.86 | 54.95 | 39.12 | 39.02 | 47.71 |
| 14 | 54.34 | 54.21 | 54.02 | 54.34 | 54.62 | 56.45 | 54.16 | 56.49 | 63.34 | 51.82 | 51.65 |
| 15 | 22.49 | 20.98 | 21.36 | 21.71 | 21.23 | 21.34 | 20.99 | 19.56 | 215.91 | 38.38 | 21.77 |
| 16 | | | | | | | | | 35.07 | 217.45 | 35.78 |
| 17 | | | | | | | | | | 55.78 | 220.61 |
| 18 | 17.74 | 17.42 | 17.41 | 17.54 | 17.46 | 17.41 | 18.18 | 17.73 | 18.31 | 18.06 | 13.83 |
| 19 | 12.10 | 12.45 | 11.22 | 13.67 | 12.95 | 11.66 | 12.10 | 11.84 | 12.17 | 12.17 | 12.22 |

[a] Relative to internal TMS. All samples were measured as 1–2 M solutions in 5-mm tubes, requiring approximately 15 min total observing time.

175

## THE SPIN-LATTICE RELAXATION TIME

Allerhand and co-workers[14] have recently shown that knowledge of the spin-lattice relaxation time $T_1$ not only provides useful information on molecular and segmental motion, but also that this parameter is well suited as an assignment tool. They have found that in large unsymmetrical molecules which are expected to reorient isotropically, under conditions of the least narrowing limit

$$(\omega_C + \omega_H)\tau_R \ll 1 \, ,$$

the relaxation mechanism even for nonprotonated carbons is predominantly dipolar. In such cases the relaxation rate of a particular carbon can be described by the following simple relationship

$$\frac{1}{T_1} = \frac{1}{T_2} = \hbar^2 \gamma_H^2 \gamma_C^2 \sum_i r_{CH_i}^{-6} \tau_R \, .$$

In this equation $\gamma_H$ and $\gamma_C$ are the magnetogyric ratios, $r_{CH_i}$ is the internuclear distance between the carbon and the $i$th hydrogen, and $\tau_R$ stands for the overall correlation time.

In the absence of internal motion, the relaxation rate is therefore proportional to the number of directly attached protons. In cholesteryl chloride, for instance, the methine carbons of the steroid backbone are found to have relaxation times twice as long as those for methylene carbons. This holds only as long as there is no internal motion involved. For a rapidly rotating methyl group, where $\tau_G$, the correlation time for group reorientation, is much smaller than $\tau_R$, $T_1$ of such a carbon is expected to increase. In the case of anisotropic motion or fast internal reorientation, the above simple relationship no longer applies. It has to be replaced[14] by

$$\frac{1}{T_1} = \frac{1}{T_2} = \hbar^2 \gamma_H^2 \gamma_C^2 \sum_i r_{CH_i}^{-6} \chi \, ,$$

where $\chi = 1 + [3(1 - \rho)/(5 + \rho)] \times \sin^2\Theta \{1 + [3(1 - 2\rho)/2(1 + \rho)] \sin^2\Theta\}$ and $\rho$ is defined as follows:

$$\rho = \frac{D_{\shortparallel}}{D_{\perp}}$$

$$D_{\perp} = (6\tau_R)^{-1}$$

$$D_{\shortparallel} = (6\tau_{\shortparallel})^{-1} = (6\tau_R)^{-1} + (6\tau_G)^{-1} \ .$$

$\tau_{\shortparallel}$ is the total correlation time for the rotating group and is related to $\tau_R$ and $\tau_G$ by

$$\frac{1}{\tau_{\shortparallel}} = \frac{1}{\tau_R} + \frac{1}{\tau_G} \ ,$$

and $\Theta$ is the angle between the C-H bond vector and the axis of internal reorientation.

If internal motion is much slower than overall reorientation ($\tau_G \gg \tau_R$), then $\chi$ becomes equal to unity, whereas in the case of fast internal reorientation ($\tau_G \ll \tau_R$), $\chi$ becomes

$$\chi = (\tfrac{3}{2}\cos^2\Theta - \tfrac{1}{2})^2 \ .$$

From this it follows for a rotating group of tetrahedral symmetry (CH$_3$) that $T_1$ will be nine times longer than in the absence of internal rotation.

Instead of 0.17 sec ($T_1$ for a methine carbon is found to be 0.52 sec[14]), 1.5 sec is observed for the relaxation time of the angular methyl groups in cholesteryl chloride. Such carbons can therefore be distinguished in a straightforward manner from those in the steroid scaffold, which is particularly useful in spectra where the lines are close together and off-resonance spectra consequently are no longer analyzable.

Quarternary carbons are shown to exhibit the longest relaxation times, although in molecules similar to cholesterol, the dipolar mechanism is still dominant; the latter can be concluded from the Overhauser enhancement factor, which is still close to its theoretical maximum.

There are basically two techniques that permit the determination of $T_1$ values of individual lines in a pulsed high-resolution spectrometer, the inversion-recovery[11, 12] and progressive saturation[13] techniques. In this chapter the latter method has been applied throughout, because it often allows a rough estimate of the spin-lattice relaxation time without investing much instrument time.

In such a pulse-excited progressive saturation experiment, a repetitive train of radio-frequency pulses is applied on the spin system and thus a steady state is established between the effect of the pulse and the relaxation. Provided the magnetization is all transverse after the pulse and, on the other hand, these components are destroyed before the next pulse elapses, the spin dynamics are entirely governed by relaxation. The magnetization $M(\tau)$ as a function of the pulse repetition time $\tau$ is then given by

$$M(\tau) = M_{\infty}[1 - \exp(-\tau/T_1)] .$$

The number of experiments to be performed depends on the accuracy required. It is therefore often sufficient to record just two spectra: one with $\tau \geqslant 4 T_{1\,\text{max}}$ yielding the equilibrium value of the magnetization (with an error due to incomplete recovery of the signal of less than 2%), and one with $\tau < T_1$.

There are a few instrumental requirements to be satisfied before a progressive saturation experiment can be performed. Owing to deviation from the ideal condition

$$\frac{\gamma B_1}{2\pi} \gg \Delta F ,$$

not all resonances within the observed frequency range $\Delta F$ are equally excited. In the rotating frame this means that the effective field $B_{\text{eff}}$, which is equal to the vector sum of the offset $\Delta B$ from resonance and the rf field $B_1$, related by

$$B_{\text{eff}} = (B_1{}^2 + \Delta B^2)^{1/2} ,$$

is tilted with respect to the $x'$ axis. It has been shown,[13] however, that for ratios $\Delta B/B_1 \leqslant 0.8$ and flip angle deviations, $\Delta\alpha$, as high as 0.13 rad (7.5°) the errors in apparent $T_1$ values do not exceed ±5%. With pulse widths $\tau_{\pi/2} \leqslant 50$ $\mu$sec, frequency offsets up to 4 kHz from the carrier can be observed at 23 kG.

It is important, furthermore, that sampling of the free induction decay does not start before the steady-state regime is

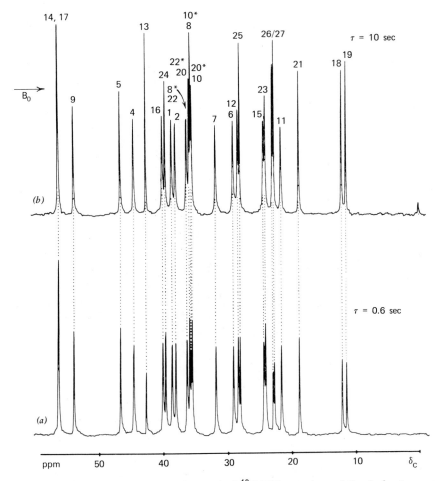

Figure 12-9. Proton noise–decoupled $^{13}C$ FT spectra of 3–cholestanone (only high–field region shown). Both traces were obtained by applying 500 pulses with $\alpha = \gamma B_1 \tau = \pi/2$ and 0.6 sec acquisition time. The employed pulse repetition time was (a) 0.6 sec; (b) 10 sec. The assignments are those given by Roberts and co–workers,[3] except for the ones marked with an asterisk, which are based on the characteristic relaxation behavior of the carbons in question. The significant peak height differences observed in the completely relaxed spectrum (b) reflect different effective spin–spin relaxation times $T_2^*$.

attained. The latter is generally reached after three precession periods.

The first example demonstrating the usefulness of the technique as an assignment tool is shown by its application to a steroid. The two spectra of 3-cholestanone, represented by Fig. 12-9, were obtained by accumulating and Fourier transforming 500 transients with a pulse repetition time of 10 and 0.6 sec, respectively. No attempt was made to calculate the relaxation times. The assignments are those given by Roberts and co-workers.[3] Since signals due to rapidly relaxing nuclei are more heavily attenuated when the repetition time is shortened, quarternary and methyl carbons and generally those exhibiting more segmental motion, such as the ones of the aliphatic side chain, can be identified readily. Between 36 and 35 ppm there are four lines which are ascribed to carbons 22, 20, 8, and 10 in order of increasing field, whereas the correct ordering should be 8, 22, 10, and 20, because these four nuclei have characteristically different spin-lattice relaxation times. The relative ordering of the corresponding relaxation times are expected to be: $T_1(10) > T_1(20) > T_1(22) > T_1(8)$. It is interesting to note that the peak heights in the completely relaxed spectrum differ significantly from each other. Since the integrated intensities are found to be equal for all signals, intrinsically narrower lines give rise to larger peak heights. From the spectrum of Fig. 12-9$b$ it is obvious that nonprotonated carbons, angular methyl, and side-chain carbons exhibit considerably higher peaks, thus reflecting their longer effective spin-spin relaxation times.

Let us now return once more to brucine, whose low-field resonances cannot all be assigned by the previously discussed techniques. The $^{13}C$ spectrum shown in Fig. 12-10, representing the low-field region, which was obtained by applying 90° pulses and a pulse repetition time of 50 sec, yields equal integrated intensities for all carbon signals. Since it can be assumed safely that in such a large molecule, protonated carbons experience full Overhauser enhancement, it can furthermore be concluded that this holds also for quarternary carbons. The relaxation mechanism is therefore of the dipolar type and the

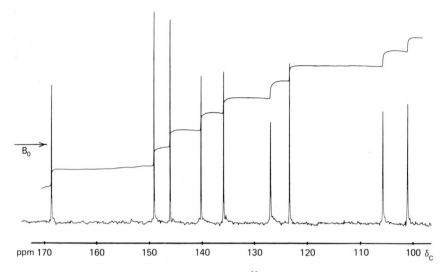

Figure 12-10. Proton noise-decoupled $^{13}$C FT spectrum of brucine (only low-field region shown). The spectrum was obtained on a de-gassed 1.2 M solution in deuterochloroform by accumulating 100 transients ($\alpha = \gamma B_1 \tau = \pi/2$) using a pulse repetition time of 50 sec in order to allow for complete relaxation between the pulses. The integrated intensities are equal within experimental error, thus indicating equal Overhauser enhancement for all carbons.

relaxation rate $1/T_1$ must thus be proportional to $\sum_i r_{CH_i}^{-6}$. Carbons 2, 3, 5, and 6 should in this case have the longest relaxation times since they have only one adjacent proton each. Signals due to carbons 2 and 3 can be distinguished easily from those due to carbons 5 and 6 because they must be deshielded as a consequence of their attachment to the electronegative oxygen. The lines at 149.1 and 146.0 ppm are therefore to be assigned to carbons 2 and 3 or vice versa, with relaxation times of 9.35 and 10.70 sec. Carbon 5 should be considerably less shielded than carbon 6 because it is bonded to nitrogen. Based on these considerations the following assignments are made: $C_5$, 136.0 ppm ($T_1 = 7.25$ sec); and $C_6$, 123.5 ppm ($T_1 = 11.75$ sec). For the carbonyl carbon which was previously sorted out for its characteristic low-field shift, a $T_1$ value of 4.3 sec is determined. This is anticipated because two $\alpha$ protons con-

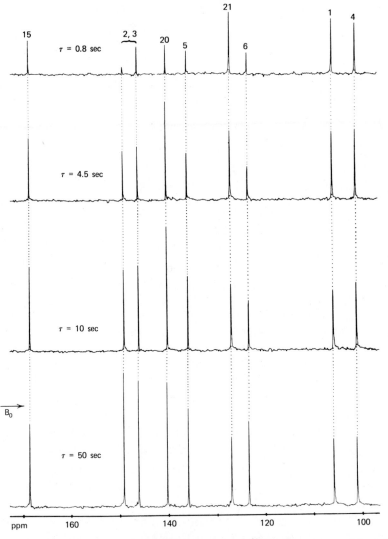

Figure 12-11. Set of progressive saturation $^{13}$C FT spectra of brucine. The instrumental parameters were the same as those employed in the spectrum of Fig. 12-10, with the exception that the pulse repetition time $\tau$ was varied as indicated. Not all traces are shown from which the relaxation times listed in Table 12-2 were deduced.

182

tribute to its relaxation.  Finally carbon 20 is relaxed by four α protons and its $T_1$ should be approximately one-fourth of those found for carbons 2, 3, 5, and 6.  The experimental value of 2.90 sec is in good agreement with these considerations.  A similar value is found for carbon 8 with $T_1 = 2.62$ sec, because this carbon is also relaxed by four α protons.  A better quantitative agreement cannot be expected because the C-H bond distance varies as a function of the carbon hybridization.

The spectral set from which these relaxation data are derived is shown in Fig. 12-11, while Fig. 12-12 represents the corresponding semilogarithmic plot of $M_0/(M_0 - M)$ versus $\tau$. The relaxation times of the quarternary carbons in brucine are listed in Table 12-2.

The relaxation times of the protonated carbons are found to be too short to be measured by the progressive saturation

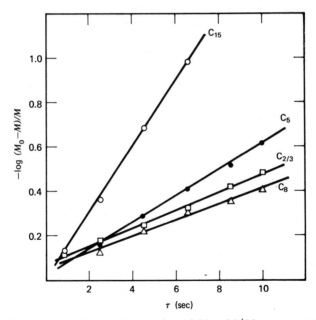

Figure 12-12.  Semilogarithmic plot of $M_0 - M/M$ versus $\tau$ from which the spin-lattice relaxation times of the quarternary carbons in brucine were derived (for the sake of clearness not all lines are shown).

TABLE 12-2.  Spin-lattice relaxation times of non-
protonated carbons in 1.2 M brucine,
determined by the progressive saturation technique

| Carbon | Chemical Shift (ppm) | $T_1$ (sec) |
|--------|----------------------|-------------|
| 15     | 168.6                | 4.30        |
| 2/3    | 149.1                | 9.35        |
| 2/3    | 146.1                | 10.70       |
| 20     | 140.3                | 2.90        |
| 5      | 136.0                | 7.25        |
| 6      | 123.5                | 11.75       |
| 8      | 51.8                 | 2.62        |

technique and were therefore determined by the inversion-recovery method.  Their values are found to be on the order of
0.20 sec for methylene and 0.40 sec for methine carbons.  Considerably longer relaxation times are observed for the methoxyl
carbons, this again being a consequence of segmental motion.
Full relaxation data on brucine and other complex alkaloids will
be reported elsewhere.

## EXPERIMENTAL

All spectra shown in the figures were recorded on a Varian
XL-100-15 NMR spectrometer equipped with a 16K 620L computer system and pulse hardware for Fourier transform operation.  Chemical shifts are reported in ppm downfield from internal TMS.  The chemical shift accuracy is better than 0.05
ppm.  The computer system allows acquisition of 8K data points
yielding 4K plot points after transformation.  If not otherwise
stated, the spectra were obtained with full proton noise decoupling by centering the synthesizer frequency at 4 ppm downfield from TMS.  Off-resonance single frequency decoupling

was achieved by turning off the noise modulation and centering the decoupler frequency at $-4$ ppm (high field) from TMS. In the selective decoupling experiments the decoupler frequency was aligned according to the observed proton frequencies and the power level was reduced to approximately one-third of its maximum value (10 W). The progressive saturation spin-lattice relaxation measurements were performed with the same program by applying 90° pulses and varying the pulse interval. The pulse width $\tau_{\pi/2} = 46 \times 10^{-6}$ sec corresponds to $\gamma B_1 / 2\pi = 5400$ Hz and thus allowed $T_1$ measurements over a range of 4000–5000 Hz with an error of less than 5%. The pulse width was calibrated by adjusting this parameter until the second null signal was observed ($2\pi$ pulse). Its accuracy is better than $\pm 5°$.

## References

1.  R. R. Ernst, *J. Chem. Phys.*, **45**, 3845 (1966).
2.  F. J. Weigert, M. Jautelat, and J. D. Roberts, *Proc. Natl. Acad. Sci. U.S.*, **60**, 1152 (1968).
3.  H. J. Reich, M. Jautelat, M. T. Messe, F. J. Weigert, and J. D. Roberts, *J. Am. Chem. Soc.*, **91**, 7445 (1969).
4.  L. F. Johnson and M. E. Tate, *Can. J. Chem.*, **47**, 63 (1969).
5.  R. A. Archer, R. D. G. Cooper, P. Demarco, and L. F. Johnson, *Chem. Commun.*, 1291 (1970).
6.  D. M. Grant and E. G. Paul, *J. Am. Chem. Soc.*, **86**, 2984 (1964).
7.  D. K. Dalling and D. M. Grant, *J. Am. Chem. Soc.*, **89**, 6612 (1967).
8.  M. Tanabe, H. Seto, and L. F. Johnson, *J. Am. Chem. Soc.*, **92**, 2157 (1970).
9.  J. W. Westley, D. L. Pruess, and R. G. Pitcher, *Chem. Commun.*, 161 (1972).
10.  R. R. Ernst and W. A. Anderson, *Rev. Sci. Instr.*, **37**, 93 (1966).
11.  R. L. Vold, J. S. Waugh, M. P. Klein, and D. E. Phelps, *J. Chem. Phys.*, **48**, 3831 (1968).

12. R. Freeman and H. D. W. Hill, *J. Chem. Phys.*, **51**, 3140 (1969); **53**, 4103 (1970).

13. R. Freeman and H. D. W. Hill, *J. Chem. Phys.*, **54**, 3367 (1971).

14. A. Allerhand, D. Doddrell, and R. Komoroski, *J. Chem. Phys.*, **55**, 189 (1971).

15. G. C. Levy and D. C. Dittmer, *Org. Magn. Resonance*, **4**, 107 (1972).

16. R. Hollenstein and W. v. Philipsborn, *Helv. Chim. Acta*, **55**, 2030 (1972).

17. M. Karplus and J. A. Pople, *J. Chem. Phys.*, **38**, 2803 (1963).

18. T. D. Alger, D. M. Grant, and E. G. Paul, *J. Am. Chem. Soc.*, **88**, 5397 (1966).

19. B. V. Cheney and D. M. Grant, *J. Am. Chem. Soc.*, **89**, 5319 (1967).

20. D. M. Grant and B. V. Cheney, *J. Am. Chem. Soc.*, **89**, 5315 (1967).

21. L. F. Johnson, private communication.

22. E. R. Malinowski, *J. Am. Chem. Soc.*, **83**, 4479 (1961).

23. N. H. Fischer, R. Wiley, and J. D. Wander, *Chem. Commun.*, 137 (1972).

24. N. H. Fischer, N. S. Bhacca, and F. W. Wehrli, to be published.

25. B. Birdsall, J. Feeney, J. A. Glasel, R. J. P. Williams, and H. V. Xavier, *Chem. Commun.*, 1473 (1971).

26. M. Christl, H. J. Reich, and J. D. Roberts, *J. Am. Chem. Soc.*, **93**, 3463 (1971).

27. O. A. Gansow, M. R. Willcott, and R. E. Lenkinski, *J. Am. Chem. Soc.*, **93**, 4295 (1971).

CHAPTER 13

# $^{13}C$-$^{13}C$ COUPLING CONSTANTS

Gary E. Maciel

*Department of Chemistry*
*Colorado State University*
*Fort Collins*

## INTRODUCTION

In the last two or three years, $^{13}C$ NMR has evolved from
a promising, but somewhat esoteric, development in a few
laboratories into a popularly acknowledged primary tool for
chemical studies. However, of the various types of high-resolution NMR parameters that most readily come to mind for organic compounds, $^{13}C$-$^{13}C$ coupling constants $J(C$-$C)$ have probably been the most neglected. The reason for this is clear.
The low natural abundance (1%) of $^{13}C$ combined with the com-
paratively unfavorable intrinsic sensitivity of the nuclide to
NMR detection result in a somewhat formidable challenge to
experimental studies, even with relatively modern equipment.
Until recent instrumental advances, $^{13}C$-enriched materials
were an absolute requirement for such studies.

We have been interested in $^{13}C$-$^{13}C$ coupling constants for
more than ten years. Our early interest in $^{13}C$-$^{13}C$ coupling
constants, and presumably that of other early workers in the
field, stemmed from a belief that such studies would constitute
an attractive and straightforward approach for elucidating de-

187

tails of bonding in carbon frameworks, especially in view of
the great deal of attention then being given to the proposed relationship between $J$(C-H) and carbon hybridization.[1-4] However, until relatively recently, the overall status of $J$(C-C)
studies was that they were of immediate interest probably to
only a very small group of participating laboratories. By many
chemists, even NMR specialists, [13]C-[13]C coupling constants
were likely to be considered spectroscopic curiosities.

Recent developments in the [13]C NMR field will transform
this restricted scope of interest in the $J$(C-C) parameter. With
the introduction of pulse Fourier transform (FT) techniques,[5]
making possible rapid repetitions of [13]C spectral observations,
it is now possible to measure [13]C-[13]C coupling constants reli-
ably on natural abundance samples. This promises to alter the
scope of interest in $J$(C-C) values for a variety of reasons.
First of all, the experimental accessibility of values of $J$(C-C)
and the promise of fruitful explorations of the fundamental nature of C-C bonding via this parameter will attract considerable
activity. Second, as $J$(C-C) data become more accessible and
more adequately represented in the literature, a realization
will emerge that these parameters can be useful in making
spectral assignments in certain [13]C NMR studies. Third, with
contemporary reductions in the cost of [13]C-enriched materials
and the concomitant popularization of [13]C NMR as a viable trac-
er technique, more and more spectra are likely to be determined on [13]C-enriched substances, especially reaction products
that contain mixtures of isotopic isomers. In such cases, the
sensitivity of the FT method is such that "satellite" peaks due
to [13]C-[13]C splittings will be prominent in the spectra obtained
under conditions that determine the "background" level of [13]C
resonances due to natural abundance. Such satellite peaks will
be recognized as bonuses in the sense that they can be used in
making assignments or confirming them. Thus within a very
short time, [13]C-[13]C coupling constants may emerge from a history of specialized curiosity and constitute the basis of frequently useful spectroscopic studies.

## EARLY $^{13}$C-$^{13}$C COUPLING STUDIES

The first literature reports of the determination of carbon-carbon coupling constants were those by Lynden-Bell and Sheppard[6] and by Graham and Holloway.[7] In both cases the reports presented data on the fundamental compounds, ethane ($J$(C-C) = 34.6 Hz), ethylene ($J$(C-C) = 67.6 Hz), and acetylene ($J$(C-C) = 171.5 Hz). The results were obtained on doubly enriched samples, using conventional single-scan, field-sweep techniques. It was demonstrated that the $J$(C-C) values could be extracted from either the $^1$H or $^{13}$C spectra. The results were discussed qualitatively in terms of percent "carbon $s$ characters" in the carbon-hybrid orbitals involved in the C-C bonding. The qualitative indications were that carbon-carbon couplings might well be dominated by the $s$ character of the carbon-hybrid orbital. At the time it was believed that there was a simple linear relationship between $J$(C-H) values and the $s$ character in the carbon-hybrid orbitals directed toward hydrogen, based on provocative empirical relationships for hydrocarbons that were reported earlier by Muller and Pritchard[1,2] and Shoolery.[3] These empirical relationships were interpreted in early theoretical terms by valence bond treatments,[8-10] and by Muller and Pritchard in terms of a localized molecular orbital picture. These interpretative approaches yield results that are essentially equivalent to those of the McConnell formula, given here in the form of equation 1:

$$J(\text{A-B}) = (\tfrac{4}{3})^2 \hbar \beta^2 \gamma_A \gamma_B (\Delta E)^{-1} s_A{}^2(0) s_B{}^2(0) P^2_{s_A s_B} , \qquad (1)$$

where $\beta$ is the Bohr magneton, $\gamma_A$ is the magnetogyric ratio of nucleus A, $\Delta E$ is a mean electronic excitation energy resulting from an approximation common to treatments at this level, $s_A{}^2(0)$ is the value of the valence-shell $s$ orbital of atom A at the nucleus, and $P_{s_A s_B}$ is the molecular orbital bond order between valence shell $s$ orbitals of atoms A and B.

The simple relationship between the formulation embodied in equation 1 and the concept of $s$ character can be seen readily for the case of a localized molecular orbital $\psi$, expressed as a combination of hybrid orbitals $\phi_A$ and $\phi_B$ centered on carbon atoms $C_A$ and $C_B$.

$$\psi = \frac{[2(1+\alpha)]^{1/2}}{2}\phi_A + \frac{[2(1-\alpha)]^{1/2}}{2}\phi_B \tag{2}$$

$$\phi_A = \lambda_A^{1/2}s_A + (1-\lambda_A)^{1/2}p_A \tag{3}$$

and

$$\phi_B = \lambda_B^{1/2}s_B + (1-\lambda_B)^{1/2}p_B . \tag{4}$$

In equation 2, $\alpha$ represents the ionic character in the bond. In equation 3, $\lambda_A$ represents the fractional $2s$ character of the hybrid orbital of atom $C_A$ forming the sigma bond to atom $C_B$. Then, using the standard formula for computing bond orders,

$$P_{s_A s_B} = 2\frac{[2(1+\alpha)\lambda_A]^{1/2}}{2} \times \frac{[2(1-\alpha)\lambda_B]^{1/2}}{2} \tag{5}$$

$$P_{s_A s_B}^2 = (1+\alpha)(1-\alpha)\lambda_A\lambda_B . \tag{6}$$

From equation 6, one sees that $P_{s_A s_B}^2$ is equivalent to the product of the $s$ characters in the sigma-bonded hybrid orbitals, times a factor that is rather close to unity except for large ionic characters in the sigma bond. Furthermore, any attempt to discuss $^{13}C$-$^{13}C$ coupling constants in terms of $s$ characters exclusively requires the assumption that the product of the other factors in an equation such as equation 1 is constant. Such assumptions have previously been subject to considerable criticism.

In 1963, Frei and Bernstein published data on $^{13}C$-$^{13}C$ coupling constants that were also based on the use of dilabeled compounds and conventional field-sweep techniques. This paper reported data on 13 compounds that represent a variety of C-C bond types.[11] These data and the earlier results on ethane, ethylene, and acetylene are collected in Table 13-1. The results of this study were also considered in terms of apparent hybridization parameters, based on a localized molecular orbital approach. In addition, Frei and Bernstein determined empirical values of carbon hybridization in compounds containing $C_A$-$C_B$ bonds, by collecting $J(C_A$-$C_B)$ data on the corresponding $C_A$-$H_B$ cases, and by utilizing the popular Muller-Pritchard[1]

TABLE 13-1. Some early $^{13}C$–$^{13}C$ spin-spin coupling data[a]

| Compound | J(C–C) (Hz) | Compound | J(C–C) (Hz) | Compound | J(C–C) (Hz) |
|---|---|---|---|---|---|
| $CH_3CH_3$ | 34.6[b] | $\overset{*}{C}H_3$–CN | 57.3 | $C_6H_5CH\overset{*}{=}\overset{*}{C}HCO_2C_2H_5$ | 76.3 |
| $\overset{*}{C}H_3CH_2C_6H_5$ | 34 ± 1 | $CH_2=\overset{*}{C}H_2$ | 67.6[b] | ⬡—$\overset{*}{C}\equiv N$ | 80.3 |
| $\overset{*}{C}H_3CHOHC_6H_5$ | 38.1 | $C_6H_5C\equiv \overset{*}{C}-\overset{*}{C}H_3$ | 68.6 | $C_6H_5\overset{*}{C}\equiv \overset{*}{C}-CO_2C_2H_5$ | 126.8 |
| $\overset{*}{C}H_3COC_6H_5$ | 43.3 | $C_6H_5\overset{*}{C}H=\overset{*}{C}H_2$ | 70 ± 3 | $C_6H_5\overset{*}{C}\equiv \overset{*}{C}-CN$ | 155.8 |
| ⬡—$\overset{*}{C}H_3$ | 44.2 | ⬡—$\overset{*}{C}OCl$ | 74.1 | $H\overset{*}{C}\equiv \overset{*}{C}H$ | 171.5[b] |
| | | | | $C_6H_5\overset{*}{C}\equiv \overset{*}{C}H$ | 175.9 |

[a]All data from reference 11, except where indicated otherwise.
[b]Taken from references 6 and 7.

191

formula relating $J$(C-H) coupling constants to carbon-hybrid $s$ characters. The results obtained by this "fitted," empirical approach are much better correlated than those based on hybridization parameters obtained simply by accounting for the carbon coordination number.

As predicted from the elementary treatments of the Fermi contact mechanism referred to above, the sign of a directly bonded $^{13}$C-$^{13}$C coupling constant is expected to be positive. This was confirmed (relative to the sign of $J$(C-H)) for acetonitrile by McLauchlan[12] and for acetic acid by Grant[13] on the basis of double resonance techniques, using $^{13}$C-enriched substances.

### MORE RECENT STUDIES ON DIRECTLY BONDED CARBON ATOMS

In 1967 Weigert and Roberts published a communication which reported $^{13}$C-$^{13}$C coupling constants obtained on natural abundance samples of substituted cyclopropanes.[14] The data were obtained using multiple-scan time averaging, with a frequency-sweep spectrometer, and were consistent with an interpretation in terms of the percent $s$ characters of the carbon-hybrid orbitals in the cyclopropane ring. Using the Muller-Pritchard formula and the known value of $J$(C-H) in cyclopropane, it was concluded that each hybrid orbital in a C-C bond should have about 16% $s$ character. The experimental data, shown in Table 13-2, are found to be consistent with this prediction, in terms of the known $J$(C-C) value for ethane, a $sp^3$-$sp^3$ case, and an assumed proportionality between $J(C_A-C_B)$ and $\lambda_A \lambda_B$.

Bertrand, Grant, Allred, Hinshaw, and Strong[15] reported $J$(C-C) data in small, fused hydrocarbon ring systems, based upon spectra obtained on natural abundance samples and cw time averaging. The results were interpreted on the basis of carbon orbital hybridizations, assuming that $J$(C-C) is proportional to the product of orbital $s$ characters, and the assumption that hybridizations are predictable from bond angle and symmetry considerations. A self-consistent picture

TABLE 13-2.   $^{13}$C-$^{13}$C coupling constants of cyclopropane derivatives

| Compound | Bond | $J$(C-C)(Hz) | Reference |
|---|---|---|---|
| Cyclopropyl chloride | 1,2 | 13.9 | 18 |
| Cyclopropyl bromide | 1,2 | 13.3 | 14 |
| Cyclopropyl iodide | 1,2 | 12.9 | 14 |
| 1,1-Dichlorocyclopropane | 1,2 | 15.5 | 14 |
| Methylcyclopropane | 1,X | 44.0 | 14 |
| Cyclopropyl cyanide | 1,2 | 10.9 | 18 |
|  | 1,X | 77.9 | 18 |
| Cyclopropanecarboxylic acid | 1,2 | 10.0 | 18 |
|  | 1,X | 72.5 | 18 |
| Dicyclopropyl ketone | 1,2 | 10.2 | 18 |
|  | 1,X | 54.0 | 18 |

emerged, lending support to the view that $^{13}$C-$^{13}$C coupling constants provide valuable information on hybridization, at least in hydrocarbon systems. Similar conclusions have been drawn from analogous data on somewhat larger polycyclic saturated hydrocarbons.[16] The experimental data from these two studies are summarized in Table 13-3.

Using the cw time-averaging technique on natural abundance samples, Litchman and Grant[17] published in 1967 a report of $^{13}$C-$^{13}$C coupling constants in a series of $t$-butyl compounds. This was the first report of $J$(C-C) data in which the emphasis was on substituent effects. The results were discussed in relationship to $J$(C-H) data, and it was concluded that bond polarization effects (e.g., variations in $s_A^2(0)s_B^2(0)$) might not be as important in the $J$(C-C) case as was believed to be the case with $J$(C-H).

A recent report by Weigert and Roberts[18] described the results of a study of $^{13}$C-$^{13}$C coupling constants for several types

TABLE 13-3. $^{13}C$–$^{13}C$ coupling constants of some small ring hydrocarbons

| Compound | Bond | J(C–C) (Hz) | Reference | Compound | Bond | J(C–C) (Hz) | Reference |
|---|---|---|---|---|---|---|---|
| | 1,2 | 12.6 | 16 | $C^1H_2=C^2=CH_2$ | 1,2 | 98.7 | 15 |
| | 1,7 | 41.5 | 16 | | | | |
| | 1,7 | 40.4 | 16 | | 1,3 | 20.2 | 15 |
| | 3,4 | 29.8 | 16 | | | | |
| | 1,2 | 33.4 | 16 | | 1,2 | 29.8 | 15 |
| | 1,7 | 32.5 | 16 | | 3,5 | 36.0 | 15 |
| | 2,3 | 18.2 | 15 | | 1,2 | 21.0 | 15 |
| | | | | | 1,2 | 36.7 | 15 |
| | | | | | 4,5 | 16.0 | 15 |

TABLE 13-4.   $J$(C-C) data for the ethyl, isopropenyl, and acetyl series[a]

|  | Ethyl[b] | Isopropenyl[c] | | Acetyl[d] |
|---|---|---|---|---|
|  |  | $CH_3$-C | $CH_2$=C |  |
| C≡N | 33.0 | 44.9 | 73.8 | — |
| $CH_3$ | 33 ± 2 | 41.8 | 72.6 | 40.1 |
| $C_6H_5$ | 34 ± 1 | 42.9 | 72.9 | 43.3 |
| H | 34.6 | 41.9 | 70.0 | 39.4 |
| COOH | 34.8 | 45.0 | 70.5 | — |
| $CONH_2$ | — | 43.8 | 70.6 | — |
| $N(R)_2$ | 35.8 (H) | — | — | 52.2 ($CH_3$) |
| OH | 37.7 | — | — | 56.7 |
| $OCOCH_3$ | 38.6 | 51.2 | 84.2 | 60.0 |
| $OCH_2CH_3$ | 38.9 | 51.8 | 80.9 | 58.8 |
| I | 35.8 | — | — | 46.5 |
| Br | 36.0 | — | — | 54.1 |
| Cl | 36.1 | 48.5 | 80.8 | 56.1 |

[a]Values in Hz.
[b]Data taken from reference 23.
[c]Data taken from reference 28.
[d]Data taken from reference 19.

of compounds.  The data were obtained by the cw time-averaging approach.  The results are discussed qualitatively in terms of the relation between structure (including substituent effects) and the assumed Fermi contact coupling mechanism, via the role of carbon hybridization and effective nuclear charge.

A study of $^{13}C$-$^{13}C$ coupling constants in a series of acetyl compounds has been reported.[19]  The data (given in Table 13-4) were obtained on di- or monolabeled compounds, utilizing frequency-sweep techniques and time averaging.  The range of

TABLE 13-5. Some calculated and experimental $J$(C–C) values

| | Calculated[a] | | | Experimental | | Reference |
|---|---|---|---|---|---|---|
| | $J(C_A C_B)$(Hz) | $P_{s_A s_B}$ [b] | $\Delta^c$(Hz) | $J(C_A C_B)$(Hz) | $\Delta^c$(Hz) | |
| Bond type $sp^3$–$sp^2$ | | | | | | |
| *CH₃CH=CH₂ | 55.4 | 0.2734 | −9.9 | | | |
| *CH₃COC₂H₅ | 63.4 | 0.2817 | −1.9 | 38.4 | −1.7 | 19 |
| *CH₃COH | 64.0 | 0.2885 | −1.3 | 39.4 | −0.7 | 19 |
| *CH₃COCH₃ | 65.3 | 0.2833 | 0 | 40.1 | 0 | 19 |
| *CH₃COC₆H₅ | 66.0 | 0.2820 | 0.7 | 43.3 | 3.2 | 19 |
| *CH₃COŌ(aq) | 71.1 | 0.3006 | 5.8 | 51.6 | 11.5 | 19 |
| *CH₃CON(CH₃)₂ | 78.7 | 0.2951 | 13.4 | 52.2 | 12.1 | 19 |
| *CH₃COOH | 80.7 | 0.3046 | 15.4 | 56.7 | 16.6 | 19 |
| *CH₃COOC₂H₅ | 82.4 | 0.3040 | 17.1 | 58.8 | 18.7 | 19 |
| *CH₃COF | 91.6 | 0.3125 | 26.3 | | | |

196

Bond type $sp^3$–$sp$

| Bond type $sp^3$–$sp$ | | | | | |
|---|---|---|---|---|---|
| *CH₃*C≡*O⁺ | 64.4 | 0.3551 | −12.3 | 46.5 | −10.0 | 22 |
| (CH₃)₃*CC*N | 71.3 | 0.3228 | −5.4 | 52.0 | −4.5 | 22 |
| (CH₃)₂*CHC*N | 74.5 | 0.3298 | −2.2 | 54.8 | −1.7 | 22 |
| CH₃*CH₂C*N | 76.2 | 0.3352 | −0.5 | 55.2 | −1.3 | 22 |
| *CH₃*CN | 76.7 | 0.3393 | 0 | 56.5 | 0 | 22 |
| *CH₃*C≡CH | 77.5 | 0.3314 | 0.8 | 67.4 | 10.9 | 18 |

[a]Calculated by finite perturbation/INDO method; results taken from reference 25.

[b]Calculated bond order between $2s_A$ and $2s_B$ orbitals.

[c]Difference between calculated (or experimental) coupling constant and that of an arbitrary reference (acetone or acetonitrile for these two bond types).

197

$J$(C-C) values observed in this series is found to be much larger (20 Hz) than can be accounted for simply in terms of arguments based on hybridization considerations and effective nuclear charges. A qualitative interpretation has been developed based on the application of the Pople-Santry[20] theory to a localized model of the C-C bond which is reminiscent of the MO theory of geminal H-H coupling constants of Pople and Bothner-By.[21] Results obtained on a series of nitriles have also been reported.[22] Similar techniques for the determination and interpretation of the data are employed. Major emphasis is placed on the effect of methyl substitution at $C_A$ on the $C_A$-$C_B$N coupling in a series $(CH_3)_n CH_{3-n}$-CN, and the results are included in Table 13-5.

Bartuska and Maciel[23] have reported data on twelve mono- or dilabeled ethyl compounds, for which the $^{13}$C-$^{13}$C coupling constants are found to vary between 33 and 39 Hz, as shown in Table 13-4. This is a much smaller range than had been observed previously for comparably substituted acetyl compounds. These data are also obtained by multiple-scan, time-averaging techniques, in the frequency-sweep mode, and have been interpreted via the finite perturbation/INDO method.[24,25] This molecular orbital approach, discussed in some detail below, is carried out at the level of intermediate neglect of differential overlap[26] (INDO) and provides a complete density matrix of the ground state with each calculation. Hence it is often of interest to search for correlations between computed coupling constants and density matrix elements from the point of view of exploring how studies of coupling constants can be used to elucidate certain features of electronic distributions.

For the ethyl compounds, variations of the computed $^{13}$C-$^{13}$C coupling constants with changes in substituents have been investigated empirically with respect to the changes that occur in the computed density matrix. Two interesting correlations emerge from this empirical evaluation of the calculated results. Approximate relationships are found to exist between the computed $^{13}$C-$^{13}$C coupling constants and ($a$) measures of the electron-withdrawing characteristics of the substituent, and

(b) the extent to which the substituent polarizes the C-C fragment of the ethyl group. These two electron-distribution characteristics are presented formally in terms of the parameters,
$\Delta$ and $\pi$, where $\Delta = 8 - (P_1 + P_2)$ and $\pi = (P_2 - P_1)$, and where the
$P$'s represent total calculated valence-shell electron densities
of the carbon atoms in question. The subscripts 1 and 2 refer
to the carbon atoms situated at the $\alpha$ and $\beta$ positions, respectively, with respect to the substituent. These parameters are
directly related to substituent characteristics that have been
classified under the term "$-I^+$" by Pople and Gordon.[27] This
label describes a substituent that withdraws electron density
from the carbon framework and polarizes it in the sense in

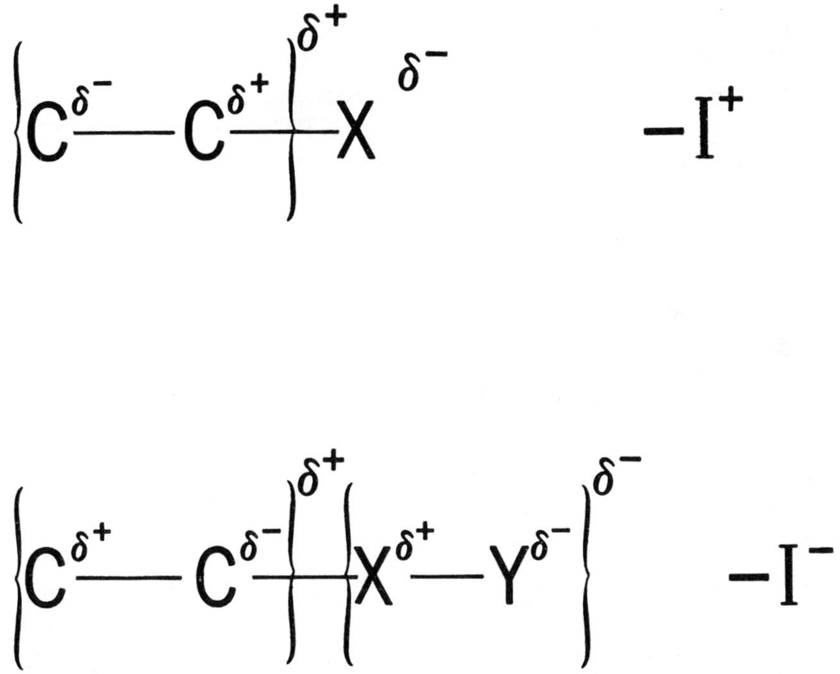

Figure 13-1. Pictorial representation of $-I^+$ and $-I^-$ substituent effects.

which the $\beta$ carbon has a higher electron density than the $\alpha$ carbon (shown pictorially in Fig. 13-1). Hence it is concluded that $^{13}C$-$^{13}C$ coupling constants in the ethyl series reflect the $-I^+$ substituent characteristics of the attached group. It is noteworthy that the lowest computed (and experimental) values of $J$(C-C) in the ethyl system, other than for hydrocarbons, occur with $-I^-$ substituents, for example, -CN and -CO$_2$H, which also withdraw electron density from the carbon framework, but which polarize it in the opposite sense to the $-I^+$ case.

The great disparity between the relatively low sensitivity to substituent variation of $J$(C-C) in the ethyl group (6 Hz range, 22% increase from lowest to highest value) compared to the great sensitivity in the acetyl group (50% increase from lowest to highest value) constituted a provocative set of results. It is of interest to discover whether the higher acetyl sensitivity is due primarily to the presence of a nominally $sp^2$ carbon at the position of substitution, or if the carbonyl group itself is specifically responsible. In order to answer this question, a study was undertaken by Bartuska and Maciel[28] to determine the $^{13}C$-$^{13}C$ coupling constants in the isopropenyl system, where the

$$CH_3-C\overset{\displaystyle /\!\!/ CH_2}{\underset{\displaystyle \backslash X}{}}$$

CH$_3$-C=CH$_2$ group is formally similar to the CH$_3$C=O group of the acetyl system. $^{13}C$ NMR spectra were obtained on the isopropenyl compounds by the same techniques employed for the ethyl compounds. The $J$(C-C) values, given in Table 13-4, exhibit a 10 Hz range (22% increase from lowest to highest value) for the CH$_3$C case, and a 14 Hz range (20% increase from lowest to highest values) for the C=CH$_2$ case. These results indicate that it is specifically the carbonyl group that is associated with the high sensitivity found in the acetyl system. The isopropenyl system was also subjected to finite perturbation/INDO

TABLE 13-6. $^{13}$C–$^{13}$C coupling constants between a ring carbon and an attached carbon ($C_x$) in X-substituted benzenes

| X | $J(C–C_x)$ (Hz) | Reference | X | $J(C–C_x)$ (Hz) | Reference |
|---|---|---|---|---|---|
| $CH_3$ | 44.2 | 11, 29 | $CO_2^-Na^+$ | 65.9 | 29 |
| $CH_2OH$ | 47.7 | 29 | $CO_2H$ | 71.9 | 29 |
| $CH_2Cl$ | 47.8 | 29 | $CO_2CH_3$ | 74.8 | 29 |
| $C(CH_3)_2OH$ | 47.3 | 42 | $CO_2C_2H_5$ | 74.3 | 42 |
| $CN$ | 80.4 | 11, 29 | $COCl$ | 74.3 | 11, 29 |
| $CO_2H$ in mesitoic acid | 71.0 | 42 | $C(CH_3){=}CH_2$ | 52.2 | 42 |

calculations to explore the relationships between coupling constants and electron distributions. Rough correlations are again found between the computed coupling constants and the parameters $\Delta$ and $\pi$ mentioned above. It is found that the acetyl case, also treated by the finite perturbation/INDO method, follows the same rough relationships.

The great sensitivity difference between the acetyl and ethyl series has been explored further at an empirical level by comparing the above-mentioned correlations within the four series of $J$(C-C) cases. This was done by determining the dependence of the coupling constant upon changes in the $\Delta$ and $\pi$ parameters, and also determining the dependence of these computed parameters upon changes in substitution. The result of this approach was to show that the uncommonly high $J$(C-C) range in the acetyl series is *not* due to an uncommonly high sensitivity of the electron distribution to substituent changes. On the other hand, it is found that the computed $J$(C-C) values for the acetyl system are uncommonly sensitive to even the small changes in the electron distribution, as reflected in the $\Delta$ and $\pi$ parameters.

Ihrig and Marshall[29] have recently communicated the results of a study of C(7)-labeled derivatives of toluene and benzoic acid, in which $^{13}$C-$^{13}$C coupling constants were observed in the cw mode. These data also demonstrate that the presence of a carbonyl group in the coupled carbons enhances the sensitivity of the couplings to substituent variation, and are included in Table 13-6.

Using natural abundance samples and the repetitive pulse Fourier transform method, Summerhays and Maciel have obtained $^{13}$C-$^{13}$C coupling constants on a series of isopropyl and $t$-butyl compounds.[30] This represents the first application of modern FT approaches to the determination of $J$(C-C) values, and demonstrates the suitability of that approach for natural abundance samples. Figure 13-2 shows a typical example of the $^{13}$C satellites from which these data are extracted.

From the data on isopropyl and $t$-butyl compounds, given in Table 13-7, one sees that the sensitivity to substituent vari-

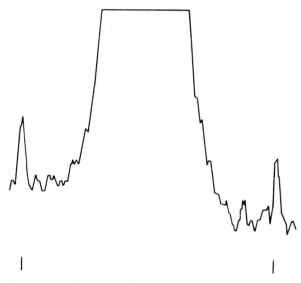

Figure 13-2.   Spectral region of the CH region of diisopropyl ether, after 55 min of observation time.   Satellite splitting is 39.9 Hz.   Taken from reference 30.

ation is approximately the same as reported earlier for ethyl and $t$-butyl compounds; however, in the latter case, a substantially different pattern of substituent effects is obtained from what had been reported earlier.[17]  For a given substituent, $J$(C-C) is found generally to increase with methyl substitution in the ethyl, isopropyl, $t$-butyl series.  Standard arguments in terms of electronic effects (e.g., inductive, hyperconjugative, and polarization) seem incapable of explaining this trend, and the finite perturbation/INDO calculations did not alter this view. It was concluded that steric influences were operative, and that the main effect was an enhancement of the R-C-R' angle in a $CH_3RR'C$-X system as R and/or R' are made $CH_3$ instead of H. The hybridization changes that one would expect with such geometry perturbations are consistent with the trend in $J$(C-C) values.  Finite perturbation/INDO calculations on $t$-butyl systems in which the $CH_3$-C-$CH_3$ angle was varied substantiate this view, and indicate that the origin of the trend is almost entirely based

TABLE 13-7.  $^{13}C-^{13}C$ coupling constants in
monosubstituted alkanes, $RX^a$

|  | R | | |
| X | Ethyl[b] | $i$-Propyl[c] | $t$-Butyl[c] |
| --- | --- | --- | --- |
| CN | 33 ± .5 | — | 33.6 |
| $C_6H_5$ | 34.2 | 34.8 | 35.7 |
| $CO_2H$ | 34.8 | 34.4 | 35.3 |
| $\overset{\overset{O}{\parallel}}{C}-CH_3$ | 35.5 | 34.7 | 34.9 |
| $CH_3$ | 33 ± 2 | — | 35.7 |
| $NO_2$ | 35.7 | 36.2 | 37.6 |
| I | 35.8 | 36.7 | 36.6 ± .5 |
| Br | 36.0 | 37.1 | 37.8 |
| Cl | 36.1 | 37.3 | 38.0 |
| $NH_2$ | 35.8 | 37.3 | 37.6 |
| OH | 37.7 | 38.6 | 39.6 |
| $OCH_3$ | 39.0 | 39.9 | 40.3 |

[a]Values in Hz.
[b]Data from reference 23.
[c]Data from reference 30.

upon hybridization changes (changes in $P^2_{s_A s_B}$).

Further support for the steric interpretation of the ethyl, isopropyl, $t$-butyl trend is obtained from recent data obtained on some $n$-propyl, isobutyl, and neopentyl systems.[31] These data are compared in Table 13-8 with the corresponding results on ethyl, isopropyl, and $t$-butyl compounds. If the steric effect is truly responsible for the trend of decreasing $J(C-C)$ values in the series $t$-butyl, isopropyl, ethyl, then this trend should continue with further methyl substitution at the $\beta$ position in the

TABLE 13-8. Relationships among $^{13}\text{C}-^{13}\text{C}$ coupling constants in monosubstituted alkanes, $RX^a$

| X | $\begin{array}{c}\text{C}\\|\\\text{C--C--X}\\|\\\text{C}\end{array}$ | $\begin{array}{c}\text{C}\\\diagdown\\\text{C--X}^b\\\diagup\\\text{C}\end{array}$ | $\text{C--C--X}^c$ | $\begin{array}{c}\text{C}\\\diagdown\\\overset{*}{\text{C}}\text{--C--X}^d\end{array}$ | $\begin{array}{c}\text{C}\\\diagdown\\\overset{*}{\text{C}}\text{--C--X}^d\\\diagup\\\text{C}\end{array}$ | $\begin{array}{c}\text{C}\\|\\\text{C--}\overset{*}{\text{C}}\text{--C--X}^d\\|\\\text{C}\end{array}$ |
|---|---|---|---|---|---|---|
| $CO_2H$ | 35.3 | 34.4 | 34.8 | 34.2 | 33.7 | |
| $C_6H_5$ | 35.7 | 34.8 | 34 ± 1 | 33.4 | 33.3 | 33.2 |
| I | 36.6 | 36.7 | 35.8 | 35.4 | 34.9 | |
| Br | 37.8 | 37.1 | 36.0 | 36.2 | 35.4 | 34.9 |
| Cl | 38.0 | 37.3 | 36.1 | 35.6 | 35.8 | 35.8 |
| OH | 39.6 | 38.6 | 37.7 | 37.6 | 37.4 | |

$^a$Values in Hz.
$^b$Data from reference 30.
$^c$Data from reference 23.
$^d$Data from reference 31.

Gary E. Maciel

series $n$-propyl, isobutyl, neopentyl. Table 13-8 shows that this trend is followed approximately.

## STUDIES OF LONG-RANGE $^{13}C$-$^{13}C$ COUPLING CONSTANTS

There have been a few studies of geminal and longer-range $^{13}C$-$^{13}C$ coupling constants reported.[18,29,32,33] The available data on geminal $J(C-C)$ values are summarized in Table 13-9. Unfortunately, some of the primary benefits to be derived from

TABLE 13-9.  Geminal $^{13}C$-$^{13}C$ coupling constants[a]

| Compound | $^3J(C-C)$ (Hz) | Reference | Compound | $^3J(C-C)$ (Hz) | Reference |
|---|---|---|---|---|---|
| CH$_3$-O-CH$_3$ | (−)2.4 | 33 | ⬡—C(=O)OCH$_3$ | 2.63 | 29 |
| CH$_3$-O-C(=O)-Cl | −2.8 | 32 | ⬡—CH$_3$ | 3.10 | 29 |
| CH$_3$-C(=O)-CH$_3$ | +16.1 | 33 | ⬡—CH$_2$OH | 3.45 | 29 |
| CH$_3$-CCl=CH$_2$ | 4.6 | 42 | ⬡—CH$_2$Cl | 3.69 | 29 |
| CH$_3$-C≡CH | 11.8 | 18 | ⬡—CO$_2$-Na$^+$ | 2.23 | 29 |
| CH$_3$COCH$_2$CH$_3$ | 15.2 | 18 | ⬡—CO$_2$H | 2.54 | 29 |
| (cyclobutanone) | 9.0 | 18 | ⬡—CO$_2$CH$_3$ | 2.38 | 29 |
| CH$_3$CH$_2$CN | 33.0 | 22 | ⬡—COCl | 3.53 | 29 |
| (CH$_3$)$_3$CCN | 33.6 | 22 | ⬡—C(CH$_3$)=CH$_2$ | 3.9 | 42 |
| CH$_3$-Hg-CH$_3$ | +22.4 | 33 | ⬡—C≡N | 2.61 | 29 |

[a]The absence of explicit sign implies that the sign was not determined.

long-range $J$(C-C) values rest upon a knowledge of the signs, and these have been determined in very few cases.

## $^{13}$C-$^{13}$C COUPLING CONSTANTS IN $^{13}$C PEAK ASSIGNMENTS

Dorn and Maciel[34] have recently demonstrated how $^{13}$C-$^{13}$C double resonance approaches can be used in making $^{13}$C peak assignments in cases where other approaches are not practicable. Such a case is the assignment of the $^{13}$C resonances of the C(2) and C(3) carbons in 1-substituted 4-methylbicyclo-[2.2.2]octanes. Figure 13-3 shows the type of $^{13}$C resonance

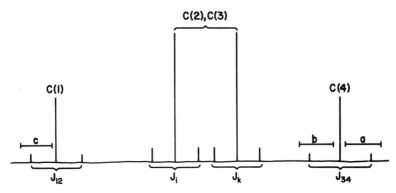

pattern ($^{1}$H-decoupled) that one expects for the bicyclo[2.2.2]-octane carbons, and one can see that if the indicated $J$(C-C) satellites from splittings of directly bonded carbons can be matched up properly, then the distinction between C(2) and C(3)

Figure 13-3. Idealized representation of the proton-decoupled $^{13}$C resonance pattern of the C(1), C(2), C(3), and C(4) carbons of a 1-substituted bicyclo [2.2.2] octane system, showing exaggerated $^{13}$C splitting satellites. Taken from reference 34.

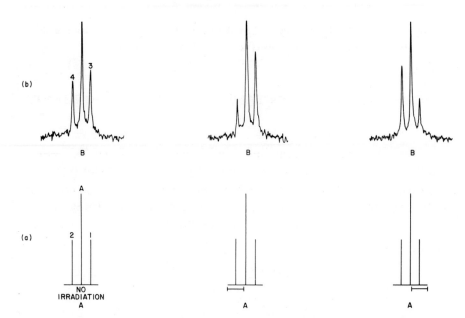

Figure 13-4.   Adiabatic rapid passage $^{13}$C-$\{^{13}$C$\}$ double resonance experiment on 25% dilabeled ethyl acetate, $C^AH_3C^BH_2OCOCH_3$ (25% $C^{A*}$-$C^B$, 25% $C^{A*}$-$C^{B*}$, 25% $C^A$-$C^{B*}$, 25% $C^A$-$C^B$), employing proton noise decoupling and 100 8-μsec pulses.   The number 1, 2, 3, and 4 refer to transition.   (a) Representation of the irradiation of resonance lines in the A region (CH$_3$), indicating the range of 36-Hz sweeps (0.9 sec). (b) Effects on the B region (CH$_2$) of the $^{13}$C spectrum under conditions of the adiabatic rapid passage over the regions of the A spectrum indicated in 3a.   Taken from reference 34.

can be made directly.   The satellites of the C(2) and C(3) carbon resonances were observed in the pulse FT mode, while the $^{13}$C satellites about the C(1) or C(4) carbon resonances are subjected to adiabatic rapid passage.   Figure 13-4 shows the type of intensity perturbations that one obtains for directly bonded carbons.   In this way, the C(2), C(3) assignments can be made on the basis of the predicted signal perturbations for irradiation of a directly bonded (and strongly coupled) carbon.   The saving in time over alternative methods of assignment (e.g., deuterium labeling) is enormous.

TABLE 13-10. $J$(C–C) and $J$(C–H) values of some simple cyclic hydrocarbons[a]

| Compound | Calculated | | | | Experimental | |
|---|---|---|---|---|---|---|
| | $J$(C–C) | $P^2_{s_Cs_C}$ | $J$(C–H) | $P^2_{s_Cs_H}$ | $J$(C–C) | $J$(C–H) |
| | | | | | | |
| $C_a$ | −1.8 | 0.02729 | 190.5 | 0.3512 | | 202 |
| $C_b$ | 32.6 | 0.05176 | 165.1 | 0.2913 (endo) | 21.0[b] | 170 |
| | | | 148.1 | 0.2775 (exo) | | 152 |
| $C_3H_6$ | 21.4 | 0.04653 | 155.2 | 0.2858 | | 161 |
| $C_4H_8$ (puckered) | 37.5 | 0.04389 | 128.3 | 0.2511 | | 134.6 |
| $C_5H_{10}$ ($C_2$ or $C_s$) | 44.2 | 0.05636 | 120.1 | 0.2344 | | 128.5 |
| $C_6H_{12}$ (chair) | 43.8 | 0.0612 | 118.1 | 0.2292 | | 124 |

[a]Calculated and experimental data for $J$(C–H) taken from reference 36. Calculated results for $J$(C–C) taken from reference 38 (same geometries used as described in reference 36). Values in Hz.
[b]Taken from reference 15.

TABLE 13-11. $J$(C-C) and $J$(C-H) values of some simple heterocyclic compounds[a]

$(CH_2)_n \overbrace{\phantom{xx}} X$

| Compound | Calculated | | | | Experimental | |
| --- | --- | --- | --- | --- | --- | --- |
| | $J$(C-C) | $P^2_{s_C s_H}$ | $J$(C-H) | $P^2_{s_C s_H}$ | $J$(C-C) (ref.) | $J$(C-H) |
| X = NH | | | | | | |
| $C_2H_5N$ | 24.9 | 0.0517 | 163.4 | 0.3002 | | 168.0 |
| $C_3H_7N$ | 47.1 | 0.0478 | 132.4α | 0.2606 | | 140.0 |
| | | | 131.6β | 0.2590 | | 134.0 |
| $C_4H_9N$ | 44.0αβ | 0.0576 | 132.5α | 0.2511 | | 139.1 |
| | 41.1βγ | 0.0550 | 122.1β | 0.2375 | | 131.4 |
| $C_4H_{11}N$ | 48.2αβ | 0.0630 | 124.9α | 0.2409 | | 133.4 |
| | 42.1βγ | 0.0583 | 119.5β | 0.2315 | | 125.6 |
| X = O | | | | | | |
| $C_2H_4O$ | 38.0 | 0.0640 | 166.1 | 0.3078 | | 175.7 |
| $C_3H_6O$ | 46.5 | 0.0500 | 142.8α | 0.2670 | | 148 |
| | | | 122.2β | 0.2387 | | 137.3 |
| $C_4H_8O$ | 44.1αβ | 0.0588 | 140.7α | 0.2650 | | 144.6 |
| | 39.6βγ | 0.0548 | 122.6β | 0.2386 | | 133.2 |
| $C_5H_{10}O$ | 50.0αβ | 0.0645 | 133.6α | 0.2537 | | 139.4 |
| | 40.1βγ | 0.0608 | 116.1β | 0.2256 | | 128.0 |

210

X = CO

| Compound | | | | | | |
|---|---|---|---|---|---|---|
| $C_3H_4O$ | 18.3αβ | 0.0335 | 152.2 | 0.2870 | | |
| | 37.2αCO | 0.0668 | | | | |
| $C_4H_6O$ | 49.4αβ | 0.0514 | 123.4α | 0.2397 | 28.5(18) | 134.8 |
| | 45.2αCO | 0.4836 | 124.6β | 0.2474 | 29.7(18) | 138.7 |
| $C_5H_8O$ | 47.1αβ | 0.0568 | 117.0α | 0.2284 | 34.4(18) | 129.4 |
| | 37.1βγ | 0.0562 | 122.3β | 0.2358 | 37.2(18) | |
| | 67.0αCO | 0.0780 | | | | |
| $C_6H_{10}O$ | 42.7αβ | 0.0577 | 115.8α | 0.2257 | | 128.0 |
| | 38.5βγ | 0.0527 | 121.5β | 0.2312 | 37.3(18) | |
| | 63.5αCO | 0.0797 | | | | |
| furan (O, α, β) | 94.5 | 0.1302 | 175.9α | 0.3339 | 69.1(18) | 201.4[b] |
| | | | 156.1β | 0.3039 | | 175.3[b] |
| pyrrole (H, N) | 89.1 | 0.1217 | 166.8α | 0.3221 | 65.6(18) | 184[c] |
| | | | 153.5β | 0.3001 | | 171[c] |

[a]Calculated and experimental data for $J(C-H)$ taken from reference 36, except where noted. Calculated results for $J(C-C)$ taken from reference 38 (same geometries used as described in reference 36). All $J(C-C)$ and $J(C-H)$ values given in Hz.
[b]Taken from reference 43.
[c]Taken from reference 44.

## THEORETICAL INTERPRETATIONS

After the early valence bond and molecular orbital treatments of directly bonded coupling constants, the next major advance in spin-spin coupling theory was the molecular orbital theory of Pople and Santry.[20] This approach can be used either at the level of the "mean excitation energy" approximation (in which case it reverts to equation 1 for Fermi contact coupling), or in the expanded form employing individual "excitation" terms of second order perturbation theory. It has been applied by Pople and Santry to the $^{13}C$-$^{13}C$ couplings in simple systems. The results, although only qualitative, clearly demonstrate the dominant importance of the Fermi contact mechanism, which has been assumed in earlier work.

The most promising and generally successful theoretical approach for calculating $^{13}C$-$^{13}C$ coupling constants has been the finite perturbation method of Pople, McIver, and Ostlund.[24,25] This conceptually simple general method for calculating second-order properties has been applied to spin-spin coupling constants at the INDO MO level. In most cases the calculations have been restricted to the Fermi contact mechanism. Basically this application involves the computation of an unrestricted, single-determinant MO wave function in the presence of a contact perturbation due to the nuclear spin $I_B$ with magnetic moment $\mu_B$. In the INDO approximation this perturbation has the form

$$h_B = \frac{8\pi}{3} \beta \mu_B s_B^2(0) .$$ (7)

Then the Fermi contact contribution to the coupling constant is given by

$$J(A-B) = h \left(\frac{4\beta}{3}\right)^2 \gamma_A \gamma_B s_A^2(0) s_B^2(0) \left\{ \frac{\partial}{\partial h_B} \rho_{s_A s_A}(h_B) \right\}_{h_B=0} .$$ (8)

Thus within the framework of the approximations made, it is the sensitivity of the spin density $\rho_{s_A s_A}$ in the orbital $s_A$ to the presence of the nuclear spin perturbation $\mu_B$ that determines the coupling constant.

TABLE 13-12. Calculated contributions to $J$(C-C)[a]

| Molecule | Experimental | Theory | | | |
|---|---|---|---|---|---|
| | | Contact | Orbital | Dipolar | Total |
| $C_6H_5\overset{*}{C}\equiv\overset{*}{C}H$ | 175.9 | 134.41 | 22.01 | 8.04 | 164.46 |
| $C_2H_2$ | 171.5 | 140.80 | 23.59 | 8.31 | 172.70 |
| $C_6H_5\overset{*}{C}\overset{*}{N}$ | 80.3 | 77.94 | −2.38 | 0.46 | 76.02 |
| $\overset{*}{C}H_2{=}\overset{*}{C}HCN$ | 70.6 | 69.66 | −18.39 | 3.91 | 55.18 |
| $C_2H_4$ | 67.6 | 70.61 | −18.58 | 3.92 | 55.95 |
| $\overset{*}{C}H_3\overset{*}{C}\equiv CH$ | 67.4 | 66.54 | −2.65 | 0.53 | 64.42 |
| $C_6H_6$ | 57.0 | 64.34 | −12.83 | 1.59 | 53.10 |
| $CH_3COOH$ | 56.7 | 74.00 | −2.79 | 0.59 | 71.81 |
| $CH_3CN$ | 56.5 | 65.96 | −2.56 | 0.57 | 63.97 |
| $CH_3\overset{*}{C}H_2\overset{*}{C}N$ | 55.2 | 65.54 | −2.76 | 0.46 | 63.24 |
| $CH_3CO^+$ | 46.5 | 54.23 | −6.65 | −0.10 | 47.48 |
| $\overset{*}{C}H_3\overset{*}{C}OCH_3$ | 40.1 | 56.15 | −2.92 | 0.73 | 53.97 |
| $CH_3CHO$ | 39.4 | 60.29 | −3.19 | 0.71 | 57.81 |
| $C_2H_6$ | 34.6 | 35.63 | −2.91 | 0.73 | 33.45 |
| $\overset{*}{C}H_3\overset{*}{C}H_2CN$ | 33.0 | 34.97 | −2.48 | 0.79 | 33.27 |

[a]Values in Hz; taken from reference 40.

The method embodied in equation 8 has been applied with considerable success to the calculation of directly bonded C-H and C-C couplings in acyclic compounds,[4,25] in small ring compounds,[36] and in carbonium ions.[37] Although most structural effects and qualitative patterns of substituent effects have been accounted for rather well by this method, some difficulties have been encountered in molecules with $-I^-$ substituents. Some typical calculated results, based only upon the Fermi contact mechanism, are shown in Tables 13-5, 13-10, and 13-11, together with the corresponding experimental data and some related $J$(C-H) results.

TABLE 13–13.  Some miscellaneous $J$(C–C) values

| Compound | $J$(C–C) (Hz) | Reference | Compound | $J$(C–C) (Hz) | Reference |
|---|---|---|---|---|---|
| (*CH₃)₂C(OC₂H₅)₂ | 47.8 | 42 | cis-CHI=CHI | 78.7 | 35 |
| C₆H₅*C(CH₃)₂OH | 39.0 | 42 | trans-CHI=CHI | 78.3 | 35 |
| *CH₃CH₂COCH₂CH₃ | 35.7 | 18 | *CH₂=CHCN | 70.6 | 18 |
| CH₃*CH₂COCH₂CH₃ | 39.7 | 18 | CH₂=*CHCO₂H | 70.4 | 18 |
| *CH₃CH₂CH(OH)CH₂CH₃ | 35.0 | 18 | (CH₃)₂C=*C=*CH₂ | 99.5 | 18 |
| CH₃*CH₂CH(OH)CH₂CH₃ | 37.9 | 18 | CH₃O₂CC≡*CCO₂CH₃ | 123 | 18 |
| *CH₂(CO₂H)₂ | 56.9 | 42 | CH₂=*C(CO₂H)CH₃ | 67.7 | 42 |
| CH₂BrCH₂Br | 38.9 | 41 | CH₂=*C(CONH₂)CH₃ | 58.9 | 42 |
| CH₃CH₂F | 38.2 | 23 | CH₂=*C(CO₂CH₃)CH₃ | 70.2 | 42 |
| Bromocyclobutane | | | Cyclopentanol | | |
| 1,2 | 29.6 | 18 | 1,2 | 36.0 | 18 |
| 2,3 | 27.1 | 18 | 2,3 | 32.6 | 18 |

| | | | | | |
|---|---|---|---|---|---|
| Tetrahydrofuran | 33.0 | 18 | Cyclohexanol | | |
| C$_6$H$_6$ | 57.0 | 18 | 1,2 | 35.8 | 18 |
| | | | CH$_3$CO$_2^-$Na$^+$(aq.) | 51.6 | 28 |
| C$_6$H$_5$NO$_2$ | | | C$_6$H$_5$NH$_2$ | | |
| 1,2 | 55.4 | 18 | 1,2 | 61.3 | 18 |
| 2,3 | 56.3 | 18 | 2,3 | 58.1 | 18 |
| 3,4 | 55.8 | 18 | 3,4 | 56.2 | 18 |
| C$_6$H$_5$I | | | Pyridine α β | 53.8 | 18 |
| 1,2 | 60.4 | 18 | β γ | 53.5 | 18 |
| 2,3 | 53.4 | 18 | | | |
| 3,4 | 58.0 | 18 | Thiophene | 64.2 | 18 |
| C$_6$H$_5$OCH$_3$ | | | | | |
| 2,3 | 58.2 | 18 | | | |
| 3,4 | 56.0 | 18 | | | |

An interesting result in Table 13-10 is the computed $J(C_a\text{-}C_c)$ value (bond across ring) for bicyclo[1.1.0]butane, $-1.8$ Hz.[38] The calculated $J(C\text{-}H)$ values and many other $J(C\text{-}C)$ cases are in qualitative agreement with experiment, but there is as yet no report of an experimental value for $J(C_a\text{-}C_c)$. This is the only negative directly bonded $J(C\text{-}C)$ value so far reported for such calculations. It is of course impossible to account for a negative Fermi contact coupling by means of simple hybridization arguments (e.g., $P^2_{s_A s_B}$ in equation 1 cannot be negative). It is interesting to note that $J(C_a\text{-}C_c)$ in this system can be viewed not only as a directly bonded $^{13}C\text{-}^{13}C$ coupling, but also as the superposition of two geminal $^{13}C\text{-}^{13}C$ couplings. Because calculated $^3J(C\text{-}C)$ values are often negative, in this light, the negative value of $J(C_a\text{-}C_c)$ in bicyclobutane may not be entirely unexpected, especially in view of the rather small $s$ character of the $C_a\text{-}C_c$ bond indicated by the $P^2_{s_a s_c}$ value,[15] and by the detailed theoretical discussion of this molecule by Newton and Schulman.[39]

In 1971 Blizzard and Santry[40] extended the finite perturbation/INDO method to include the orbital and spin dipolar mechanisms. Their results put on a more quantitative basis the earlier conclusions[20] regarding the relative importance of the other two mechanisms in directly bonded $^{13}C\text{-}^{13}C$ coupling constants. Their results, some of which are given in Table 13-12, show that these contributions are nonnegligible between multiply bonded carbon atoms. Some additional miscellaneous $J(C\text{-}C)$ values are summarized in Table 13-13.

## References

1.  N. Muller and D. E. Pritchard, *J. Chem. Phys.*, **31**, 768 (1959).
2.  N. Muller and D. E. Pritchard, *J. Chem. Phys.*, **31**, 1471 (1959).
3.  J. N. Shoolery, *J. Chem. Phys.*, **31**, 1427 (1959).
4.  G. E. Maciel, J. W. McIver, Jr., N. S. Ostlund, and J. A. Pople, *J. Am. Chem. Soc.*, **92**, 1 (1970).

5.  T. C. Farrar and E. D. Becker, *Pulse and Fourier Transform NMR*, Academic Press, New York, 1971.

6.  R. M. Lynden-Bell and N. Sheppard, *Proc. Roy. Soc. London Ser. A*, **269**, 385 (1962).

7.  D. M. Graham and C. E. Holloway, *Can. J. Chem.*, **41**, 2114 (1963).

8.  M. Karplus and D. M. Grant, *Proc. Natl. Acad. Sci. U. S.*, **45**, 1269 (1959).

9.  H. S. Gutowsky and C. Juan, *J. Am. Chem. Soc.*, **84**, 307 (1962).

10. C. Juan and H. S. Gutowsky, *J. Chem. Phys.*, **37**, 2198 (1962).

11. K. Frei and H. J. Bernstein, *J. Chem. Phys.*, **38**, 1216 (1963).

12. K. A. McLauchlan, *Chem. Commun.*, **105** (1965).

13. D. M. Grant, *J. Am. Chem. Soc.*, **89**, 2228 (1967).

14. F. J. Weigert and J. D. Roberts, *J. Am. Chem. Soc.*, **89**, 5962 (1967).

15. R. D. Bertrand, D. M. Grant, E. L. Allred, J. C. Hinshaw, and A. B. Strong, *J. Am. Chem. Soc.*, **94**, 997 (1972).

16. J. B. Grutzner, M. Jautelat, J. B. Dence, R. A. Smith, and J. D. Roberts, *J. Am. Chem. Soc.*, **92**, 7107 (1970).

17. W. M. Litchman and D. M. Grant, *J. Am. Chem. Soc.*, **89**, 6775 (1967).

18. F. J. Weigert and J. D. Roberts, *J. Am. Chem. Soc.*, **94**, 6021 (1972).

19. G. A. Gray, P. D. Ellis, D. D. Traficante, and G. E. Maciel, *J. Magn. Resonance*, **1**, 41 (1969).

20. J. A. Pople and D. P. Santry, *Mol. Phys.*, **8**, 1 (1964).

21. J. A. Pople and A. A. Bothner-By, *J. Chem. Phys.*, **42**, 1339 (1965).

22. G. A. Gray, G. E. Maciel, and P. D. Ellis, *J. Magn. Resonance*, **1**, 407 (1969).

23. V. J. Bartuska and G. E. Maciel, *J. Magn. Resonance*, **5**, 211 (1971).

24. J. A. Pople, J. W. McIver, Jr., and N. S. Ostlund, *J. Chem. Phys.*, **49**, 2960, 2965 (1968).

25. G. E. Maciel, J. W. McIver, Jr., N. S. Ostlund, and J. A. Pople, *J. Am. Chem. Soc.*, **92**, 11 (1970).

26. J. A. Pople, D. L. Beveridge, and P. A. Dobosh, *J. Chem. Phys.*, **47**, 2026 (1967).

27. J. A. Pople and M. Gordon, *J. Am. Chem. Soc.*, **89**, 4253 (1967).

28. V. J. Bartuska and G. E. Maciel, *J. Magn. Resonance*, **7**, 36 (1972).

29. A. M. Ihrig and J. L. Marshall, *J. Am. Chem. Soc.*, **94**, 1757 (1972).

30. K. D. Summerhays and G. E. Maciel, *J. Am. Chem. Soc.*, **94**, 8348 (1972).

31. R. L. Elliott and G. E. Maciel, unpublished results.

32. D. Ziessow, *J. Chem. Phys.*, **55**, 984 (1971).

33. H. Dreeskamp, K. Hilderbrand, and G. Pfisterer, *Mol. Phys.*, **17**, 429 (1969).

34. H. C. Dorn and G. E. Maciel, *J. Phys. Chem.*, **76**, 2972 (1972).

35. G. E. Maciel, P. D. Ellis, J. J. Natterstad, and G. B. Savitsky, *J. Magn. Resonance*, **1**, 589 (1969).

36. P. D. Ellis and G. E. Maciel, *J. Am. Chem. Soc.*, **92**, 5829 (1970).

37. G. E. Maciel, *J. Am. Chem. Soc.*, **93**, 4375 (1971).

38. P. D. Ellis and G. E. Maciel, unpublished results.

39. M. D. Newton and J. M. Schulman, *J. Am. Chem. Soc.*, **94**, 767 (1972).

40. A. C. Blizzard and D. P. Santry, *J. Chem. Phys.*, **55**, 950 (1971).

41. R. E. Carharts and J. D. Roberts, *Org. Magn. Resonance*, **3**, 139 (1971).

42. V. J. Bartuska and G. E. Maciel, unpublished results.

43. G. S. Reddy and J. H. Goldstein, *J. Am. Chem. Soc.*, **84**, 583 (1962).

44. B. Dishler, *Z. Naturforsch.*, **19a**, 887 (1964).

## CHAPTER 14

## USES AND MISUSES OF $^{13}$C-H COUPLING CONSTANTS BETWEEN DIRECTLY BONDED NUCLEI*

Victor M. S. Gil and (in part) Carlos F. G. C. Geraldes

*The Chemical Laboratory*
*University of Coimbra*
*Portugal*

The interpretation of coupling constants between directly bonded $^{13}$C and H nuclei $J$(C-H)[1] has been dominatingly influenced by the excellent agreement between experimental $J$(C-H) values for simple hydrocarbons and the well-known theoretical model of Muller and Pritchard.[2] We refer, in particular, to the almost exact proportionality of $J$(C-H) (observed for simple hydrocarbons) to the inverse of the carbon coordination number $n - J$(C-H) $\simeq 500/n$ Hz—which is exactly reproduced by the Muller and Pritchard proportionality relationship between $J$(C-H) and the $s$ character ($\rho_{CH}$) of the usual $sp^{n-1}$ carbon hybrid orbitals:

$$J(\text{C-H}) = A\rho_{CH} , \qquad (1)$$

where $A = 500$ Hz and $\rho_{CH} = 1/n$.

This success, together with the apparent simplicity of the Muller and Pritchard molecular orbital model (and of its va-

*This work is included in the Project of Molecular Structure (CQ-2) of the Instituto de Alta Cultura of Portugal.

219

lence bond counterpart[3]), has attracted too much uncritical en-
thusiasm, which is largely responsible for incorrect interpretations, applications, and extensions of equation 1; some of
them are unfortunately already in textbooks. In fact, quite often proper attention is not paid to the various approximations,
assumptions, and pitfalls associated with the theory. In this
chapter we try to give a critical evaluation of some of these.

## UNFORTUNATE STATEMENTS CONCERNING HYBRIDIZATION

One frequently encounters unfortunate statements related
to the Muller-Pritchard relationship (equation 1) which reveal
or at least perpetuate a common misunderstanding concerning
hybridization and bent bonds. For hydrocarbons, when $A$ in
equation 1 is taken as practically constant, the $J(C-H)$ values
are often regarded as reflecting "the hybridization state" of
carbon atoms, or even "the *true* hybridization state" of C
atoms; by this it is implied that "the *true* hybrid orbitals" do
not have to point to the neighboring nuclei; that is, interorbital
angles may be different from bond angles, "bent bonds" thus
arising. The same is said of $J(C-H)$ for substituted hydrocarbons after due account is taken of changes of $A$ in equation 1
caused mainly by variations in the carbon effective nuclear
charge.[4,5] Here, "rehybridization of C brought about by a substituent" or "transfer of $s$ character from one bond to another"
are even taken as being able to minimize the molecular energy.
Such views wrongly imply that the hybridization of an atom is
something unique and even a real phenomenon. To state that
$J(C-H)$ for $C_2H_2$ is twice $J(C-H)$ for $CH_4$ *because* the carbon $s$
character for the former $(\frac{1}{2})$ is twice that for the latter $(\frac{1}{4})$ is
just as incorrect as to say that $C_2H_2$ is linear and $CH_4$ tetrahedral *because* the carbon atoms are, respectively, $sp$ and $sp^3$
hybridized.

In fact, no physical property related to molecular wave
functions and energies—coupling constants, bond angles, electric dipole moments, etc.—can rigorously (within a theory

based on orbitals) be made dependent on the hybridization of a given atom. There is, in fact, an infinite number of ways of defining atomic hybrid orbitals that lead to the same molecular wave functions and energies as the set of "pure" atomic orbitals. This is seen clearly in molecular orbital theory where it is immaterial whether (delocalized) molecular orbitals (MO's) are expressed as linear combinations of pure atomic orbitals (AO's) or as linear combinations of linear combinations of these. It is not so obvious in valence bond (VB) theory where full delocalization, which must be invoked, requires a large number of resonant functions to be considered.

Quantities that may depend on the nature of the hybrid orbitals are those *calculated assuming a localized bond approximation*. In such cases, therefore, hybrids should be defined by some criterion that assures that the most localized MO or VB functions or good approximations of these are obtained.* Even so, if the values so calculated agree with experiment one can only conclude that either residual delocalization effects are unimportant,[†] or they are canceled out by other neglected factors; in no case is one allowed to take differences of hybridization so defined as the *cause* of differences in observed values. The true cause is the same, that is, carbon coordination number, bond angles, bond lengths and some features of atoms, that require the hybridization as defined above to vary.

It is thus less misleading to say that, in the absence of substituent effects, the value $J(C-H) \simeq 125$ Hz, for example, is characteristic of tetracoordinated C atoms in tetrahedral arrangements, than to state that it is characteristic of $sp^3$-hy-

---

*Little investigation has been done on correlating $J(C-H)$ to such hybrids, through a localized bond approach. When good MO's are known, a full delocalization procedure has usually been preferred. This is the case in the use of approximate SCF MO's by Pople et al.[6] in their finite perturbation theory for coupling constants.

[†]These may be of no relevance for a given property (e.g., molecular energy) but important for another (e.g., coupling constants).

bridized C atoms. Moreover, the hybrid orbitals appropriate
to the construction of the best most localized MO's (or VB
functions) are not necessarily the usual $sp^3$, $sp^2$, and $sp$ ones.
In general, different bonds require different hybrids, even
though the bond angles are the common 109°. 5°, 120° and 180°
(e.g., the CH and CC bonds in $C_2H_2$). Also equivalent bonds,
as in $CH_4$, do not necessarily imply the usual $s$ to $p$ content
ratio, namely, 1 : 3 in $CH_4$.

## HYDROCARBONS—THE QUANTITATIVE SUCCESS OF THE MULLER-PRITCHARD FORMULA IS ACCIDENTAL

From the success of the Muller-Pritchard formula in the
case of simple hydrocarbons, it is concluded that, at least in
these cases, the basic assumptions of the theory are essentially valid, namely: (a) the various electronic excitation energies from the ground to all triplet states may be replaced, in
Ramsey's expression for the contact contribution, by an appropriate average which is approximately constant; (b) the delocalized MO's may be replaced by a set of nonpolar localized
MO's defined in terms of the usual hybrid orbitals $sp^{n-1}$, any
residual delocalization between them being unimportant for
C-H coupling constants. In the VB formalism, assumption (b)
corresponds to the description of each C-H bond in terms of a
localized perfect-pairing VB function, using the same hybrid
atomic orbitals.[3]

These conclusions have received some support from the
more general theory due to Pople and Santry.[7] They have accepted the $\Delta E$ approximation on the grounds of the small difference between the $2s$ and $2p$ orbital energies for carbon, and
indirectly, also implied that $\Delta E$ does not change appreciably.
In fact, if a localized MO description is valid, then in the simplified Pople-Santry expression for the contact term, there is
only one $\Delta E$ value, equal to the difference between the energies
of the antibonding and bonding MO's localized in CH, which may
correctly be taken as almost constant. As a consequence, the
Pople-Santry formula leads to the Muller-Pritchard relation:

$$J(C\text{-}H) = A'\pi_{sh} = \frac{A'}{\Delta E}\rho_{CH} = A\rho_{CH} \qquad (2)$$

($\pi_{sh}$ is the mutual polarizability between the $2s$ carbon orbital and the $1s$ hydrogen orbital; $A' = 7.89 \times 10^3$ Hz-eV[7]).

Noting that $J(C\text{-}H)$ for $CH_4$ calculated using a fully delocalized MO description depends on a molecular parameter $\frac{1}{4}$ in just the same manner as Muller-Pritchard's value depends on the $s$ character $\frac{1}{4}$ of a carbon $sp^3$ hybrid orbital, Pople and Santry conclude that the localized MO scheme is valid. Indeed it is possible to obtain fair agreement between the localized scheme $J(C\text{-}H)$ values for the series $CH_4$, $C_2H_6$, $C_2H_4$, and $C_2H_2$ (practically, the experimental values) and those obtained by numerical Pople-Santry calculations based on full delocalization and on a reasonable set of parameters.[8]

In spite of all this evidence in favor of the validity of the Muller-Pritchard model for hydrocarbons, there are in the literature two apparently distinct views opposing it. Maksić and co-workers,[9] although keeping to a localized bond description together with a constant $\Delta E$, question the use of the same type of hybrid orbitals for nonequivalent bonds. By using hybrid orbitals defined by the maximum overlap method and including overlap in the normalization factor of a localized VB function, they obtain a satisfactory linear relationship between observed $J(C\text{-}H)$ values and the maximum overlap $s$ characters, which is still far from the simple proportionality relation of Muller and Pritchard.

On the other hand, Van Duijneveldt et al.[8] and Gil et al.,[10] by making use of the more general Pople-Santry theory, have questioned the basic assumptions of the Muller-Pritchard model, namely, the neglect of residual delocalization and of changes in $\Delta E$, and suggest that the success of the localized bond models is due to a fortuitous cancellation of these neglected factors. The present authors have recently resumed and developed these arguments and, in view of the paper by Maksić et al.,[9] have compared the two approaches and investigated the possible connections between them.[12]

The Importance of Residual Delocalization Between Localized
Molecular Orbitals.  A $\Delta E$ Varying With the Carbon
Coordination Number

Van Duijneveldt et al.[8] have shown that the same Pople-
Santry calculations, referred to above for simple hydrocarbons,
which seem to indicate that the Muller-Pritchard localized
bond model is valid, do in fact show on more detailed analysis
that residual delocalization between the localized MO's defined
in terms of $sp^{n-1}$ hybrids cannot be neglected. Indeed it is found
that it is because of residual delocalization that the values of
$J(C-H)$ for hydrocarbons calculated by the localized and delocalized MO approaches do not differ greatly.

The delocalization effects on $\pi_{sh}$ were investigated analytically for the following highly symmetrical $CH_n$ molecular systems: tetrahedral $CH_4$, trigonal planar $CH_3$ (or $CH_3^+$), and linear $CH_2$ (or $CH_2^{2+}$).[8] Whereas the localized MO approach gives
a $1/n$ dependence,

$$\pi_{sh} = \frac{\rho_{CH}}{\Delta E} = - \frac{1}{2n\beta_{t_i h_i}} , \tag{3}$$

a parallel delocalized MO treatment leads to

$$\pi_{sh} = - \frac{1}{2n^{3/2}\beta_{sh}} \tag{4}$$

($\beta_{t_i h_i}$ and $\beta_{sh}$, approximately constant, are resonance integrals
between the $1s$ orbital of H and the hybrid orbital $t$ of C pointing to H and the $2s$ orbital of C, respectively). This provides
a big difference, as can be seen from the calculated $J(C-H)$
values below ($\beta_{t_i h_i}$ and $\beta_{sh}$ are chosen so as to reproduce the
experimental value of $J(C-H)$ for $CH_4$):

|                  | $CH_4$ | $CH_3$ | $CH_2$ |      |
| ---------------- | ------ | ------ | ------ | ---- |
| Localized MO's   | (125)  | 167    | 250    | Hz   |
| Delocalized MO's | (125)  | 192    | 354    |      |

The difference between equations 3 and 4 arises from different $\Delta E$ values. In the localized MO approach, ignoring overlap,

$$\Delta E = -2\beta_{t_i h_i} \tag{5}$$

is the energy gap for the localized antibonding and bonding C-H MO's, whereas in the delocalized MO scheme, ignoring overlap,

$$\Delta E = -2\sqrt{n}\,\beta_{sh} \tag{6}$$

is the energy difference for the totally symmetric antibonding and bonding delocalized MO's.  Gil and Teixeira-Dias[10] have found that, in the latter case, $\Delta E$ can also be written

$$\Delta E = -2\beta_{t_i h_i} - 2(n-1)\beta_{t_j h_i}\,, \tag{7}$$

which shows that among the various resonance integrals leading to delocalization, $\beta_{t_j h_i}$ is the important one and that its contribution depends on the coordination number $n$ of the carbon atom.

If we take the reasonable value $\beta_{sh} = -4.0$ eV in order to approximately reproduce the $J(C\text{-}H)$ value for $CH_4$, then the various $\Delta E$ values are 16.0, 14.1, and 11.3 eV, respectively, for $CH_4$, $CH_3$, and $CH_2$ .

We may therefore write, in reasonable approximation,

$$\pi_{sh} = \frac{\rho_{CH}}{\Delta E} \tag{8}$$

if we allow for changes of $\Delta E$.

### Changes of Delocalization Effects (Namely, $\Delta E$) on Replacing H by C Atoms, as an Explanation of the Success of the Muller-Pritchard Formula

Obviously, the delocalization effects discussed above change on going from $CH_4$ to $C_2H_6$, from $CH_3$ to $C_2H_4$, and from $CH_2$ to $C_2H_2$ .  The differences result mainly from changes in overlap integrals, and hence in resonance integrals, on replacing a $1s$ H orbital by a carbon-hybrid orbital.  This has been investigated numerically[10, 11] and analytically[12] for $CH_n$ molecules.  For example, it is found that an increase of $\beta_{sh_2}$ increases $\pi_{sh_1}$, the effect being larger for smaller $n$.

It is mainly this factor that makes the delocalized MO $\pi_{sh}$ values for $C_2H_4$ and $C_2H_2$ much smaller than those for $CH_3$ and $CH_2$, respectively, and thus in reasonable agreement with the observed $J(C-H)$ values. [8, 10]  This is because the overlap integral between a $2s$ C orbital and an adjacent $1s$ H orbital is appreciably less than that involving the same $2s$ orbital and an adjacent $sp^2$ (for $C_2H_4$) or $sp$ (for $C_2H_2$) C orbital. [10]  We may say that the resulting decrease of the corresponding resonance integrals, particularly noticeable in the case of $C_2H_2$, is equivalent to an increase in $\Delta E$ relative to the corresponding $CH_n$ species (see equation 6).  By the use of Slater orbitals, [10] the increase in $\Delta E$ on going from $CH_2$ to $C_2H_2$ is estimated to be 15%, [12] which leads to a reduction of 40 Hz, about 80% of the difference between the numerically calculated $J(C-H)$ values for linear $CH_2$ and $C_2H_2$. [8]

We thus note that the effective $\Delta E$ values for simple hydrocarbons, when equation 8 is used with $\rho_{CH} = 1/n$, turn out to be closer than those for the $CH_n$ molecules.  This is essentially why the delocalized MO values of $J(C-H)$ for the simple hydrocarbons do not differ so much from the Muller-Pritchard values as they do for the $CH_n$ series.

We may therefore conclude that the observed $1/n$ dependence of $J(C-H)$ arises in part because of special fortuitous delocalization effects, and not because residual delocalization is fundamentally unimportant.

Similar arguments enable a better understanding of $J(C-H)$ in strained hydrocarbons. [12]


## HYDROCARBONS—USE OF MAXIMUM OVERLAP HYBRIDS. RESIDUAL DELOCALIZATION IS STILL IMPORTANT (CHANGES IN $\Delta E$ WITH C COORDINATION NUMBER)

We now note that the changes of overlap integrals that are responsible for the variations in $\Delta E$ and $\pi_{sh}$ are the same changes that cause variations of the hybrid orbitals as defined by the maximum overlap method, irrespective of changes in the geometry. The following question thus arises: is the "re-hybridization" so defined equivalent to the $\Delta E$ changes? In other words, for the pair $CH_2$, $C_2H_2$, for example, if we want to use equation 8 is there no difference between keeping $\rho_{CH}$ constant while varying $\Delta E$ and keeping $\Delta E$ constant while allowing $\rho_{CH}$ to vary according to the maximum overlap method?

The maximum overlap hybrids for $CH_2$ and $C_2H_2$ are $sp$ and $sp^{1.25}$, respectively.[9] (The corresponding hybrids in the CC bond are $sp^{0.8}$.) Therefore the decrease in $\rho_{CH}$ has the right sign to explain the decrease of $\pi_{sh}$ in $C_2H_2$ with respect to $CH_2$, assuming $\Delta E$ constant, in the same manner as the increase of $\Delta E$ explains it when assuming $\rho_{CH}$ to be constant. It also has the correct magnitude. In fact, from equations 2 and 8 it is found that one obtains almost the same decrease of 40 Hz in $J(C\text{-}H)$, using either method.

It thus appears that, on going from the simple $CH_n$ systems to (nonstrained) hydrocarbons, keeping the coordination number of C the same, we may take $\Delta E$ as a constant and calculate changes of $\pi_{sh}$ and $J(C\text{-}H)$ by using the $\rho_{CH}$ values given by the maximum overlap method. However, $\Delta E$ will be different for different coordination numbers. According to our discussion of the $CH_n$ systems, the values will be about 16, 14, and 11 eV, respectively, for $n = 4$, 3, and 2. That is, we can only hope to find a proportionality of $J(C\text{-}H)$ to $\rho_{CH}$ within the same $n$ value. If a general proportionality to maximum overlap $\rho_{CH}$ value is assumed which then reproduces the $J(C\text{-}H)$ values for $n = 4$, the calculated $J(C\text{-}H)$ for $n = 2$ will be appreciably lower than observed. This is exactly what Maksić et al.[9] have found.

This discussion shows that, even when using maximum overlap hybrids to define localized MO's, the residual delocalization cannot be ignored. Indeed, it seems to lead to differences in $\Delta E$ for different carbon coordination numbers that are, for simple hydrocarbons, even more pronounced than when the conventional $sp^3$, $sp^2$, and $sp$ hybrids are used. Those are the differences found for the symmetrical $CH_n$ systems, the previous discussion of which still applies because the hybrids used are exactly those required in the usual form of the maximum overlap method, on grounds of symmetry.

This, we suggest, is a fundamental reason why the linear correlation found by Maksić et al.[9] between $J$(C-H) values and $\rho_{CH}$ given by the maximum overlap method requires a large additive constant. The alternative explanation in terms of ionic character of the C-H bonds[9] is not quite satisfactory.[12]

## SUBSTITUENT EFFECTS. INDUCTIVE, HYPER-CONJUGATIVE, AND OTHER CONTRIBUTIONS

In view of the previous considerations about hydrocarbons, it is not surprising that simple proportionality relationships between $J$(C-H) and $s$ characters fail when substituent effects are present. This is certainly so, not only when the usual $sp^{n-1}$ hybrid orbitals ($n$ being the C coordination number) are assumed, but also when modifications of these to follow geometry changes or to conform to the maximum overlap criterion are considered.[1]

The variation of $J$(C-H) on going from a hydrocarbon to a corresponding derivative may be interpreted in terms of several effects: ($a$) changes in molecular geometry, ($b$) variations of the carbon effective nuclear charge and hence of the density of the C $2s$ orbital at the nucleus,[4,5] and ($c$) inductive and hyper-conjugative effects.

The contribution of geometry variations has not been investigated thoroughly. The relative importance of factor ($b$) has been the object of some controversy[1]; if the Slater screening rule is used, then a decrease of 0.2 in the total electron density of C relative to a neutral C atom gives rise to an increase in $J$(C-H) of about 7%, through an increase of factor $A'$ in equation 2. Associated with effects ($c$) are changes in resonance integrals, as we have considered before on replacing H by C. Special attention has been devoted, however, to inductive effects related to changes of orbital coulomb integrals and to hyperconjugative effects.

## Inductive Effects

Inductive effects have been studied using the Pople-Santry theory, both numerically[11] and analytically.[10] For the fragment H-C-X, the more negative the coulomb integral ($\alpha$) of the X atomic orbital in the C-X bond, the larger is $\pi_{sh}$. This is in accordance with the observation that an electronegative X atom leads to a bigger $J$(C-H). For perturbed $CH_n$ molecules[10] we have

$$\frac{\Delta\pi_{sh_1}}{\Delta\alpha_{h_2}} = -\frac{5\beta_{t_it_j} - \beta_{h_ih_j}}{8n\,\beta^3_{h_it_i}} \tag{9}$$

(providing $\Delta\alpha_{h_2}$, $\beta_{h_ih_j}$, and $\beta_{t_it_j}$ are small compared to $\beta_{h_it_i}$). Since $\beta_{t_it_j} = (\alpha_s - \alpha_p)/n$, $\Delta\pi_{sh_1}/\Delta\alpha_{h_2}$ has essentially a $1/n^2$ dependence.

## Hyperconjugative Effects. The Effect of Lone Pairs and of $\beta$ Substitution

Gil and Teixeira-Dias,[10] using the Pople-Santry theory, have obtained an expression for the contribution to $\pi_{sh}$ of an X lone pair in conjugation with the CH MO's. The effect is strongly stereospecific, as is seen in the following expression:

$$\Delta\pi_{sh_i} = \frac{O}{n} + \frac{P}{n^2} + \frac{Q}{n}\beta_{lt_i}. \tag{10}$$

In equation 10, $O$, $P$, and $Q$ are functions of several resonance integrals independent of the $CH_i$ bond taken; $\beta_{lt_i}$ is the integral between the lone-pair orbital $l$ and the $t_i$ hybrid orbital (appropriate to approximately localized $CH_i$ MO's), a type of delocalization mechanism similar to that ($\beta_{h_it_j}$) responsible for $\Delta E$ changes in $CH_n$. The last term will be, in general, the largest one in absolute value. Since $Q < 0$, $\Delta\pi_{sh_i}$ has got the sign of the overlap integral $S_{lt_i}$, opposite to that of $\beta_{lt_i}$. For the case of

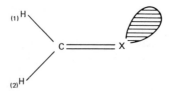

this means an increase of $J(C\text{-}H_1)$ and a decrease of $J(C\text{-}H_2)$, the calculated effects being quite large (of the order of 20 Hz). The numerical calculations of Yonezawa et al.[14] also predict similar effects. The decrease of $J(C\text{-}H)$ predicted for a C-H bond *trans* to a lone pair has been unambiguously confirmed experimentally by Gil and Alves.[13] Similar although less pronounced effects have also been found by Yonezawa and Morishima.[15]

If instead of a lone pair there is a C-Y bond, for example,

$$\underset{H}{\overset{H}{\diagdown}}C = C\underset{H}{\overset{Y}{\diagup}}$$

Y being more electronegative than H, then conjugation of the C-Y with the C-H MO's will have the opposite effect of that due to an $sp^2$ lone pair. This is again supported by numerical calculations.[16] In general, however, there will be additionally the Y lone pairs to be taken into account.

### References

1.   W. McFarlane, *Quart. Rev.*, 23, 187 (1969); J. H. Goldstein, V. S. Watts and L. S. Rattet, *Prog. Nucl. Magn. Res. Spectrosc.*, 8, 103 (1971); and references therein.

2.   N. Muller and D. E. Pritchard, *J. Chem. Phys.*, 31, 758, 1471 (1959).

3.   H. S. Gutowsky and C. Juan, *Discussions Faraday Soc.*, 34, 52 (1962); *J. Chem. Phys.*, 37, 2198 (1962).

4.   J. N. Shoolery, *J. Chem. Phys.*, 31, 1427 (1959).

5.  D. M. Grant and W. M. Litchman, *J. Am. Chem. Soc.*, **87**, 3394 (1964).

6.  J. A. Pople, J. W. McIver, and N. S. Ostlund, *Chem. Phys. Lett.*, **1**, 465 (1967); *J. Chem. Phys.*, **49**, 2965 (1968).

7.  J. A. Pople and D. P. Santry, *Mol. Phys.*, **8**, 1 (1964).

8.  F. B. VanDuijneveldt, V. M. S. Gil, and J. N. Murrell, *Theoret. Chim. Acta*, **4**, 85 (1966).

9.  Z. B. Maksić, M. Eckert-Maksić, and M. Randić, *Theoret. Chim. Acta*, **22**, 70 (1971).

10. V. M. S. Gil and J. J. C. Teixeira-Dias, *Mol. Phys.*, **15**, 47 (1968).

11. R. Ditchfield, M. A. Jensen, and J. N. Murrell, *J. Chem. Soc. A*, 1674 (1967).

12. V. M. S. Gil and C. F. G. C. Geraldes, *J. Magn. Resonance*, **11**, 268 (1973).

13. V. M. S. Gil and A. C. P. Alves, *Mol. Phys.*, **16**, 527 (1969).

14. T. Yonezawa, I. Morishima, K. Fukuta, and Y. Ohmori, *J. Mol. Spect.*, **31**, 341 (1969).

15. T. Yonezawa and I. Morishima, *J. Mol. Spect.*, **27**, 210 (1968).

16. T. Yonezawa, I. Morishima, M. Fujii, and H. Kato, *Bull. Chem. Soc. Japan*, **42**, 1248 (1969).

# CHAPTER 15

# $^{13}$C NMR STUDIES OF PROLINE PEPTIDES AND CARBOHYDRATES

O. Oster, E. Breitmaier and W. Voelter

*Chemisches Institut der Universität Tübingen*

## PROLINE PEPTIDES

It has been estimated that the energy difference between the *cis* and the *trans* forms of a peptide unit is larger than 2 kcal/mole/residue.[1,2] Energetically it is therefore not possible to have a series of *cis* peptide units in an open polypeptide chain. Further, *cis* bonds might occur, if at all, only at isolated points.[1] So far, however, they have never been observed in proteins.

For imino acid residues, the energy difference between the *cis* and the *trans* forms is much smaller, [1,2] and therefore *cis* peptide units are observed in proline[3] and sarcosine[4] peptides (for the assignment of *cis/trans* isomers see reference 1 and structure 1).

Because of its importance concerning the secondary structure of polypeptides and proteins, the *cis/trans* isomerism of monomer, oligomer, and polymer proline peptides has been investigated by several spectroscopic methods such as circular dichroism, [5,6] IR, [7,8] ORD, [9] $^1$H NMR, [10] and Raman spectroscopy. [11]

We have begun our $^{13}$C NMR studies on the *cis/trans* isomerism of the peptide bond with the simplest models for peptides, the formamide derivatives. The $^{13}$C NMR spectra of $N$-methylformamide, $N$-isopropylformamide, and $N$-*tert*-butyl-formamide have confirmed the interpretation of the $^1$H NMR spectra of LaPlanche and Rogers.[12] At room temperature the concentration of the *cis* isomer increases as the bulkiness of the N residue increases. The C atoms of the alkyl residues of the *cis* conformation appear about 2–3 ppm to lower field of those in the *trans* conformation.[13] This is expected since they are less shielded by the oxygen atom of the carbonyl group.

An increase of the bulkiness of the residues attached to the carbonyl carbon favors the *trans* isomer as is seen from the measurements on acetamide, which exists at room tempera-ture only in the *trans* form.

In the $^{13}$C NMR spectra of Boc-proline (2), measured in several solvents, the signals of the $C_\beta$, $C_\gamma$, and $C_\delta$ atoms of the pyrrolidine moiety of the proline[14, 15, 17] appear at high field. The $\delta$ and $\gamma$ carbons of the *cis* isomer are more shielded than those of the *trans* isomer; this is reversed for the $\beta$ carbons. Therefore the $C_\beta$ and $C_\gamma$ atoms give rise to two signals per atom, if the Boc-proline (2) is measured in DMSO, CDCl$_3$, or CD$_3$OD.[15] Surprisingly, in these solvents the *cis* isomer of Boc-proline predominates.[16] When Boc-proline is measured in aqueous solution (pH 7) signals for only one isomer are found.

From the $^{13}$C NMR spectra of Boc-Gly-Pro (3), dissolved in CDCl$_3$, DMSO, or H$_2$O (pH 7), it is evident that two isomers are present in all solutions, contrary to Boc-proline[15] and Boc-Pro-Pro (5b).[17] For these latter two compounds only one iso-mer is found in aqueous solutions.

Boc-Ala-Pro (4), measured in solvents such as CDCl$_3$, DMSO, or H$_2$O (pH 7), exists in only one conformation. Thomas and Williams[18] have stated that in proline dipeptides the bulk of

BOC – GLYCYL– PROLINE

BOC- ALANYL-PROLINE

**3**

**4**

the side chain in the amino acid preceding proline should have almost no influence on the *cis/trans* ratio. Our results clearly show, however, that the bulk of the side chain of N-substituted amino acids preceding proline has a remarkable influence on the *cis/trans* isomerism of proline peptides. The [13]C NMR spectra of Boc-Ala-Pro (4) in $CDCl_3$, DMSO, or water solutions indicate that only one isomer, most probably the *trans* form, is present.

At room temperature the [13]C NMR spectrum of Boc-Pro-Pro-OBzl (5a) shows six resonances from two carbons in the region where the signals of the $C_\gamma$ carbons appear, indicating that three conformations are present. On cooling down to − 23 °C (in $CDCl_3$) eight resonances are observed in the $C_\gamma$ region, and in the region where the signals of the urethane carbonyl carbons occur four resonances appear. Boc- Pro- Pro-OBzl (5a) has two amide bonds in which two proline residues are involved. Therefore four conformations should be possible: *trans/trans*, *cis/trans*, *trans/cis*, and *cis/cis*. Tonelli[19] has calculated the energies of the possible conformations of N- and carboxyl-protected proline peptides as a function of the *cis/trans* character of the amide bond and the rotation of the pyrrolidine ring around the $C_\alpha$-C=O bond and found that the *cis/cis* conformation is

BOC –PROLYL–PROLINE - O – R

5    a  R = BENZYL

5    b  R = H

sterically less favored.  Our measurements show that at low temperatures four isomers exist in solution.  Measurements on Boc-Pro-Pro (5b) in DMSO and $CDCl_3$ show that two conformations are present, whereas in aqueous solution (pH 7) only a single isomer exists.  This result provides strong evidence for the large influence of a free carboxyl group on the conformation of proline dipeptides.

In order to study the interaction of a $C_\beta$-containing amino acid with the pyrrolidine ring we have synthesized model compounds containing a pyrrolidine ring.  Our investigations of the pyrrolidine ring show two signals for the four ring C atoms; this result may be interpreted to show that the five-membered ring exists in an envelope form.  Low-temperature measurements ($-90\,^{\circ}$C) do not show a change in the spectrum.

In the pyrrolidine ring the nitrogen is tetrahedral.  If there is an amide bond it has to be planar, such as in the proline peptides.  $N$-formyl- (6), $N$-acetyl- (7), and $N$-propionylpyrrolidine (8) were synthesized and investigated by $^{13}$C NMR. For $N$-formylpyrrolidine (6) and $N$-acetylpyrrolidine (7) in the region of the carbonyl carbon signals two peaks are found in the ratio 1:20, indicating that more than one conformation is present.  Low-temperature measurements ($-90\,^{\circ}$C) do not change the spectrum.  Propionylpyrrolidine (8), which has a $C_\beta$ carbon similar to alanine, exists in two isomers.  The energy

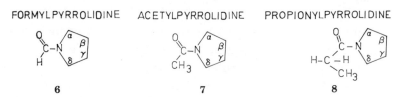

FORMYLPYRROLIDINE       ACETYLPYRROLIDINE       PROPIONYLPYRROLIDINE

6                              7                              8

barriers between the possible conformations seem to become higher with increasing alkyl chain length.  As demonstrated by models there is a strong interaction between the $C_\beta$ carbon of the propionyl residue and the pyrrolidine ring.  The $^{13}$C NMR spectra give evidence that Boc-Ala-pyrrolidine (9) occurs in three conformations, if dissolved in DMSO or $CDCl_3$; this has been predicted by calculations.[20-23]

BOC - ALANINPYRROLIDINE

## CARBOHYDRATE CHEMISTRY

The signal assignments of the $^{13}$C NMR spectra of carbo-
hydrates or their moieties in composite molecules are performed on the following basis.[24-37]   All chemical shifts are
given with respect to TMS; low-field shifts are reported with
positive values.

1.  The anomeric carbons resonate at lowest field (+90 to
+98 ppm for pyranoses) because two oxygen atoms withdraw
electrons.[24,27,33,36]

2.  The signals of the CH$_2$OH groups (C-6) in hexapyranoses
and of the ring CH$_2$ groups (C-5) in pentopyranoses generally
absorb at highest field (+60 to +62 ppm and +61 to +68 ppm,
respectively).[24,27,30,33,36]

3.  The resonances of hydroxyl groups bearing ring carbons
and those of C-5 in hexapyranoses occur in the range +65 to
+76 ppm.[24,27,30,36]

4.  In 6-deoxy sugars the CH$_3$ carbons resonate at the highest field positions (+16 to +18 ppm); the resonances of ring
deoxy carbons occur at somewhat lower field (+30 to +40
ppm).[24,30,33]

5.  Methylation of a sugar hydroxyl group causes the resonance of the neighboring ring carbon atom to shift 6−10 ppm
downfield.

6.  Carbons bearing axial hydroxyl groups generally absorb at higher field than corresponding ones with equatorial OH
substituents.

7.  The change of an equatorial pyranose hydroxyl group
to axial orientation causes, due to 1,3-diaxial interaction, the
resonances of carbons in $\gamma$ positions to shift upfield.[24-37]

8. Most signal assignments are possible only by spectral comparison of different analogues.

9. Proton off-resonance spectroscopy allows the differentiation between primary (quartet), secondary (triplet), tertiary (doublet), and quaternary (singlet) carbons.

10. The signals of different anomers can easily be assigned by recording the $^{13}C$ NMR spectrum both before and after mutarotational equilibrium has occurred. The more abundant isomer gives rise to stronger signal intensities in the spectrum.[36]

11. The carbon resonances of adjacent hydroxyl groups in *cis* positions are identified by a drastic downfield shift in the presence of boric acid.[37]

β-D-GLUCOSE (C1)      α-D-GLUCOSE (C1)

10 a                 10 b

Glucose (10a, 10b), for instance, measured immediately after dissolving the sugar, shows five resonances: at highest field the $CH_2OH$ group, at lowest field the signal of the anomeric carbon. The residual resonances can be assigned by spectral comparison with other pyranose sugars. The spectrum, recorded from a solution two days old (the mutarotational equilibrium has now occurred), shows the following characteristics: the resonances of the α anomer (10b) are still found but in addition to the signals of the β anomer (10a). The carbons of the β anomer (10a) resonate always at lower field than the corresponding ones of the α anomer. At lowest field, well separated from the signals of the other carbons, the resonances of the anomeric ones appear and the line of the α-anomeric (10b) carbon occurs at higher field than the corresponding one of the β-anomeric (10a) carbon. This confirms the applicability of the above-mentioned, general rule: ring carbons bearing axial substituents usually resonate at higher field than the corresponding carbons with equatorial ones. From the intensity of the lines it is further seen that in mutarotated solutions of glucose the $\beta$ anomer (**10a**) predominates. Integration of the signals shows that the solution contains more than 60% of the $\beta$ anomer (**10a**).

A similar rule applies for the methoxy resonances and can be used to establish the conformations of pyranoses. Most pyranoses occur in the so-called C-1 conformation. The $\alpha$ anomer of a pyranose C-1 conformation bears an axial substituent at C-1; the $\beta$ anomer bears an equatorial oriented substituent at C-1. This is reversed in the 1-C conformation of a pyranose as is demonstrated for arabinose (**11a, 11b, 11c, 11d**).

β -D-ARABINOSE (C1)

11 a

∝ -D- ARABINOSE (C1)

11 b

β-D-ARABINOSE (1C)

11 c

∝-D-ARABINOSE (1C)

11 d

From the resonance position of an anomeric carbon it is possible to decide whether C-1 bears an axial or equatorial hydroxyl group, the latter absorbing at lower field. As C-1 in methyl $\alpha$-$d$-arabino-pyranoside (**11b**) resonates at lower field than C-1 in methyl $\beta$-$d$-arabino-pyranoside we must draw the conclusion that arabinose exists in the 1-C and not the C-1 conformation.

## References

1. G. N. Ramachandran and C. M. Venkatachalam, *Biopolymers*, **6**, 1255 (1968).
2. G. N. Ramachandran and U. Sasisekharan, *Advan. Protein Chem.*, **23**, 287 (1968).
3. W. Traub and U. Shmueli, in G. N. Ramachandran, *Aspects of Protein Structures*, Academic Press, New York, 1963, p. 83.
4. J. P. Carver and E. R. Blout, in G. N. Ramachandran, *Treatise on Collagen*, Academic Press, New York, 1967, p. 441.
5. M. Rothe, R. Theyson, K.-D. Steffen, M. Kostrzewa, and M. Zamani, *Peptides*, E. Scoffone, Ed., North-Holland, Amsterdam, 1969, p. 179.
6. E. S. Pysh, *J. Mol. Biol.*, **23**, 587 (1967).
7. H. Okabayashi, H. Isemura, and S. Sakakibara, *Biopolymers*, **6**, 323 (1968).
8. H. Isemura, H. Okabayashi, and S. Sakakibara, *Biopolymers*, **6**, 307 (1968).
9. F. A. Bovey and F. P. Hood, *J. Am. Chem. Soc.*, **88**, 2326 (1966).
10. C. M. Deber, F. A. Bovey, J. P. Carver, and E. R. Blout, *J. Am. Chem. Soc.*, **92**, 6191 (1970).
11. W. B. Rippon, J. L. König, and A. G. Walton, *J. Am. Chem. Soc.*, **92**, 7455 (1970).
12. L. A. LaPlanche and M. T. Rogers, *J. Am. Chem. Soc.*, **86**, 337 (1963).
13. G. N. Ramachandran and R. M. Vankatachalam, *Biopolymers*, **6**, 1255 (1968).
14. W. Voelter, G. Jung, E. Breitmaier, and E. Bayer, *Z. Naturforsch.*, **26b**, 213 (1971).
15. W. Voelter and O. Oster, *Chemikerztg.*, **10**, 586 (1972).
16. V. Madison and J. Schellman, *Biopolymers*, **9**, 511 (1970).
17. W. Voelter, O. Oster, and E. Breitmaier, *Z. Naturforsch.*, in press.

18. W. A. Thomas and M. K. Williams, *Chem. Commun.*, 994 (1972).
19. A. E. Tonelli, *J. Am. Chem. Soc.*, **92**, 6187 (1970).
20. M. Maigret, B. Pullman, and J. Caillet, *Biochem. Biophys. Res. Commun.*, **40**, 808 (1970).
21. P. R. Schimmel and P. J. Flory, *J. Mol. Biol.*, **34**, 105 (1968).
22. A. Damiani, P. deSantis and A. Pizzi, *Nature*, **226**, 542 (1970).
23. G. N. Ramachandran, A. V. Lakahminarayanan, R. Balasubramanian, and G. Tegoni, *Biochim. Biophys. Acta*, **221**, 165 (1970).
24. D. E. Dorman and J. D. Roberts, *J. Am. Chem. Soc.*, **92**, 1355 (1970).
25. E. Breitmaier and W. Voelter, *Tetrahedron*, **29**, 227 (1973).
26. E. Breitmaier, G. Jung, and W. Voelter, *Chimia*, **26**, 136 (1972).
27. A. S. Perlin, B. Casu, and H. J. Koch, *Can. J. Chem.*, **48**, 2596 (1970).
28. W. Voelter, E. Breitmaier, G. Jung, T. Keller, and D. Hiss, *Angew. Chem.*, **82**, 812 (1970); *Angew. Chem. Intern. Ed.*, **9**, 803 (1970).
29. D. E. Dorman, S. J. Angyal, and J. D. Roberts, *J. Am. Chem. Soc.*, **92**, 1351 (1970).
30. E. Breitmaier, W. Voelter, G. Jung, and D. Tänzer, *Chem. Ber.*, **104**, 1147 (1971).
31. D. E. Dorman and J. D. Roberts, *J. Am. Chem. Soc.*, **93**, 4463 (1971).
32. E. Breitmaier, G. Jung, and W. Voelter, *Chimia*, **25**, 362 (1971).
33. W. Voelter, V. Bilĭk, and E. Breitmaier, *Collection Czech. Chem. Commun.*, **38**, 2054 (1973).
34. W. Voelter and E. Breitmaier, *Org. Magn. Resonance*, **5**, 311 (1973).
35. W. Voelter, E. Breitmaier, R. Price, and G. Jung, *Chimia*, **25**, 168 (1971).

36. W. Voelter, E. Breitmaier, and G. Jung, *Angew. Chem.*, **83**, 1011 (1971); *Angew. Chem. Intern. Ed.*, **10**, 935 (1971).

37. W. Voelter, C. Bürvenich, and E. Breitmaier, *Angew. Chem.*, **84**, 589 (1972); *Angew. Chem. Intern. Ed.*, **11**, 539 (1972).

CHAPTER 16

# APPLICATIONS OF [13]C NMR SPECTROSCOPY TO THE SOLUTION OF STRUCTURAL PROBLEMS IN SOME SANTONIN DERIVATIVES

P. S. Pregosin

*Department of Chemistry*
*University of Delaware*
*Newark*

and

E. W. Randall

*Department of Chemistry*
*Queen Mary College*
*London*

The study of relative configuration in natural products, via spectroscopic techniques, remains an area of active research.[1] In santonin derivatives of type 1, which have been tested as combatants against certain intestinal parasites and as potential oral contraceptives, both nuclear magnetic resonance[2] and mass spectrometric[3] techniques have been employed in an effort to obtain information concerning the relative configuration at C-11 and the nature of the C-6, C-7 lactone fusion. In view of the relatively large range of carbon-13 chemical shifts (about 400 ppm) and the recognized[4] sensitivity of carbon chemical shifts to small changes in both electronic and molecular structure, we have measured the carbon NMR spectra of a series of san-

1

2

tonin derivatives with the aim of establishing a facile method for determining relative configuration in these terpene deriva- tives.

Since carbon bound to oxygen is known[4] to appear at rela- tively low fields, it is expected that C-6 will appear in a sparsely populated region of the carbon spectrum, to low field of most $sp^3$ carbon resonances but to high field of $sp^2$ carbons.[4] Thus changes in the position of C-6 as a function of the lactone fusion may be monitored quite readily. Although C-13 might well appear in a more "crowded" region of the spectrum, as- signment of its resonance position is facilitated by the use of known aliphatic substituent effects.[5] These effects, usually designated by the Greek letters $\alpha$, $\beta$, and $\gamma$, refer to the effects at carbons one, two, and three bonds away, respectively, of

substituting methyl for hydrogen at the $\alpha$-carbon. In partic- ular, carbon-13 NMR studies on methylcyclohexanes[6] and cy- clohexanols[7] have demonstrated a configurational dependence of these substituent effects on the orientation of the methyl or hydroxyl group. Thus substitution of either methyl or hydroxyl in an equatorial position produces a larger *deshielding* at *both* the $\alpha$ and $\beta$ positions than does the corresponding axial substi- tution. The effect of equatorial substitution on the $\gamma$ carbon, however, is small while the effect of axial substitution at this

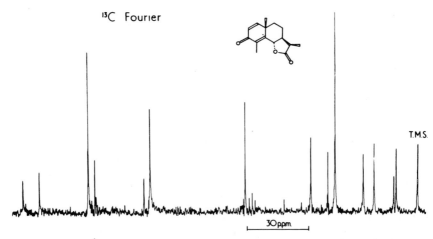

Figure 16-1. Natural abundance, proton decoupled carbon-13 NMR spectrum of $\beta$-santonin.

same position is relatively large (about 5-7 ppm) and in the opposite direction. Since the $\gamma$ effect is believed to arise from steric interactions, it should prove a useful tool for ascertaining the configuration at C-11.

The proton decoupled natural abundance carbon-13 NMR spectrum of $\beta$-santonin is shown in Fig. 16-1, and the resonance positions are listed in Table 16-1. For purposes of interpretation, the spectrum may be divided into two sections, six $sp^2$ resonances which appear between 125 and 155 ppm and nine $sp^3$ carbons which fall between 10 and 80 ppm. A discussion of the factors affecting the carbon chemical shifts in these molecules may be found elsewhere.[8] In keeping with our expectation C-6 resonates at considerably lower field (about 80 ppm) than the remaining aliphatic carbons. The assignments for these latter resonances were made using the previously mentioned substituent effects in conjunction with single frequency off-resonance (SFOR) techniques. The use of SFOR techniques permit the detection of the multiplicity arising from residual one-bond carbon-hydrogen coupling (two- and three-bond interactions are not usually observed), thus readily differentiating methyl, methylene, methine, and quaternary carbons.

TABLE 16-1.  Carbon-13 chemical shifts[a] in santonin derivatives

| Compound | 1 | 2 | 3 | 4 | 5 | 6 |
|---|---|---|---|---|---|---|
| α-Santonin | 155.1 | 125.9 | 186.0 | 128.4 | 151.5 | 81.5 |
| β-Santonin | 155.0 | 126.0 | 186.1 | 128.8 | 151.9 | 80.8 |
| α-epi-Santonin | 157.5 | 126.1 | 186.2 | 137.8 | 149.0 | 76.5 |
| β-epi-Santonin | 157.6 | 126.1 | 186.3 | 137.6 | 149.3 | 76.9 |
| 1,2-Dihydrosantonin | 42.1 | 33.9 | 198.5 | 128.5 | 152.9 | 82.1 |
| Desoxypseudosantonin | 213.9 | 30.8[b] | 35.5 | 139.3 | 128.1 | 76.6 |
| Pseudosantonin | 214.3 | 29.6 | 35.0 | 139.4 | 127.0 | 77.5 |
| α-Santonin oxime | 145.6 | 112.5 | 150.7 | 122.5 | 139.5 | 82.3 |
| | | | | OH | | |
| | | | | C=N | | |
| Cyclohexanone oxime | | | | 160.8 | | |

[a]Estimated to be correct to ± 0.2 ppm.  Values are downfield from internal TMS.
[b]Assignment tentative.
[c]With respect to nitrogen lone pair.

The carbon chemical shifts for the isomers α, β, α-epi, and β-epi are grouped together in Table 16-1.  Since the primary differences in these molecules are centered in the lactone ring, it might be expected that the dienone ring carbon resonance positions should be relatively unchanged, and to a large extent this is observed.  It is readily seen that the absorption of C-6 is dependent upon the stereochemistry of the C-6, C-7 junction; a high field set at about 76.5 ppm for the epi (lactone cis fused) derivatives and another set at about 81 ppm (lactone trans fused).  The exact differences, 5.0 ppm for the α isomers (C-13 methyl pseudoequatorial) and 3.9 ppm for the β isomers (C-13 methyl pseudoaxial) are in good agreement with the configurational dependence observed in cyclohexanols.[7]  Specifically, the difference, Δδ, at C-1 between cis- and trans-2-methylcyclohexanol is 5.8 ppm with the trans isomer at lower field.  The configurational dependence of the lactone fusion may

| Carbon Number | | | | | | | | |
| --- | --- | --- | --- | --- | --- | --- | --- | --- |
| 7 | 8 | 9 | 10 | 11 | 12 | 13 | 14 | 15 |
| 54.0 | 23.3 | 39.3 | 41.7 | 41.2 | 177.4 | 12.5 | 10.9 | 25.3 |
| 49.5 | 20.3 | 38.2 | 41.5 | 38.2 | 178.3 | 9.9 | 11.0 | 25.2 |
| 43.8 | 23.4 | 34.9 | 39.4 | 44.2 | 179.6 | 14.9 | 11.0 | 25.2 |
| 41.8 | 18.3 | 34.7 | 39.5 | 41.3 | 179.0 | 9.6 | 11.0 | 24.9 |
| 53.4 | 23.7 | 38.7[b] | 38.5[b] | 41.4 | 177.6 | 12.5 | 11.3 | 23.4 |
| 42.8 | 23.6 | 31.1[b] | 46.1 | 44.4 | 180.0 | 15.2 | 19.5 | 24.4 |
| 51.0 | 67.3 | 39.2 | 46.7 | 41.5 | 180.4 | 15.6 | 19.8 | 25.1 |
| 53.7 | 23.8 | 38.5 | 41.1 | 41.4 | 178.0 | 12.5[b] | 12.2[b] | 26.0 |

| $\alpha$-$CH_2$ | |
| --- | --- |
| syn[c] | anti |
| 32.4 | 24.8 |

also be observed via the changes in the resonance position of C-9. Change from a *trans* fusion to a *cis* fusion, in which the lactone ring is raised significantly above the plane described by carbons 5, 6, and 7, causes C-9 to move upfield by approximately 4 ppm. This increased shielding may be likened to a $\gamma$-substituent effect which has been induced at C-9 by the change in configuration at C-7.

If a similar line of reasoning is employed, inversion of configuration of the methyl group at C-11, from a pseudoequatorial ($\alpha$ isomer) to a pseudoaxial orientation ($\beta$ isomer), should effect an *increase* in shielding at both C-13 and C-8. In the $\beta$ isomer, C-8 and C-13 should exert reciprocal $\gamma$ effects due to the increased nonbonded interactions between the atoms attached to these carbons when the methyl is in an axial position. The observed differences ($\delta_\alpha - \delta_\beta = 3$ ppm at C-13; 3–5 ppm at C-8) are in agreement with this expectation.

Recently, Gough et al.[9] have demonstrated a similar configurational dependence of the carbon-13 shielding of the angular methyl group in a series of 5α and 5β steroids.

The utility of this type of NMR fingerprint is readily demonstrated. Catalytic hydrogenation of α-santonin affords a product whose structure may be formulated as **2**. The carbon-13 NMR spectrum readily shows that the 1, 2 double bond has been reduced. Additionally, inspection of the resonance positions of C-6 and C-13 reveals that the reaction has proceeded with retention of configuration at both sites. Thus from a single measurement, which may be performed in less than 30 min,[10] one obtains precise information concerning both the methyl configuration and the nature of the lactone fusion.

The carbon-13 NMR spectra of the pseudosantonin derivatives **3** and **4** are revealing in a similar sense. In desoxypseudosantonin **3** we note the excellent agreement of the positions of the C-6 and C-13 resonances with those in α-*epi*-santonin, suggesting that the lactone junction in the former is *cis* and that the C-13 methyl group is pseudoequatorial. In the parent pseudosantonin **4**, one can assign not only the lactone

3　　R═H
4　　R═OH

fusion (*cis*) and the C-13 methyl configuration (pseudoequatorial) but the configuration of the C-8 hydroxyl group as well. If the cyclohexanols are used as models, substitution of an equatorial hydroxyl is expected to produce a low-field shift of about 43.2 ppm at C-8 (α effect) and concomitant low-field shifts of about 7.9 ppm at carbons 7 and 9 (β effect), whereas the values for axial hydroxyl substitution should be 37.8 ppm and 5.5 ppm, respectively. The observed low-field shifts of 43.9 ppm at C-8 and 8.1 ppm at C-9 in **4** relative to **3**, strongly suggest equatorial hydroxyl orientation. Thus from one measurement we can make three configurational assignments.

In addition to providing information about the lactone ring, the carbon spectrum of $\alpha$-santonin oxime (5) allows us to distinguish between the two possible orientations of the hydroxyl group. Although unsymmetrical oximes are known[11] to exist in two configurations, 6 and 7, the carbon-13 NMR spectrum of 5 shows only one set of resonances, thus indicating the predominance of one isomer. From Table 16-1, we observe that

both the C-2 and C-4 resonances are shifted upfield when the dienone (1) is converted to the oxime; however, the upfield shift of C-2 is larger by 7.5 ppm. To consider the possibility that this difference may be related to the orientation of the nitrogen lone pair, we have examined the carbon spectrum of cyclohexanone oxime. The data (Table 16-1) for this molecule reveal that the $\alpha$-methylene group *anti* to the nitrogen lone pair is shielded by 7.6 ppm relative to the corresponding *syn* methylene group. A similar lone-pair dependence for acetone oxime[12] and N-nitrosamines[13] has been reported. Thus we conclude that for $\alpha$-santonin oxime the most abundant isomer is that one having the nitrogen lone pair *syn* to C-4.

In conclusion, we have shown carbon-13 NMR to be an excellent probe for detecting subtle differences in molecular structure in some santonin natural products.

## References

1. Sir John Simmonsen and D. H. R. Barton, *The Terpenes*, Vol. 3, Cambridge University Press, 1952.

2.  J. T. Pinhey and S. Sternhell, *Aust. J. Chem.*, **18**, 543 (1965).

3.  N. Woseda, T. Tsuchiya, E. Yoshii, and E. Watanabe, *Tetrahedron*, **23**, 4623 (1967).

4.  G. C. Levy and G. L. Nelson, *Carbon-13 Nuclear Magnetic Resonance for Organic Chemists*, Wiley-Interscience, New York, 1972.

5.  D. M. Grant and E. G. Paul, *J. Am. Chem. Soc.*, **86**, 2984 (1964).

6.  D. K. Dalling and D. M. Grant, *J. Am. Chem.*, **89**, 6612 (1967).

7.  J. D. Roberts, F. J. Weigert, J. I. Kroschwitz, and H. J. Reich, *J. Am. Chem. Soc.*, **92**, 1338 (1970).

8.  P. S. Pregosin, E. W. Randall, and T. B. H. McMurry, *J. Chem. Soc. Perkin Trans.*, I, 299 (1972).

9.  J. L. Gough, J. P. Guthrie, and J. B. Stothers, *Chem. Commun.*, 979 (1972).

10. P. S. Pregosin and E. W. Randall, in *The Determination of Organic Structures by Physical Methods*, Vol. 4, Academic Press, New York, 1972, for a discussion of techniques used in carbon-13 NMR. See also Ref. 8.

11. G. J. Karabotsos and R. N. Taller, *Tetrahedron*, **24**, 3347 (1968).

12. N. Gurudata, *Can. J. Chem.*, **50**, 1956 (1972); G. C. Levy and G. L. Nelson, *J. Am. Chem. Soc.*, **94**, 4897 (1972).

13. P. S. Pregosin and E. W. Randall, *Chem. Commun.*, 399 (1971).

# CHAPTER 17

# $^{29}$Si FOURIER TRANSFORM NMR

George C. Levy* and Joseph D. Cargioli

*General Electric Corporate Research and Development
Schenectady, New York*

Organosilicon compounds and silicon-containing inorganic compounds may be studied by direct NMR observation of the 4.7% naturally abundant $^{29}$Si isotope.[1-10] The degree of difficulty encountered in obtaining $^{29}$Si NMR spectra is comparable to that for $^{13}$C NMR. Although the natural abundance of $^{29}$Si is about four times that of $^{13}$C (1.1%), the relative sensitivity for equal numbers of nuclei, at constant field strength, is about $\frac{1}{2}$ that for carbon ($7.84 \times 10^{-3}$ for $^{29}$Si and $1.59 \times 10^{-2}$ for $^{13}$C both relative to $^{1}$H $\equiv 1.0$).

Taking into account the isotopic abundances and NMR sensitivities, $^{29}$Si experiments should be about twice as sensitive as $^{13}$C experiments on comparable samples.[11] The situation is more complex[5] and $^{29}$Si NMR may be considerably less efficient than predicted from these two factors.[9,10] First, the magnetogyric ratio, $\gamma$, for $^{29}$Si is negative. The nuclear Overhauser effect (NOE) which increases integrated $^{13}$C peak areas as much as threefold in $^{1}$H decoupled $^{13}$C experiments is negative in proton decoupled $^{29}$Si spectra. $^{29}$Si$\{^{1}$H$\}$ spectra may have resonance

*Present address: Department of Chemistry, Florida State University, Tallahassee, Florida.

lines of greatly reduced intensity, *negative* peaks, or the signals may be *nulled*, depending on the mixing of relaxation mechanisms resulting in $^{29}$Si spin-lattice relaxation.

The second complication results from the fact that $^{29}$Si nuclei generally have very long $T_1$'s and thus are easy to saturate.[9, 10] Fourier transform (FT) $^{29}$Si experiments cannot in general achieve sensitivity improvements comparable to those in FT $^{13}$C NMR. However, with the use of several techniques it is possible to efficiently obtain $^{29}$Si FT NMR data on a wide variety of compounds.[10]

The information obtained from NMR spectroscopy may be separated into chemical shift data, peak areas, spin-spin (scalar) coupling constants, and spin-relaxation parameters. Here attention is focused primarily on spin-lattice relaxation parameters, since they affect operation most significantly in FT NMR studies (especially in the case of $^{29}$Si nuclei). Furthermore, the $^{29}$Si relaxation data can be used to characterize molecular structure and dynamics, in the same way that $^{13}$C relaxation data have been used.[12]

## $^{29}$Si CHEMICAL SHIFTS

Most known $^{29}$Si chemical shifts occur within a range of just over 100 ppm. Figure 17-1 summarizes $^{29}$Si chemical shifts for a number of organosilicon compounds. Highly shielded $^{29}$Si nuclei generally have multiple oxygen substitution, whereas

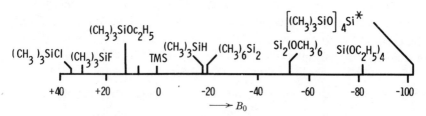

Figure 17-1. $^{29}$Si chemical shifts for some organosilicon compounds (relative to tetramethylsilane, TMS).

TABLE 17-1.[13]  $^{29}$Si chemical shifts relative to a $(CH_3CH_2O)_4$ Si standard[a]

| Compound | Solvent | Concentration (% v/v) | δ (ppm) |
|---|---|---|---|
| $[(CH_3)_3Si]_2O$  (MM) | $CCl_4$ | 15 | 89.48 |
| $(CH_3)_4Si$  (TMS) | $CCl_4$ | 15 | 82.58 |
| $[(CH_3)_2SiO]_4$ | $CCl_4$ | 15 | 63.06 |
| $(CH_3)_3SiO[(CH_3)_2SiO]_xSi(CH_3)_3$ <br> $x \simeq 50$ | $CCl_4$ | 15 | 60.39 |
| $(CH_3O)_4Si$ | $CCl_4$ | 15 | 3.36 |
| TMS <br> MM | Cyclohexane | 15% each | 82.4 <br> 89.2 |
| TMS <br> MM | Benzene | 15% each | 81.7 <br> 89.1 |
| TMS <br> MM | Acetone | 15% each | 82.0 <br> 89.2 |
| TMS <br> MM | Methylene chloride | 15% each | 82.2 <br> 89.3 |
| TMS <br> MM | Dioxane | 15% each | 82.3 <br> 89.4 |

[a]At ~30 °C.  Chemical shifts determined on several instruments, ± 0.1 ppm (actual deviations $\lesssim 0.05$ ppm).  *Positive* shifts downfield.  $(CH_3CH_2O)_4Si$ concentration = 15%.

substitution by aliphatic groups or halogens results in deshielding.

The $^{29}$Si chemical shifts given in Fig. 17-1 are reported relative to the reference standard, tetramethylsilane (TMS). TMS is used both for convenience and because it is now the generally accepted standard for both $^{1}$H and $^{13}$C NMR studies. However, any number of common organosilicon compounds could be used as an internal NMR reference standard. Proton and $^{13}$C NMR scales based on TMS report shifts to lower field (higher frequency) as *positive* chemical shifts. With this convention a compound such as $(MeO)_4Si$ or $(EtO)_4Si$ could be set equal to zero; almost all observed $^{29}$Si shifts are then positive (Fig. 17-1). Tetramethylorthosilicate has been used as a reference,[6] but its high toxicity[13] makes the tetraethylorthosilicate a better choice. Early $^{29}$Si workers used polydimethylsiloxane as a standard, reporting upfield shifts as *positive*. It is fair to say that a universally accepted $^{29}$Si chemical shift scale is not yet available.

To facilitate the reporting of $^{29}$Si chemical shift data, accurate internal chemical shifts have been determined for the many organosilicon compounds used as references.[13] Table 17-1 lists these shifts. As indicated in Table 17-1 solvent-induced shifts for these compounds are small, confirming work by Maciel et al.[14]

Scholl, Maciel, and Musker have reported the chemical shifts of 120 organosilicon compounds covering a range of about 70 ppm.[6] They report that the effects of $\alpha$, $\beta$, and $\gamma$-saturated carbons on $^{29}$Si shielding are substantial, and roughly the same as the effects observed in $^{13}$C, $^{31}$P, and $^{15}$N shieldings.

Although the chemical shift range for $^{29}$Si resonances is smaller than that observed in $^{13}$C NMR, the *spectral dispersion* for many organosilicon compounds may be considerably greater in the $^{29}$Si spectra. For example, $^{29}$Si shifts for some cyclic and acyclic polydimethylsiloxanes are reported in Table 17-2.

In Table 17-2 the M-D siloxane nomenclature is used, where M represents a trimethylsiloxy end unit, and D represents an internal dimethylsiloxy group.[15] In the linear siloxane oligomers each different silicon nucleus can be distinguished.

TABLE 17-2. [29]Si chemical shifts for some poly(dimethyl)-siloxanes[a]

| Compound | M | D$^1$ | D$^2$ | D$^3$ | D$^4$ | D$^5$ |
|---|---|---|---|---|---|---|
| | | | [29]Si | | | |
| MM | 6.79 | | | | | |
| MDM | 6.70 | − 21.5 | | | | |
| MD$_2$M | 6.80 | − 22.0 | | | | |
| MD$_3$M | 6.90 | − 21.8 | − 22.6 | | | |
| MD$_4$M | 7.0 | − 21.8 | − 23.4 | | | |
| MD$_5$M | 7.0 | − 21.8 | − 22.4 | − 22.3 | | |
| MD$_6$M | 7.0 | − 21.8 | − 22.3 | − 22.2 | | |
| MD$_7$M | 7.0 | − 21.89 | − 22.49 | − 22.33 | − 22.29 | |
| MD$_8$M | 6.93 | − 21.86 | − 22.45 | − 22.30 | − 22.20 | |
| D$_3$ cyclic | | − 9.12 | | | | |
| D$_4$ cyclic | | − 19.51 | | | | |
| D$_5$ cyclic | | − 21.93 | | | | |
| D$_6$ cyclic | | − 22.48 | | | | |

[a]Relative to internal TMS. Positive numbers, downfield. Accuracy of reported shifts better than 0.05 ppm (1 Hz). Solutions in acetone-$d_6$.

The [13]C chemical shifts of the CH$_3$ carbons in these compounds are not as well separated. To illustrate this, the [13]C spectrum of octadecamethyloctasiloxane (MD$_6$M) and the [29]Si spectrum of the undecasiloxane MD$_9$M, shown in Figs. 17-2 and 17-3, can be compared. The [13]C spectrum of the smaller compound resolves only the M and outer D units (D$^1$), while the [29]Si spectrum easily resolves the M unit from the D units and further separates the four outermost D units (D$^1$→ D$^4$), failing to resolve only the central D unit (D$^5$). For smaller chains, all D units are resolved in the [29]Si spectra, as in MD$_7$M shown in Fig. 17-4.

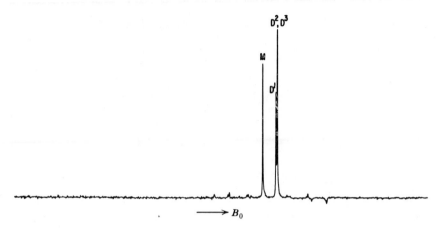

Figure 17-2.   $^{13}$C FT NMR spectrum of $MD_6M$.   Spectral width, 500 Hz.

In the cyclic polydimethylsiloxanes there is a decrease in $^{29}$Si shielding as the ring size gets smaller.   For the six-membered ring, $D_3$, this effect is particularly significant (see Table 17-2).

Proton chemical shifts of silicone polymer solutions have been used to study the effect of chain length on NMR signals.[16]

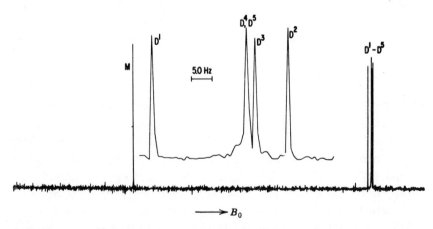

Figure 17-3.   $^{29}$Si FT NMR spectrum of $MD_9M$.   Spectral width, 1000 Hz.   Experiment used pulse-modulated $^1$H decoupling (see section on experimental methods).

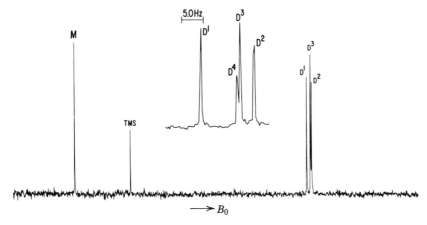

Figure 17-4. $^{29}$Si FT NMR spectrum of MD$_7$M (40% in ethylacetate-acetone); spectral width, 1000 Hz; 250 90° pulses, pulse interval 4 sec; total acquisition time 17 min. Added paramagnetic relaxation reagent, Cr(acac)$_3$ at $3 \times 10^{-2}$ $M$ (see section on experimental methods).

The observed shift of the end unit CH$_3$ protons have been correlated with chain length in different solvent media, and information obtained about configurational environments of chain segments. This kind of information should be more easily obtainable from $^{29}$Si NMR studies, but no results have yet been reported.

## COUPLING CONSTANTS

Directly bonded and long-range $^{29}$Si-$^1$H coupling constants have been measured for a few silicon compounds. Those one-bond $^{29}$Si-$^1$H couplings are, in general, larger than $^{13}$C-$^1$H couplings for structurally similar compounds. Figure 17-5 shows the undecoupled $^{29}$Si spectrum of diphenylsilane. The one-bond $^{29}$Si-$^1$H coupling constant is 200 Hz, compared with 127 Hz for the analogous coupling in diphenylmethane. Long-range $^{29}$Si-$^1$H coupling with the *ortho* and *meta* protons can also be measured in the $^{29}$Si spectrum of diphenylsilane.

When spin-spin splitting patterns are well resolved and

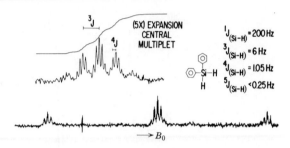

Figure 17-5.  Undecoupled $^{29}$Si FT NMR spectrum of diphenylsilane; spectral width, 1000 Hz.  Insert: expanded scale.

easily interpreted, they can be useful for confirming the structure of organosilicon compounds.  In most molecules however, long-range $^{29}$Si-$^{1}$H coupling patterns are too complex for manual interpretation, sometimes even taxing available computer spectral analysis programs.  For example, Fig. 17-6 shows the undecoupled $^{29}$Si spectrum of dimethylvinylsilane.

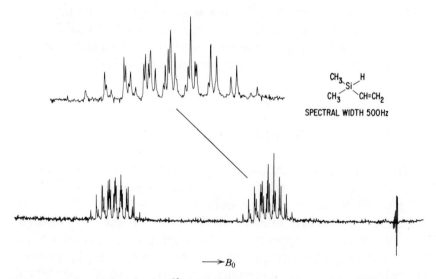

Figure 17-6.  Undecoupled $^{29}$Si FT NMR spectrum of dimethylvinylsilane.  Insert: expanded scale.

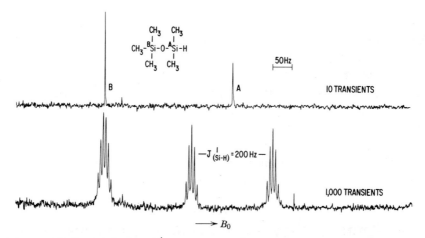

Figure 17-7. Wide-band $^1$H decoupled (above) and undecoupled (below) $^{29}$Si FT NMR spectra of pentamethyldisiloxane.

TABLE 17-3. Some $^{13}$C–$^{29}$Si coupling constants

| Compound | $^1J$(C–Si) (Hz)[a] | $^2J$(C–Si) (Hz)[b] | Hybridization at Carbon |
|---|---|---|---|
| $(CH_3)_4Si$ | 50.3 | | $sp^3$ |
| $(CH_3CH_2)_4Si$ | 50.2 | | $sp^3$ |
| $\phi$-Si$(CH_3)_3$ | 66.5 | | $sp^2$ |
| $(CH_2{=}\overset{*}{C}H)_4Si$ | 70 | $<10$[c] | $sp^2$ |
| $(CH_2{=}\overset{*}{C}H)_2Si(CH_3)_2$ | 66 | | $sp^2$ |
| $CH_2{=}\overset{*}{C}H{-}Si(CH_3)_3$ | 64 | | $sp^2$ |
| $\phi{-}\overset{*}{C}{\equiv}\overset{*}{C}{-}Si(CH_3)_3$ | 83.6 | 16.1 | $sp$ |

[a]To (*) carbon.
[b]To (≠) carbon.
[c]Not observed.

259

The acquisition time required to obtain $^{29}$Si-$^1$H coupling information can be very much longer than that required to record a proton decoupled spectrum (Fig. 17-7).

A few $^{29}$Si-$^{13}$C coupling constants have been measured from the 2.3% $^{29}$Si satellites in $^{13}$C spectra[17] and from multiple resonance experiments.[18] One-bond $^{13}$C-$^{29}$Si couplings are found to be roughly proportional to the amount of $s$ character on the carbon atom, ranging from 50.2 to 83.6 Hz (Table 17-3).[17]

In two recent communications[7,8] comparisons were made between $^{19}$F-$^{13}$C and $^{19}$F-$^{29}$Si couplings in analogous compounds.

## SPIN-LATTICE RELAXATION PARAMETERS

The process of nuclear spin-lattice energy exchange (*spin-lattice relaxation*) that attempts to maintain "thermal equilibrium" populations in the available spin-energy levels is relatively inefficient because of the small energy differences between nuclear spin states. It is not difficult to override the spin-lattice relaxation process by excitation of the spins, thus *saturating* the spin system.

All available $^{29}$Si spin-lattice relaxation mechanistic pathways have one characteristic in common. They all depend on the presence of fluctuating local magnetic fields produced at or near the $^{29}$Si nuclei undergoing relaxation. Four spin-lattice relaxation mechanisms are usually considered for nuclei of spin $\frac{1}{2}$ such as $^{29}$Si: dipole-dipole (DD) interactions, the spin-rotation (SR) interaction, scalar (SC) interaction, and chemical shift anisotropy (CSA). As a result of a lower magnetogyric ratio and changes in molecular and nuclear characteristics (bond lengths, molecular geometry, shift anisotropy, etc.) the relative efficiencies of the four mechanisms change in comparisons of $^{29}$Si relaxation with $^{13}$C relaxation.[12] Of the four mechanisms, scalar spin-lattice relaxation is a special case, possible only in specific situations for $^{13}$C or $^{29}$Si nuclei.[10,19]

Table 17-4 compares generalized contributions to $^{29}$Si and $^{13}$C relaxation for four cases: protonated nuclei (silicon or carbon atoms with one or more directly attached protons) and non-protonated nuclei (having no directly bonded protons), in small

TABLE 17-4.  Contributions of $^{29}$Si and $^{13}$C spin-lattice relaxation mechanisms[a]

|  | DD | SR | CSA | DD (inter) |
|---|---|---|---|---|
| **Very small molecules[b]** | | | | |
| Protonated | | | | |
| $^{29}$Si | Minor → + | + → ++ | Neg | Neg → Minor |
| $^{13}$C | + | + → ++ | Neg | Neg |
| Nonprotonated | | | | |
| $^{29}$Si | Minor → + | ++ | Neg | Neg → Minor |
| $^{13}$C | + → ++ | + → ++ | Neg | Neg |
| **Medium and large molecules** | | | | |
| Protonated | | | | |
| $^{29}$Si | + → ++ | Neg → ++ | Neg | Neg → Minor |
| $^{13}$C | ++ → 100% | Neg → Minor | Neg | Neg |
| Nonprotonated | | | | |
| $^{29}$Si | + → ++ | Neg → + | Minor → + | Minor → + |
| $^{13}$C | Usually ++ | Neg → Minor | Neg → + | Neg → Minor |

[a]All contributions refer to relaxation near room temperature at ∼23 kG magnetic field; $O_2$ excluded.
[b]Assumes no intermolecular association.

and large molecules.  In Table 17-4, + and ++ indicate appreciable and nearly exclusive contributions from the indicated relaxation mechanism.  Minor and negligible contributions indicate 5–10% and ∼1–5% contributions, respectively.  The last column in Table 17-4 lists typical contributions from *intermolecular* dipole-dipole interactions with protons.

Since the $^{29}$Si-$^1$H dipole-dipole mechanism is directly responsible for observed nuclear Overhauser effects in proton decoupled $^{29}$Si FT spectra we treat that mechanism in some detail.

The dipole-dipole spin-lattice relaxation rate $(R_1^{DD} \equiv 1/T_1^{DD})$ for a $^{13}$C or $^{29}$Si nucleus in a molecule that is not subject to severe motional constraints (within the so-called region of extreme spectral narrowing) is described by equation 1, which assumes that no high $\gamma$ nuclei other than protons are present.

$$1/T_1^{DD} \equiv R_1^{DD} = \sum_{i=1}^{n} \hbar^2 \gamma_X^2 \gamma_H^2 r^{-6} \tau_c \,, \tag{1}$$

where $\hbar$ is Planck's constant divided by $2\pi$, $\gamma_X$ is the magneto-gyric ratio for the $^{13}C$ or $^{29}Si$ nucleus being relaxed, $\tau_c$ is the molecular correlation time, and $r$ is the distance between the $i$th proton and the X nucleus. In practice, the number of protons, $n$, used for the summations may be restricted, but in principle all nearby protons are included. The molecular correlation time, $\tau_c$, describes molecular tumbling in solution. It can be thought of as the time required for the rotation of the molecule through 1 rad. $\tau_c$ is strictly defined only for rigid molecules that tumble isotropically. When these conditions do not hold it is sometimes still possible to speak of an effective correlation time, $\tau_c^{eff}$, that may be used to describe molecular motional characteristics.

It has been shown that for protonated $^{13}C$ nuclei only directly attached $^1H$ nuclei need to be considered, eliminating the summation in equation 1:

$$1/T_1^{DD} = R_1^{DD} = N\hbar^2 \gamma_X^2 \gamma_H^2 r_{XH}^{-6} \tau_c \ (X = {}^{13}C) \,. \tag{2}$$

In equation 2, $N$ is the number of directly attached protons and $r_{XH}$ is equal to $1.09 \times 10^{-8}$ cm, the C–H bond length. Equation 2 shows that the combined effect of the smaller $\gamma$ for $^{29}Si$ and the longer Si–H bond length (1.48 Å versus 1.09 Å for a C–H bond) is a decrease of the DD relaxation rate, $R_1^{DD}$, about tenfold—that is, $T_1^{DD}$ will be 10 times longer for the $^{29}Si$ nucleus.

The $^{13}C/^{29}Si$ DD relaxation differential is not as great for nonprotonated nuclei. Long-range intra- and intermolecular DD interactions, usually unimportant for $^{13}C$ nuclei, can contribute significantly to $^{29}Si$ relaxation. Van der Waals' radii, $d$, for silicon, carbon, and proton atoms are approximately 1.97, 1.57, and 1.2 Å, respectively. The efficiency of non-nearest neighbor protons in causing DD relaxation of a nucleus, X, *relative to DD relaxation by a directly bonded proton*, is given by equation 3[10]:

$$\% \text{ efficiency} = 100N \left[\frac{d_X + d_H}{r_{XH}}\right]^{-6}, \tag{3}$$

where $N$ is the number of nuclei at a distance not significantly greater than $(d_X + d_H)$ from the X nucleus, $d_X$ and $d_H$ are the van der Waals' radii for the X and H nuclei, and $r_{XH}$ is the X-H bond length (1.48 Å for silicon; 1.09 Å for carbon). For $^{13}$C and $^{29}$Si nuclei, the relative efficiency of a single long-range $^1$H DD interaction operating at the distance $(d_X + d_H)$ is 0.3% and 1.0%, respectively. As a result of the larger nonbonded distance between silicon and hydrogen nuclei, a larger number of protons may be accommodated in the van der Waals' "sphere" around a silicon than around a carbon nucleus. Relaxation contributions from long-range DD interactions are ~ 3–7% for $^{13}$C nuclei (corresponding to about 8–18 nearby nonbonded protons) and $\lesssim$ 10–30% for $^{29}$Si nuclei ($N$ = 10–30), both relative to the contributions from a single, *bonded* proton. Long-range *intramolecular* interactions may be substantially higher in those molecules where steric compression forces the protons close to the carbon or silicon nuclei.

$^{29}$Si dipole-dipole relaxation times are very long, the shortest measured in this study being 26 sec. Thus for $^{29}$Si nuclei relaxation from other mechanisms can contribute significantly.

Spin-rotation relaxation can occur with rapid overall molecular tumbling (*spin overall rotation*) or from free internal spinning of symmetrical groups (*spin internal rotation*). Spin-rotation relaxation can be particularly important for $^{29}$Si nuclei where $^1$H DD relaxation is relatively inefficient.

The contributions of DD and SR relaxation were calculated by standard methods.[6a, 13] The $^{29}$Si-$^1$H dipole-dipole $T_1$ ($T_1^{DD}$) can be determined directly from equation 4:

$$T_1^{DD} = \frac{2.52\, T_1^{obs}}{\eta}, \tag{4}$$

where $T_1^{obs}$ is the observed $T_1$ and $\eta$ is the observed NOE ($\eta$ = –2.52 is the NOE corresponding to 100% $^{29}$Si-$^1$H DD relaxation). The spin-rotation relaxation time $T_1^{SR}$ can be calculated from equation 5:

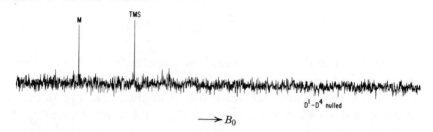

Figure 17-8. $^{29}$Si FT NMR spectrum of MD$_7$M. Same sample and experimental conditions as in Fig. 17-4 but no added paramagnetic relaxation reagent.

TABLE 17-5. $^{29}$Si spin-lattice relaxation times and nuclear Overhauser effects in TMS[a]

| Temperature (°C)[b] | $T_1$ (sec)[c] | NOE $(\eta)$[d] | $T_1^{DD}$ (sec)[e] | $T_1^{SR}$ (sec)[e] |
|---|---|---|---|---|
| +25 | 19 | −0.09 | | |
| 0 | 23.3 | −0.205 | 286 | 25.4 |
| −20 | 32.5 | −0.35 | | |
| −50 | 42.0 | −0.495 | 213.8 | 52 |
| −62.5[f] | 55[g] | −1.0 | 139[g] | 91[g] |
| −64[f] | | −1.03 | | |
| −83[f] | 37 | −1.59 | 58.6 | 100 |

[a]85% TMS, 15% acetone-$d_6$. Oxygen removed with $N_2$. Experiments performed on a Varian XL-100-FT system at 19.9 MHz.
[b]± 2−3°.
[c]Inversion-recovery pulse sequence. Estimated errors <10%.
[d]Theoretical maximum, $\eta - 2.52$. Determined directly from the decoupled and undecoupled spectra, estimated maximum error < 0.1.
[e]Estimated error in $T_1^{SR} \lesssim 20\%$. Chemical shift anisotropy relaxation assumed insignificant.
[f]Sample: 75% TMS, 25% $CD_2Cl_2$.
[g]$^{29}$Si $\{^1H\}$ signal nulled. $T_1$ determined from undecoupled pulse Fourier transform resonance spectra.

$$\frac{1}{T_1^{obs}} = \frac{1}{T_1^{DD}} + \frac{1}{T_1^{SR}} + \frac{1}{T_1^{CSA}} . \tag{5}$$

For most calculations the CSA term is assumed to be negligible.

When the contribution of DD relaxation is 40%, the observed NOE in [29]Si$\{$[1]H$\}$ experiments is $\eta = -1.0$ and the signal is *nulled*. In situations where the DD contribution is close to 40%, observation of the [29]Si resonance is very difficult.

It is not uncommon for different [29]Si nuclei in a single compound to relax by different pathways. Thus in MD$_7$M (Fig. 17-8) the M silicons are dominantly relaxed by the SR mechanism (resulting from internal $(CH_3)$ Si group rotation) and give rise to positive spectral peaks, whereas DD and SR relaxation are more competitive for the four kinds of D unit [29]Si nuclei, resulting in peaks that are essentially nulled. When it is important in routine modes of operation to be able to observe all [29]Si nuclei with equal facility one or more of the techniques mentioned in the section on experimental methods must be used.

The competition between SR and DD relaxation is illustrated by the [29]Si relaxation behavior of TMS given in Table 17-5. At the higher temperatures given in Table 17-5 SR relaxation is strongly favored, probably occurring in the vapor phase near room temperature. At very low temperatures, as molecular motion decreases, DD relaxation dominates while SR relaxation becomes inefficient.[20] Figure 17-9 shows plots of the dipole-dipole $T_1$ ($T_1^{DD}$) and spin-rotation $T_1$ ($T_1^{SR}$) for TMS as a function of inverse temperature. Such a plot allows determination of the activation energy, $E_a$, for the relaxation process. For both $T_1^{DD}$ and $T_1^{SR}$ $E_a$ is ~1.8 kcal/mole, in qualitative agreement with the activation energies expected for diffusion-controlled processes. The relaxation times for several other organosilicon compounds studied[10] have higher activation energies, ranging from 2.6 to 3.7 kcal/mole.

Table 17-6 summarizes the [29]Si relaxation behavior of some typical organosilicon compounds near room temperature. The relatively large molecule diphenylsilane (1) has two protons di-

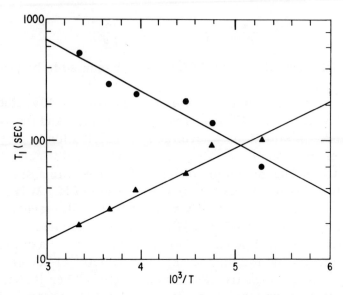

Figure 17-9.   The dependence of $T_1^{DD}$ and $T_1^{SR}$ on temperature$^{-1}$ for TMS (data plotted from Table 17-5).[10]

rectly attached to the $^{29}$Si nucleus.  The observed NOE indicates that within the limits of experimental accuracy ($\pm 0.10\,\eta$) all the relaxation arises from $^{29}$Si-$^1$H DD interactions.  For diphenylsilane $T_1^{DD}$ is therefore ~26 sec, the shortest $^{29}$Si-$^1$H DD $T_1$ so far observed.  The $^{29}$Si $T_1$ for diphenylsilane can be compared with $^{13}$C $T_1$'s obtained for the phenyl ring carbons in the same sample at 25.2 MHz.  Those $T_1$'s should be about 5.3 times shorter than the $^{29}$Si $T_1$ (allowing for the differences between $^{13}$C and $^{29}$Si nuclear characteristics and the *two* protons bonded to the silicon—see equation 2).  The average ring carbon (dipole-dipole) $T_1$ is about 5 sec (5.5, 5.5, and 3.1 sec for the *ortho*, *meta*, and *para* carbons, respectively).  Multiplying 5 sec by 5.3 predicts $T_1^{DD} = 26.5$ sec, in excellent agreement with the observed value.  (This is partly accidental since anisotropic molecular motion affects the $^{13}$C and $^{29}$Si values differently.)  The direct NOE determination for diphenylsilane obtained from continuous wide-band proton decoupled and pulse-

TABLE 17-6.  $^{29}$Si relaxation in some representative organosilicon compounds[a]

| Compound | | $T_1$ (sec) | NOE $(-\eta)$ | $T_1^{DD}$ (sec) | $T_1^{SR}$ (sec) |
|---|---|---|---|---|---|
| PhSiH$_2$ | 1 | 26 | 2.5 | 26 | >500 |
| PhSi(CH$_3$)$_2$H | 2 | 46 | 0.62 | 187 | 61 |
| [(CH$_3$)$_3$Si]$_2$O | 3 | 39.5 | 0.31 | 320 | 45 |
| [(CH$_3$)$_2$SiO]$_3$ | 4 | 80 | 1.1 | 183 | 142 |
| [(CH$_3$)$_2$SiO]$_6$ | 5 | 91 | 2.24 | 102 | 820 |
| Si(OCH$_2$CH$_3$)$_4$ | 6 | 135 | 1.5 | 227 | 334 |
| | 7  A | 31 | 1.8 | 43.4 | 108 |
| | B | 66 | 1.45 | 115 | 155 |
| | C | 67 | 1.40 | 121 | 151 |

[a]Data from Reference 10.  $T_1$'s at 19.9 MHz and 38°.  Samples run neat or with 15% acetone-$d_6$; degassed with N$_2$.  Derived quantities ($T_1^{DD}$ and $T_1^{SR}$) ±15–30%.

267

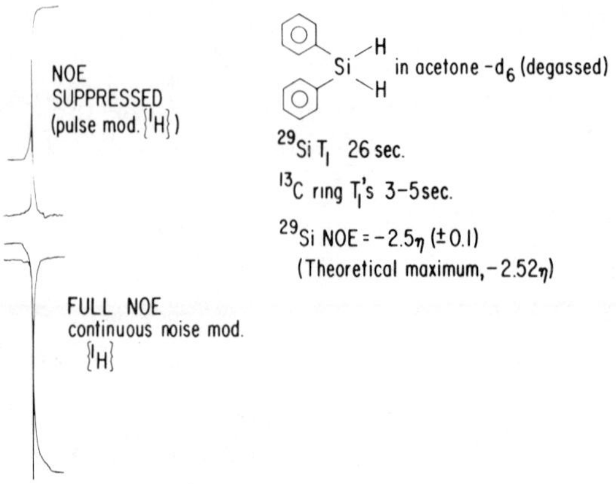

Figure 17-10. Direct $^{29}$Si-$^1$H nuclear Overhauser effect determination for diphenylsilane.

modulated $^1$H decoupled spectra is shown in Fig. 17-10.

For the $^{29}$Si nucleus in phenyldimethylsilane (2) $T_1^{DD}$ might be expected to be approximately twice that observed for the silicon in 1. Instead, $T_1^{DD}$ is about 187 sec while $T_1^{SR}$ is about 61 sec. The phenyldimethylsilane would appear to be tumbling more rapidly, $\tau_c$ being about $\frac{1}{3}$ $\tau_c$ for 1. The $^{13}$C $T_1$'s for the ring carbons in 2 largely confirm this: $T_1$ for the *ortho* and *meta* carbons is 13 sec, and $T_1$ for the *para* is 8 sec. (Exact comparisons of the $T_1$ data cannot be made since the two compounds have different anisotropic tumbling characteristics.) In hexamethyldisiloxane (3) both M units are able to spin rapidly. Also, the overall size of the molecule is small. Both these characteristics favor SR relaxation, which is dominant for this compound (Table 17-6).

In the "large" cyclic dimethylsiloxane $D_6$ (5) DD relaxation accounts for almost all observed $^{29}$Si relaxation (NOE = $-2.24$, Table 17-6) whereas in the smaller cyclic molecule, $D_3$ (4), SR and DD relaxation are competitive. In organosilicon compounds containing $^{29}$Si nuclei without nearby protons and with molecular motion characterized by slow overall tumbling, the observed $T_1$'s are very long with generally high NOE's (compare ethylorthosilicate (6), Table 17-6).

The relaxation behavior of **7** is an experimental verification of the increased importance of *long-range* $^{29}$Si-$^1$H dipole-dipole interactions, relative to any analogous $^{13}$C-$^1$H relaxation situation. In this molecule molecular tumbling is more or less isotropic, as evidenced by the comparable $T_1$'s for the two non-protonated $^{29}$Si nuclei (B and C). $T_1^{DD}$ for the *protonated* silicon, A, is only ~2.7 times shorter than $T_1^{DD}$ for silicon atoms B and C. In $^{13}$C relaxation, the ratio between $T_1^{DD}$ for nonprotonated and protonated carbons in a single molecule would be $\gtrsim 10:1$.[12] *Intermolecular* NOE data obtained from hexachlorodisilane and octamethylcyclotetrasiloxane solutions in protio solvents[10] allows estimation of long-range *intramolecular* and *intermolecular* contributions to $T_1^{DD}$ for the $^{29}$Si nuclei in **7**: $T_1^{DD}$ (long-range intramolecular) $\approx 140$ sec; $T_1^{DD}$ (intermolecular) $\approx 700-1000$ sec.

## EXPERIMENTAL METHODS

The spectra of MD$_7$M shown in Figs. 17-4 and 17-8 were run at the same sample concentration and with similar spectrometer settings. The total data acquisition time used to obtain each spectrum was ~17 min. The increased efficiency of the former experiment resulted from addition of a paramagnetic additive, Cr(acac)$_3$, to the sample. This technique helps in two ways. First, the $^{29}$Si $T_1$'s are shortened more than tenfold, by replacing $^{29}$Si-$^1$H DD and SR relaxation by electron-nuclear DD relaxation.[9, 10] This results in more complete relaxation between successive pulses even when the pulse repetition rate is high ($\lesssim 10$ sec). Second, by effectively replacing $^{29}$Si-$^1$H DD relaxation, electron-nuclear DD relaxation eliminates the undesirable $^{29}$Si-$^1$H NOE.[21]

With the addition of a paramagnetic additive $^{29}$Si $T_1$'s normally are equivalent in a given sample, being functions only of the solution viscosity and the concentration of the paramagnetic additive (assuming no "complexing" occurs between the additive and the organosilicon substrate).[22] In the case of Cr(acac)$_3$ additive concentrations of about $5 \times 10^{-2}$ $M$ result in $^{29}$Si $T_1$'s of a few seconds. In the absence of complex formation between the relaxation reagent and the substrate, quantitative $^{29}$Si NMR information becomes available: peak integrations become directly proportional to the *number* of nuclei making up a resonance signal.

Cr(acac)$_3$ has been shown to be a particularly useful paramagnetic relaxation reagent for $^{13}$C and $^{29}$Si NMR. [9,10,23] At typical concentrations ($10^{-2}$ to $10^{-3}$ $M$) the reagent has no effect on the substrate chemical shifts. Cr(acac)$_3$ is soluble and chemically inert in a large variety of solvent systems.

For highest efficiency operation in $^{29}$Si FT NMR experiments it is necessary to dope samples with relaxation agents such as Cr(acac)$_3$. Sometimes it is not desirable to introduce extraneous compounds, even at low concentrations. For these situations there are several alternate techniques that can facilitate $^{29}$Si FT NMR studies. Dissolved oxygen gas in the sample solution can replace normal $^{29}$Si relaxation pathways in the same way as with paramagnetic chemical additives. Unfortunately, the pressure of oxygen required to completely replace $^{29}$Si-$^{1}$H DD relaxation and result in $^{29}$Si $T_1$'s of a few seconds is $\gtrsim 10$ atm. Addition of this amount of O$_2$ to the sample would necessitate special handling and the use of heavy-wall NMR tubes. Saturating the sample with O$_2$ at 1 atm does result in some experimental improvement, reducing $T_1$'s to less than 30 sec[24] and largely suppressing the $^{29}$Si-$^{1}$H NOE, at least to some extent. [25]

Interrupted or pulse-modulated proton decoupling[26] does not affect $^{29}$Si $T_1$'s but does effectively suppress the negative $^{29}$Si-$^{1}$H NOE. This instrumental technique (which does not require *any* manipulation of the NMR sample) was developed for use in $^{13}$C NMR[26] but it is even more useful in $^{29}$Si NMR[9,10] where $T_1$'s are generally longer. In this experiment the wideband $^{1}$H decoupling power is gated on *only* during the periods of data acquisition; throughout the remaining time between $^{29}$Si excitation pulses the decoupler is gated off. The duty cycle for $^{1}$H decoupling is kept below 10–20%, typically using data acquisition periods of 1–2 sec and total pulse intervals of 10–20 sec. The long $^{29}$Si $T_1$'s prevent buildup of significant NOE during the brief $^1$H irradiation periods and yet the spectrum is completely decoupled (the presence of spin-spin coupling is instantaneously affected by the $^1$H irradiation, whereas the NOE grows as a function of the $^{29}$Si $T_1$'s).

A variation of the pulse-modulated decoupling scheme has been used to improve the efficiency of $^{29}$Si $T_1$ studies.[10] A typical set of *inversion-recovery* spectra obtained with pulse-modulated decoupling is shown in Fig. 17-11. These data allow calculation of the $^{29}$Si $T_1$'s for all the silicons in MD$_6$M. The $T_1$'s obtained (M, 42 sec; D$^1$, 64 sec; D$^2$, 55 sec; D$^3$, 59 sec) coupled with the observed NOE's from other pulse-modulated

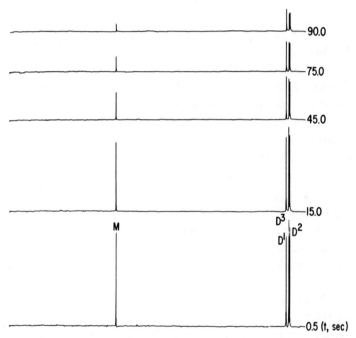

Figure 17-11. Inversion-recovery $T_1$ spectral set for MD$_6$M obtained with pulse-modulated proton decoupling. $^{29}$Si pulse sequence:

$$(T - 90^\circ_\infty - \tau' - 180^\circ - t - 90^\circ_t)_x.$$

The spectra plot $(S_\infty - S_t)$ where $S_\infty$ and $S_t$ are the Fourier transformed signals resulting from the $90^\circ_\infty$ and $90^\circ_t$ pulses (see reference 12 for details). For this experiment, $T = 220$ sec and $x = 15$; $t$ varies as shown. The spectral width was 1000 Hz and the sample was degassed.

$^{29}Si\{^{1}H\}$ experiments indicated that segmental motion was significant along this oligomer chain.[10]

## ACKNOWLEDGMENTS

We thank Dr. Peter C. Juliano and Tyrone D. Mitchell for their participation in some of the reported work. We also acknowledge the support and encouragement of the General Electric Corporate Research and Development management.

## References

1. G. R. Holzmann, P. C. Lauterbur, J. H. Anderson, and W. Koch, *J. Chem. Phys.*, **25**, 172 (1956).

2. P. C. Lauterbur, *Determination of Organic Structures by Physical Methods*, F. Nachod, Ed., Vol. 2, Academic Press, New York, 1962, p. 465.

3. B. K. Hunter and L. W. Reeves, *Can. J. Chem.*, **46**, 1399 (1968).

4. G. Englehardt, H. Jancke, M. Mägi, T. Pehk, and E. Lippmaa, *J. Organomet. Chem.*, **28**, 29 (1971).

5. G. C. Levy, *J. Am. Chem. Soc.*, **94**, 4793 (1972).

6. R. E. Scholl, G. E. Maciel, and W. K. Musker, *J. Am. Chem. Soc.*, **94**, 6376 (1972).

7. H. C. Marsmann, *Chemiker Ztg.*, **96**, 288 (1972).

8. H. C. Marsmann and H. G. Horn, *Chemiker Ztg.*, **96**, 456 (1972).

9. G. C. Levy, J. D. Cargioli, P. C. Juliano, and T. D. Mitchell, *J. Magn. Resonance*, **8**, 399 (1972).

10. G. C. Levy, J. D. Cargioli, P. C. Juliano, and T. D. Mitchell, *J. Am. Chem. Soc.*, **95**, 3445 (1973).

11. Not 20 times as sensitive, as given by Wells (P. R. Wells, *Determination of Organic Structures by Physical Methods*, F. Nachod and J. Zuckerman, Eds., Vol. 4, Academic Press, New York, 1971, p. 237.

12. G. C. Levy, *Acc. Chem. Res.*, **6**, 161 (1973).

13. E. B. Whipple, unpublished results; G. C. Levy, unpublished results.

14. M. Bacon, G. E. Maciel, W. K. Musker, and R. Scholl, *J. Am. Chem. Soc.*, **93**, 2537 (1971).

15. D units are numbered from the outside: $MD_9M$ $= MD^1D^2D^3D^4D^5D^4D^3D^2D^1M$.

16. Kang-Jen Lui, *Makromol. Chem.*, **126**, 189 (1969).

17. G. C. Levy, D. M. White, and J. D. Cargioli, *J. Magn. Resonance*, **8**, 280 (1972).

18. R. R. Dean and W. McFarlane, *Mol. Phys.*, **12**, 289 (1967).

19. For $^{29}$Si nuclei spin-spin coupled to $^{127}$I, scalar spin-lattice relaxation may be significant.

20. For further details on the SR relaxation mechanism see T. C. Farrar and E. D. Becker, *"Pulse and Fourier Transform NMR,"* Academic Press, New York, 1971, Chapter 4.

21. The identical spectrometer settings used to obtain the spectra shown in Figs. 17-4 and 17-8 were not optimized. The increase in spectral sensitivity apparent in Fig. 17-4 is thus not a quantitative measure of the effect of the paramagnetic additive.

22. G. C. Levy and J. D. Cargioli, *J. Magn. Resonance*, **10**, 231 (1973).

23. R. Freeman, K. G. R. Pachler, and G. N. LaMar, *J. Chem. Phys.*, **55**, 4586 (1971); O. A. Gansow, A. R. Burke, and W. D. Vernon, *J. Am. Chem. Soc.*, **94**, 2550 (1972).

24. At the solution concentration of oxygen under 1 atm of $O_2$, $T_1^e$ due to electron-nuclear dipole-dipole interactions is about 40 sec.[10]

25. The NOE will not of course be effectively suppressed for $^{29}$Si nuclei undergoing relatively rapid $^{29}$Si-$^1$H DD relaxation, such as in diphenylsilane.

26. R. Freeman, H. D. W. Hill, and R. Kaptein, *J. Magn. Resonance*, **7**, 327 (1972).

*Additional* $^{29}$*Si NMR References*

H. G. Horn and H. C. Marsmann, *Die Makromol. Chem.*, **162,** 255 (1972).
H. C. Marsmann and H. G. Horn, *Z. Naturforsch.*, 27B, 1448 (1972).

CHAPTER 18

# A SURVEY OF RECENT ADVANCES IN $^2$D AND $^3$T NMR

P. Diehl

*Department of Physics*
*University of Basel*

## INTRODUCTION

Although the hydrogen isotopes proton, deuteron, and triton have magnetic moments, the research in high-resolution hydrogen NMR was limited for a long period exclusively to proton spectroscopy. One reason for this limitation is the low natural abundance of deuterons (0.0156%) and the radioactivity of tritium; in addition the necessary fixed-frequency NMR spectrometers are available in only a few laboratories. There is a second, more fundamental reason for the slow progress in these fields which is related to the question whether anything can be learned from $^2$D or $^3$T spectroscopy that cannot be obtained by the use of $^1$H spectroscopy.

We began trying to answer this question in our NMR group at Basel exactly 10 years ago. The results were reported for the first time at the Pittsburgh Conference in 1963 and the data were published in 1964.[1] Although our interest soon shifted to other fields we were again asked in 1969 to report on recent advances in $^2$D NMR. A careful search of the literature showed, that in five years less than 10 papers had been published which

made use of high-resolution $^2$D NMR, and only five of these were
of general interest.   Again asked to give a survey on recent advances in 1972, we decided to look for all papers on $^2$D NMR in
the last 10 years.   Here is the statistical result of this search.

We have found 90 papers of which only 10 deal with general
chemical high-resolution studies,  10 are papers on oriented
molecules,  20 treat relaxation,  10 have a mixed character but
are predominantly of a biological nature,  and 40 treat problems
of the solid state.   Another 20 papers are related to deuterons
in that they discuss the influence of deuterium on the spectra of
different nuclei.

To our disappointment we find that there are still not very
many real advances.   Nevertheless there are some,  and we report on them in this chapter.

In our summary of 1964 we reached 10 conclusions about
$^2$D NMR and we have organized this chapter into 10 sections,
each beginning with the original conclusion followed by a report
on recent advances and,  if possible,  an indication of future
progress.


## ORIGINAL RESULTS AND RECENT ADVANCES

### Sensitivity

The relevant part of the summary[1] in 1964 is as follows:
"in order to be detectable, the volume concentration of chemically equivalent deuterons has to exceed a certain minimum
limit.   As this volume concentration decreases with increasing
molecular volume,  deuteron NMR at natural abundance is limited to small molecules and therefore of no practical importance.   (At a density of 1 the molecular weight has to be less
than 20 per chemically equivalent deuteron.)  In fully deuterated samples the signal-to-noise ratio is of the same order of
magnitude as for protons,  because the loss in sensitivity can
be compensated by an increased volume of the sample."

This statement is probably no longer true.   With the new
pulse—and computer—methods as well as the available strong

magnetic fields, $^2$D NMR at natural abundance should be possible and presumably quite interesting. With simultaneous decoupling of protons, if necessary, one should obtain simple first-order stick-spectra similar in nature to modern $^{13}$C spectroscopy. There is, of course, the line-width problem (discussed below), but we still believe in future progress.


## Chemical Shifts

The summary[1] states: "deuteron chemical shifts are very similar to proton chemical shifts. In pure compounds no deviations larger than 0.06 ppm have been detected. It is possible to measure deuteron isotope effects which are of the order of $+2.10^{-8}$ per deuteron for each proton neighbor that is replaced by a deuteron."

Here we have advances: recent measurements[2] on the chemical shifts between the two proton and deuteron functional groups in ethyl bromide show that, measured in the same sample, the deuteron and proton chemical shifts are identical. Consequently our observed deviations must have been caused by differences in susceptibilities.


## Analysis of Spectra

Our conclusion[1] was as follows: "the analysis of deuteron spectra is straightforward, as the ratio of coupling constant to chemical shift is reduced by a factor of about 6.5 compared with the proton."

This is still true; in no paper on high-resolution $^2$D NMR has the use of a computer been necessary for the analysis.[3] The situation is well illustrated in a study of axial and equatorial 4 positions of 1-$t$-butycyclohexyl derivatives.[4] Because of the simplicity of the spectra, the chemical shifts are measured directly from the observed singlets and they prove to be stereochemically sensitive.

## Indirect Spin–Spin Coupling

"Spin coupling between deuterons and $^{31}$P or $^{13}$C as well as isotope effects in the $^{31}$P-$^{2}$D coupling constants have been observed."[1]

Not very much has been added here. The most widely studied coupling is still the proton-deuteron one, but usually it is observed by $^1$H resonance and used in order to determine an otherwise undetectable proton-proton coupling in equivalent groups.

## Line Width

*Quadrupole Relaxation.* Our summary[1] states: "line widths of deuteron resonances are determined by quadrupole relaxation. In deuterated hydroxyl groups of alcohols the line widths are found to increase, in agreement with theory, with increasing viscosity and molar volume. From line width measurements it is possible to calculate approximate quadrupole coupling constants. It is found that for deuterons in carbon-deuteron bonds this constant is about a factor of two less than in oxygen-deuteron bonds."

Here we would like to report on three more recent studies: in the first[5] the relaxation of *n*-dodecane was measured and analyzed according to the relation

$$\frac{1}{T_1} = \left(\frac{3}{8}\right)\left(\frac{e^2 qQ}{\hbar}\right)^2 \tau_c$$

where

$$\tau_c = \frac{\eta M}{\rho N k T}$$

and

$$\frac{M}{\rho N} = \text{molecular volume}.$$

A comparison with known quadrupole coupling data has shown that a microviscosity $\eta_{eff}$ has to be introduced:

$$\eta_{\text{eff}} = \eta \cdot f_r$$

with

$$f_r = 6 \left(\frac{r_e}{r}\right) \sum_{m=0}^{\infty} \left(1 + \frac{2m r_e}{r}\right)^{-4}$$

and $r$ = radius of the solute molecule, $r_e$ = radius of the solvent molecule, and $m$ = number of layers.

Whereas $f_r$ is found to be 0.16 in the pure liquid, values of 0.25 and 0.22 have to be used at infinite dilution in $CS_2$ and $C_6H_6$, respectively.

In the second study[6] the relaxation of deuterons, this time at various positions in organic molecules, was measured. Here the microviscosity was kept constant at $f_r = \frac{1}{12}$ and the resulting differences between $T_{1\,\text{obs}}$ and $T_{1\,\text{calc}}$ are attributed to additional relaxation processes. So, for instance, the fact that $(T_{1\,\text{obs}}/T_{1\,\text{calc}})$ was found to be practically constant at 1.0 for all three positions in propyl iodide was interpreted as a missing internal rotation in such molecules. On the other hand, the difference between ring deuterons of toluene $(T_{1\,\text{obs}}/T_{1\,\text{calc}} = 0.81)$ and methyl deuterons $(T_{1\,\text{obs}}/T_{1\,\text{calc}} = 2.81)$ has been used in favor of the internal rotation argument.

Similarly more qualitative measurements on specifically labeled polynucleotides[7] have provided some data on their motional characteristics.

*Dipole Relaxation by Paramagnetic Ions.* "Due to their small gyromagnetic ratio, deuterons are less sensitive to relaxation by paramagnetic ions than are protons. The longer relaxation time, compared to that of protons at the same concentration of paramagnetic ions, has been found in agreement with theory."[1]

In our early measurements we compared the water line width at various concentrations of $Fe^{3+}$ ions.

Recently measurements[8] have been performed directly on some paramagnetic transition metal acetylacetonates. In these the deuteron line widths are smaller than the corresponding proton line widths by factors between 40 and 9 (theoretical factor, 42.5). It is not obvious from the study whether the influence of viscosity was accounted for, nor is the variation explained. On the other hand, the authors have concluded that deuterons are ideal probes in that they provide a gain in resolution and, by means of selective deuteration, help in signal assignment; furthermore, owing to the small value of $J$(H-D), the spectra are simplified.

In a more quantitative study[9, 10] the proton and deuteron relaxations were compared in aqueous solutions of vanadyl(IV). In this system an excess relaxation ($T_{2p}$) arises which is governed by the relaxation time of the nuclei in the first hydration shell ($T_{2M}$) and their mean residence time ($\tau_M$) in this shell:

$$1/T_{2p} = \frac{P_M}{T_{2M} + \tau_M} \, ,$$

where $P_M$ = fraction of nuclei in the first coordination sphere.

By combination of the results for protons $T_{2p}{}^H$ and deuterons $T_{2p}{}^D$, taking into account the relation

$$T_{2M}{}^D / T_{2M}{}^H = \left( \frac{\gamma_H}{\gamma_D} \right)^2 \simeq 42.5 \, ,$$

$T_{2M}$ and $\tau_M$ may be derived separately. The temperature dependence of $T_{2M}$ which is related to the variation in the electron relaxation rate can thus be studied.

*Relaxation Decoupling.* "By adding paramagnetic ions to liquids of partially deuterated molecules, the difference in relaxation times between deuterons and protons can be used to decouple the deuterons from the protons."[1] Unfortunately our suggested, and demonstrated, poor man's heteronuclear decoupler has not found much practical application.

*Solvation.* We found the following applications[1]: "as there is no interference from solvent signals, deuteron spectroscopy is an ideal method for studying selective solvation of cations by water in mixed solvents. The concentration of water in the hydration sphere of $Mn(ClO_4)_2$ at high dilution in acetone has been found to be 180 times higher than the average concentration in the acetone solvent. Selective solvation of $Mn^{+++}$ ions in water-acetone mixtures depends upon the anions."

There are only three more recent studies in this very promising application of deuteron resonance. In the first[11] $^2$D relaxation times are used to understand protein solvation in $D_2O$ solution. It is found that, whereas the correlation time in free water is of the order of $2 \times 10^{-12}$ sec, in the solvation water molecules it is reduced to $8 \times 10^{-11}$ sec, and to $3 \times 10^{-9}$ sec in the protein.

A related study[12] deals with the interaction of water with organic molecules. Quite generally, $T_1$ values for the deuterons in the mixtures decrease. This is interpreted as binding, association, or hydration. The data indicate the existence of a complex of three water molecules with DMSO.

Finally, the participation of $D_2O$ in conformational changes of biopolymers is discussed[13] and the following rules are obtained. The groups

$$C=O, \quad >NH, \quad -\overset{\overset{\displaystyle O}{\|}}{C}-OM, \quad \text{and} \quad C-\overset{-+}{NH_3}X^-$$

do not interact strongly with water, whereas the following groups do interact:

$$\overset{\overset{\displaystyle O}{\|}}{C}-OH \quad \text{and} \quad C-N\overset{\displaystyle /\,H}{\underset{\displaystyle \backslash\,H}{}} \quad .$$

## Solvent Effects

Our summary[1] was as follows: "we have observed differences in solvent effects between deuteron and proton spectroscopy. These are ascribed qualitatively to differences between hydrogen and deuteron bridges."

In our original paper we mentioned differences in chemical shifts between $D_2O$ and $H_2O$ at infinite dilution in various solvents. A more drastic effect showed up later on when we studied deuteroacetylacetone. We found that the deuteron in the bridge of the enol form is shifted 0.5 ppm to higher fields with respect to the corresponding proton.

A similarly large shift was detected[14] in chlorophyll-$a$-$d_{72}$ in which the resonance of the phytyl chain is shifted by 0.63 ppm to higher field with respect to the proton case.

## Kinetic Studies

The last point of our summary[1] in 1964 was as follows: "deuteron spectroscopy can be used to study kinetic problems. For weak deuteration the precision will be better than that obtained from proton spectroscopy. The sensitivity range for studying fast exchange is, compared to protons, shifted to longer lifetimes. This is due to the decreased chemical shift of deuterons, measured in Hz and at the same magnetic field."

Here again we have progress. Still, there are only three papers which have made use of $^2$D NMR for kinetic studies in the last few years.

In the first[15] the coupling reactions of 1-chlorocycloheptene (-2, 7, 7-$d_3$) with phenyllithium were investigated and the reaction path determined (elimination-addition reaction with cycloheptyne intermediate). The authors state that, due to the $^2$D line width of 0.6 Hz, their resolution is limited to 0.1 ppm, but they conclude that many things can be done at such a resolution.

In a second group of two papers[16, 17] the addition reaction of formaldehyde to aromatic olefins is studied (Prins reaction) and the stereochemistry of the products (*cis* or *trans*) is elucidated. The advantage of $^2$D NMR as a spectral simplification is pointed out.

In a third study[18] the kinetics of the slow deuteration of decalon-1-*trans* is described. The spectrum is again very simple and the slow reaction can be followed easily.

## NEW FIELDS

### Oriented Molecules

$^2$D NMR spectra of oriented molecules allow either the measurement of quadrupole coupling constants, if the degree of orientation is known, or the measurement of the degree of orientation if the quadrupole coupling constant is known.[19,20] Obviously such a field has an exciting potential of new applications, particularly in biology where oriented molecules play a very important role. The details are presented in chapter 24.

### $^3$T Resonance

Our report on $^2$D NMR at the Pittsburgh meeting in 1963 stirred some interest in $^3$T resonance, and in 1964 the first paper[21] on high-resolution tritium NMR appeared. It described the $^3$T spectra of the methyl and ethyl groups of ethylbenzene. Perhaps the most surprising result is that the value of $J$(T-C-C-H) is not exactly as large as predicted on the basis of gyromagnetic ratios. It is extrapolated as $7.8 \pm 0.1$ Hz against the theoretical value of $8.16 \pm 0.02$ Hz. The paper has warned potential future $^3$T spectroscopists of the inherent dangers of the method. A sample of 10 Ci of the $^3$T-enriched compounds may produce 1 ml of $H_2$ gas per day as a consequence of self-radiolysis.

It is not surprising that after this warning only one more paper has since appeared on this subject.[22] In this study a microbulb assembly was used in order to keep the total radioactivity low and to avoid the danger of breakage due to the pressure buildup. The results presented are quite promising. For example, the self-radiolysis of complex compounds may be followed easily. The simple spectra indicate the position at which the $^3$T label is fixed, in contrast to the corresponding complex proton spectra which may obscure the relevant data. This method is obviously of particular interest to the radiation chemist.

In the second study it was again found that the couplings to tritons can differ appreciably from the corresponding couplings to protons, and that there is no simple relationship for these deviations. On the contrary, the chemical shifts follow closely the [1]H shifts. As far as couplings in magnetically equivalent groups are concerned, the deviations may perhaps be attributed to the inaccuracy of $J$(H-D), which is usually determined from broad and overlapping lines. The error is subsequently multiplied by a factor of 6.5 in deriving $J$(H-H).

## CONCLUSIONS

As a conclusion we would like to repeat what was stated 10 years ago and which is also true for [3]T resonances. These methods not only give information about isotope effects but also introduce exciting new possibilities which could not have been obtained with [1]H resonance, so that besides the disadvantages there are obviously some very promising future applications as well.

## References

1.  P. Diehl and Th. Leipert, *Helv. Chim. Acta,* **47,** 545 (1964).
2.  H. Jensen and K. Schaumburg, *Acta Chem. Scand.,* **25,** 663 (1971).
3.  A. Habich et al., *Helv. Chim. Acta,* **48,** 1297 (1965).
4.  R. Wilde and J. Grimaud, *Compt. Rend. Ser. C.,* **271,** 597 (1970).
5.  D. E. Woessner et al., *J. Magn. Resonance,* **1,** 105 (1969).
6.  J. A. Glasel, *J. Am. Chem. Soc.,* **91,** 4569 (1969).
7.  J. A. Glasel, *Proc. Natl. Acad. Sci. U.S.,* **60,** 1038 (1968).
8.  A. Johnson and G. W. Everett, Jr., *J. Am. Chem. Soc.,* **92,** 6705 (1970).

9.  J. Reuben and D. Fiat, *J. Am. Chem. Soc.*, **91**, 1242 (1969).

10. J. Reuben and D. Fiat, *J. Am. Chem. Soc.*, **91**, 4652 (1969).

11. J. A. Glasel, *Nature*, **220**, 1124 (1968).

12. J. A. Glasel, *J. Am. Chem. Soc.*, **92**, 372 (1970).

13. J. A. Glasel, *J. Am. Chem. Soc.*, **92**, 375 (1970).

14. R. C. Dougherty et al., *J. Am. Chem. Soc.*, **87**, 5801 (1966).

15. L. K. Montgomery et al., *J. Am. Chem. Soc.*, **89**, 3453 (1967).

16. C. L. Wilkins and R. S. Marianelli, *Tetrahedron*, **26**, 4131 (1970).

17. C. L. Wilkins et al., *Tetrahedron Lett.*, 5109 (1968).

18. E. Casadevall and P. Metzger, *Tetrahedron Lett.*, 4199 (1970).

19. P. Diehl and C. L. Khetrapal, *NMR, Basic Principles and Progress*, P. Diehl, E. Fluck, and R. Kosfeld, Eds., Vol. 1, Springer, Berlin-Heidelberg-New York, 1969.

20. P. Diehl and P. M. Henrichs, *Specialist Periodical Reports: NMR*, Chemical Society, London, 1972.

21. G. V. D. Tiers et al., *J. Am. Chem. Soc.*, **86**, 2526 (1964).

22. J. Bloxsidge et al., *Org. Magn. Resonance*, **3**, 127 (1971).

# CHAPTER 19

# FLUORINE CHEMICAL SHIFTS AND F-F COUPLING CONSTANTS

L. Cavalli

*Montedison*
*Centro Ricerche*
*Bollate, Italy*

The fluorine-19 isotope has 100% natural abundance and has a spin of $\frac{1}{2}$; the spin response is about 20% weaker than that of the proton. Consequently fluorine resonance is almost as easily detected as proton resonance. The range of fluorine chemical shifts (about 1000 ppm) is bigger than that of protons (about 15 ppm). Because the magnitude of chemical shifts is determined almost entirely by the behavior of valence electrons, the factors which are most important in determining [19]F shifts are more interesting than those for proton resonance and are also likely to be predominant for other nuclei such as [13]C and [14]N.

## FLUORINE SHIELDING CONSTANT, $\sigma$[1,2]

Because nuclear shielding constants are not measurable directly, it is necessary to measure shielding, $\sigma_c$, relative to some reference resonance, $\sigma_R$; the differences are known as chemical shifts, $\delta$:

|  | $\delta$ (ppm) |  | $\delta$ (ppm) |
| --- | --- | --- | --- |
| $CFCl_3$ | 0.0 | $C_4F_8$ | $-135.15$ |
| $CF_3COOH$ (int) | $-76.55$ | 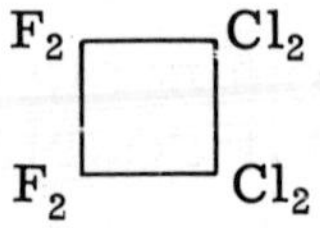 | $-114.20$ |
| $CF_3COOH$ (ext) | $-78.90$ |  |  |
| $CF_3CCl_3$ | $-82.2$ | 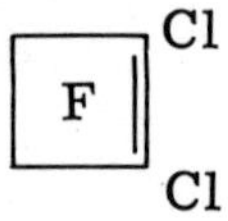 | $-118.47$ |
| $CFCl_2 \cdot CFCl_2$ | $-67.80$ | $C_6F_{12}$ | $-133.23$ |
| $CF_2Cl \cdot CCl_3$ | $-65.10$ | $C_6F_6$ | $-164.9$ |

Figure 19-1. Some $^{19}F$ NMR reference standards.[3] Shifts upfield from $CFCl_3$ are negative.

$$\delta = \sigma_R - \sigma_c \ .$$

$\delta$ is negative if the reference compound is less shielded than the sample. Both internal and external referencing are used with different reference compounds. Trifluoroacetic acid (TFA) has been used extensively, particularly as external reference. However, TFA is not satisfactory for correlating chemical shift measurements due to the bulk susceptibility corrections that must be applied. Because of the magnitude of $^{19}F$ shifts, these corrections are not very important. Several other reference standards are often used[3] (Fig. 19-1).

$$\sigma = \sigma_d(\text{local}) + \sigma_p(\text{local}) + \sigma_m + \sigma_r + \sigma_e + \sigma_s$$

where $\sigma_m$ = neighbor anisotropy effect

$\sigma_r$ = ring current effect

$\sigma_e$ = electric field effect

$\sigma_s$ = solvent effect

Figure 19-2.   Shielding constant, $\sigma$.

The most common reference compound is $CFCl_3$.  Evaluating $\sigma$ requires some detailed knowledge of the excited-state wave functions of the molecule, which is rarely available.  Following an oversimplified picture it is possible and useful to write $\sigma$ as a sum of local effects and long-range or molecular effects (Fig. 19-2).  This scheme is arbitrary but useful.  $\sigma_d$ (local) and $\sigma_p$ (local) are the diamagnetic and paramagnetic local terms of Ramsey's equation for nuclear shielding due to electrons closely associated with the nucleus.  The chemical shift of the fluorine nucleus is dominated by $\sigma_p$ (local).  For instance the $^{19}F$ resonance of $F_2$ occurs about 625 ppm to low field of that for HF.  The difference in $\sigma_d$ can be estimated to be about 20 ppm.  $\sigma_p$ is zero for spherically symmetric species such as an atom or ion, like $F^-$, whereas it is appreciable for an asymmetric distribution of $p$ electrons such as for covalently bonded fluorine.  The discrepancy between experimental and calculated values is due to the partial ionic character of the bond in $F_2$ which reduces $\sigma_p$ considerably from its calculated value.  $\sigma_m$ is related to the anisotropy in the diamagnetic susceptibilities, $\Delta x$, of a certain group of electrons.  The magnitude of this contribution depends only on the magnitude of $\Delta x$ and the geometry of the molecule.

This means that $\sigma_m$ should be equal for $^{19}F$ and $^1H$ resonance, for example, and typically of the order of 1 ppm.  The magnitude of $\sigma_r$ and its effect on nuclear shielding should again be independent of the nucleus and hence the magnitude of this term should be typically about 1 ppm.  The $\sigma_e$ term is related

$$\sigma_e = -AE_z - BE^2 - B_1 \langle E^2 \rangle$$

$$\langle E^2 \rangle = \sum_i \frac{3 P_i \, l_i}{R_i^{\,6}}$$

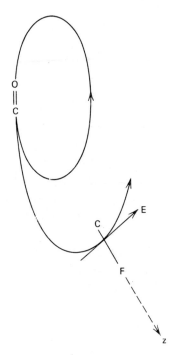

Figure 19-3. Electric field effect $\sigma_e$.

to the intramolecular electric fields arising from charges remote from the shielded nucleus. The $\sigma_e$ term for fluorine resonance is more important than $\sigma_m$ and $\sigma_r$, and can be considered
to be made up of three contributions (Fig. 19-3). When the fluorine is in an axially symmetric environment the first contribution to $\sigma_e$ may simply be related to drift of electrons along the
bond direction, and the second contribution to the distortion of
the electronic charge distribution in a direction perpendicular
to the bond direction. The third contribution is due to time-
dependent electric dipoles mutually induced between bonds in
the molecule and the bond to the shielded nucleus. Such dipoles
give a nonzero average value of $E^2$, written $\langle E^2 \rangle$. This is the

so-called van der Waals contribution, $\sigma_w$, for which $P_i$ and $I_i$ are the polarizability and first ionization potential of the electron group (bond). Experimental estimates of the constants $A$, $B$, and $B'$ have shown that they are quite large for fluorine and that $A$ is negative. Typical values are $A =$ about $-10 \times 10^{-12}$ e.s.u., $B \simeq B'$ $20 \div 40 \times 10^{-18}$ e.s.u. $\sigma_s$ contains contributions of the same type as $\sigma_m$, $\sigma_r$, and $\sigma_e$, but they are of intermolecular origin.

## Aromatic Fluoro Compounds

The chemical shifts of the aromatic fluorine atoms occur within the range $\delta = -130$ to $-180$ ppm (upfield from $CFCl_3$). $\delta_F$ of substituted compounds generally decreases in the order *ortho* > *para* > *meta*. The shifts of fluorine nuclei *meta* or *para* to a substituent can be correlated successfully with $\pi$-electron charge densities and bond orders as calculated by the Hückel MO method. Thus, for example, substituents such as $-NH_2$ and -OH, which produce an increase in $\pi$-electron density at the *para* carbon atom, produce an increase in shielding of the *para* fluorine atom and those such as $-NO_2$ and -CHO, which decrease the $\pi$-electron density, similarly decrease the fluorine shielding. For fluorine nuclei in a position *ortho* to a substituent, large deviations are always found. This is the so-called "ortho effect": whenever two fluorine atoms are in *ortho* positions to one another an upfield shift of about 20 ppm is produced. This effect is found to be a feature of all the fluorobenzene shifts and is additive. Such ortho effects can be accounted for satisfactorily in terms of intramolecular electric field contributions. Differences in $\delta_F$ in substituted aromatic compounds can be reasonably predicted according to[4]

$$\Delta\sigma = \Delta\sigma_{electronic} + \Delta\sigma_{electric} \, ,$$

where $\Delta\sigma_{electronic}$ takes into account differences in $\pi$-electron charge distribution and $\Delta\sigma_{electric}$ is the difference in the resultant electric fields at fluorine nuclei, as calculated according to the expression given in Fig. 19-3.

## Aliphatic Fluorocarbons

The interpretation of the effect of substituents on the shielding constant in terms of changes in electric fields and van der Waals interactions should be more successful than in aromatic compounds, because in such a case we have $\Delta\sigma_{electronic} \approx 0$. In fact, satisfactory results have been obtained in a variety of fluoro compounds, for example, ethanes,[5] cyclopropanes,[6] cyclobutanes,[5] cyclopentenes,[7] cyclohexanes,[8] and bicyclo(2,2,1)-heptane.[9] Qualitatively, the approach to be applied to elucidate configurational and/or conformational problems is to try to fit calculated to observed shifts, adopting the electric field and van der Waals terms and using different empirical coefficients for the $E^2$ and $\langle E^2 \rangle$ terms. In the case of cyclohexanes the method predicts the chemical shift difference, $\Delta\delta_{ae}$, between axial and equatorial fluorines with the correct sign, but with too small a magnitude; for example, for $C_6F_{12}$ at $-60\,°C$ in $CFCl_3$ $\Delta\delta_{ae}$ is experimentally equal to 18.2 ppm, with the axial fluorine at low field. The method works satisfactorily only when halogen atoms are involved in the substitution. In such a case the electric field at the fluorine nucleus under observation does not change significantly and the van der Waals effect, $\sigma_w$, can be considered to control the shielding of the fluorine nucleus. When other substituents such as hydrogen are involved, for example, the method breaks down completely. Other difficulties of application of the method are known, and they have been summarized and discussed.[1] Recently a simple method of calculating the effects of geminal and vicinal substituents upon the fluorine chemical shifts has been presented.[10,11] The suggestion is that the shielding of the fluorine nucleus is proportional to the difference in the *effective* electronegativities of F and the C to which F is directly bonded. The effective electronegativities are the Huggins electronegativities of F and C, modified by a variety of mutual interactions between all groups in the molecule. The recently developed method makes it possible to calculate the fluorine chemical shifts in a variety of molecules containing the elements C, H, N, O, F, Cl, Br, and I and to rationalize several puzzling fluorine shift trends known in the liter-

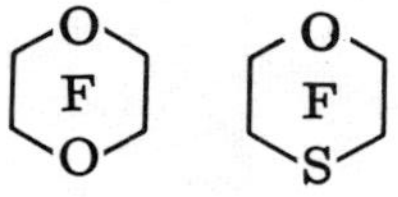

| | F (Axial) | | | | | F (Equatorial) | | | | |
| | $\alpha H_{ax}$ | $\alpha H_{eq}$ | $\beta H_{ax}$ | $\beta H_{eq}$ | Base Value | $\alpha H_{ax}$ | $\alpha H_{eq}$ | $\beta H_{ax}$ | $\beta H_{eq}$ | Base Value |
|---|---|---|---|---|---|---|---|---|---|---|
| $\alpha$ to oxygen in I | +5 | −7 | +5 | 0 | 82 | −9 | −4 | 0 | +4 | 90 |
| $\alpha$ to oxygen in II | +4 | −5 | — | 0 | 78 | −17 | −12 | 0 | +5 | 88 |
| $\alpha$ to sulfur in II | +12 | −9 | +9 | 0 | 82 | −6 | −5 | 0 | +7 | 97 |
| In fluorocyclohexanes | +2.3 | −4.3 | — | — | 124.2 | −15 | −10.5 | +4.0 | — | 142.1 |

Figure 19–4.  Parameter scheme for $>CF_2$.[15]

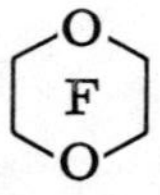

I     II

| | F (Axial) | | | | | F (Equatorial) | | | | |
|---|---|---|---|---|---|---|---|---|---|---|
| | $\alpha H_{ax}$ | $\alpha H_{eq}$ | $\beta H_{ax}$ | $\beta H_{eq}$ | Base Value | $\alpha H_{ax}$ | $\alpha H_{eq}$ | $\beta H_{ax}$ | $\beta H_{eq}$ | Base Value |
| $\alpha$ to oxygen in I | +11 | −6 | +5 | 0 | 140 | −16 | −2 | 0 | +2 | 158 |
| $\alpha$ to oxygen in II | +5 | −6 | +5 | 0 | 139 | −9 | −10 | — | +3 | 153 |
| $\alpha$ to sulfur in II | +3 | −4 | 0 | 0 | 166 | — | −11 | — | +3 | 178 |
| In fluorocyclohexanes | +4.5 | — | — | — | 212.2 | −11 | −7.6 | — | — | 233.2 |

Figure 19-5. Parameter scheme for $>$CHF.[15]

ature.[12] In a few cases the fluorine chemical shifts can also be calculated in a totally empirical way. This is possible when a large number of compounds of the same type is available. A few examples of parameter schemes are known for fluorocyclo-hexanes[13] and for some fluoro sugars.[14] A more recent chemical shift parameter scheme has been deduced for fluorodioxans and fluorooxathians.[15] It is obtained from all the compounds available by taking the best, internally most consistent, fit of experimental shifts to parameters; this is normally done by trial and error. The final, successful parameter scheme of the fluorodioxans and fluorooxathians is shown in Figs. 19-4 and 19-5 for $>CF_2$ and $>CFH$ groups.

## F-F COUPLING CONSTANTS

This discussion will be limited to F-F couplings through carbon, obtained from the literature for short-chain aliphatic fluorocarbons and small-ring cyclic compounds.

### Geminal F-F Coupling Constant, $J_{gem}$

The values for $J_{gem}$ (positive relative sign) have been obtained for a series of ethanes of the type $CF_2Q \cdot CFHCl$, for

|  |  | $J_{gem}$ (Hz) |
|---|---|---|
| $CF_2Cl \cdot CF_2Br$ | $CF_2Cl$ | 170.2 |
|  | $CF_2Br$ | 177.1 |
| $CF_2I \cdot CF_2Br$ | $CF_2I$ | 203.2 |
|  | $CF_2Br$ | 170.5 |
| $CF_2I \cdot CF_2Cl$ | $CF_2I$ | 213.0 |
|  | $CF_2Cl$ | 171.0 |

Figure 19-6. Some geminal F-F couplings in fluoroethanes.[17]

which a relationship between $J_{gem}$ and the electronegativity, $\chi$, of $Q$ has been found[16]:

$$J_{gem} = 730\chi^{-1} - 70$$

The above expression predicts a progressive increase of $J_{gem}$ upon going from $>CF_2Cl$ to $>CF_2Br$ and $>CF_2I$. This is shown by some recent results of simple fluoroethanes (Fig. 19-6), [17] for which analogous $J_{gem}$ results have been obtained by liquid crystal work. [18] The dependence of $J_{gem}$ on substituent electronegativity has been shown by some members of the cyclopropane series (Fig. 19-7). [19,20] In these cases variations of $J_{gem}$ with the electronegativity of X in the fragment $-CF_2CFX-$ have been demonstrated. Values for $J_{gem}$ in fluorinated cyclobutenes[21-25] and cyclobutanes[22,23,25-27] are known. They are difficult to interpret because few molecules that form a series, where only one substituent at a time is changed, have been studied. The

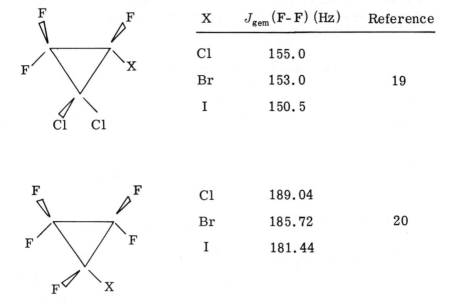

| X | $J_{gem}$ (F-F) (Hz) | Reference |
|---|---|---|
| Cl | 155.0 | |
| Br | 153.0 | 19 |
| I | 150.5 | |
| Cl | 189.04 | |
| Br | 185.72 | 20 |
| I | 181.44 | |

Figure 19-7. Some geminal F-F couplings in fluorocyclopropanes.

References

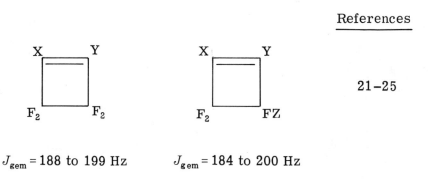

$J_{gem}$ = 188 to 199 Hz          $J_{gem}$ = 184 to 200 Hz

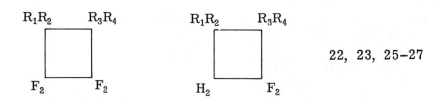

$J_{gem}$ = 190 to 230 Hz          $J_{gem}$ = 182 to 210 Hz

Figure 19-8.   Geminal F-F couplings in fluorocyclobutenes and fluoro-cyclobutanes.

series, which can be picked out for cyclobutenes and cyclobu-tanes, are listed in Fig. 19-8.

Examples of $J_{gem}$ in other ring compounds are shown in Figs. 19-9 and 19-10.   A correlation between $J_{gem}$ and the value of the internal C-Ĉ-C angle of some cyclic systems has been shown (Fig. 19-11).[35]  Because a direct interdependence between the appropriate F-Ĉ-F angle and the internal ring angle C-Ĉ-C is expected, one can predict a definitive trend toward larger cou-plings with decrease in the F-Ĉ-F angle.   In summary: (1) in -CF$_2$X groups, $J_{gem}$ shows a progressive increase as the elec-tronegativity of the X substituent is decreased; (2) in the frag-ments -CF$_2$ĊX and CF$_2$ĊX, $J_{gem}$ is likely to increase as the elec-

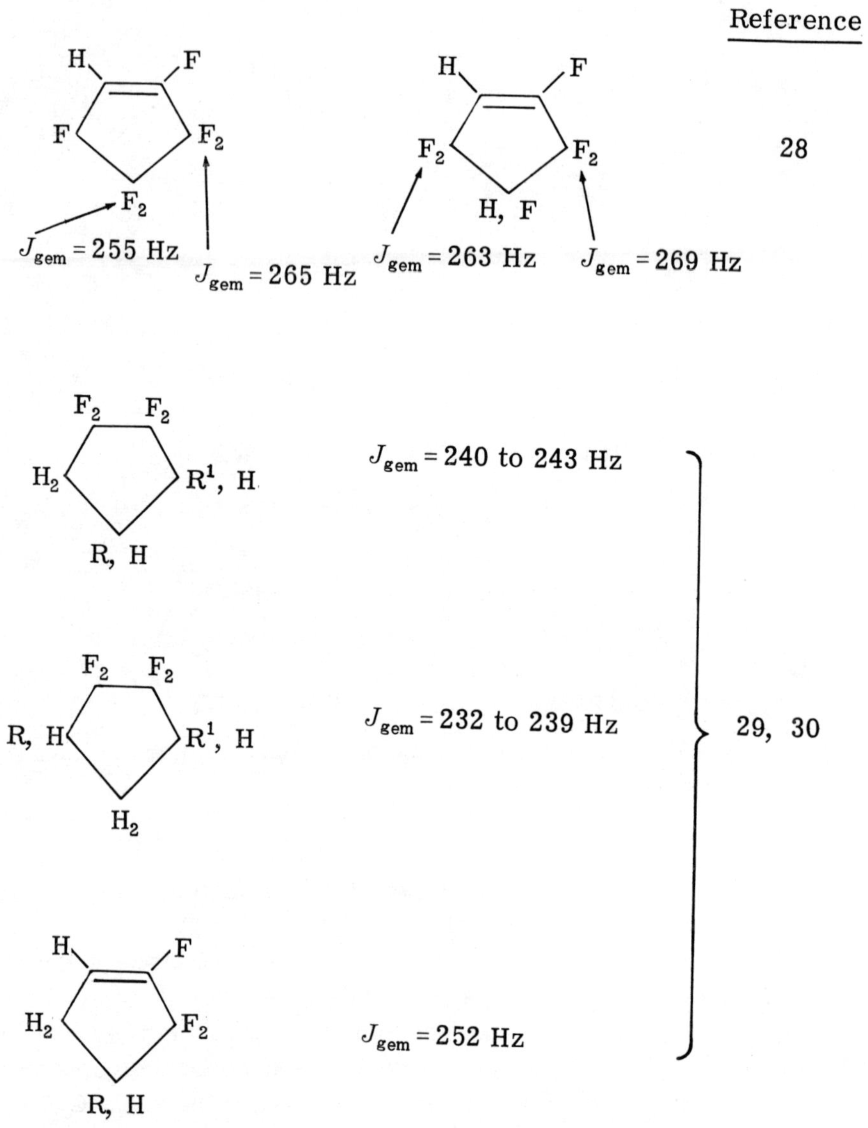
Reference
H
F
F
F2
F2
Jgem = 255 Hz
Jgem = 265 Hz
H
F
F2
F2
H, F
Jgem = 263 Hz
Jgem = 269 Hz
28
F2
F2
H2
R1, H
R, H
Jgem = 240 to 243 Hz
F2
F2
R, H
R1, H
H2
Jgem = 232 to 239 Hz
29, 30
H
F
H2
F2
R, H
Jgem = 252 Hz

Figure 19-9. Geminal F-F couplings in cyclic compounds.

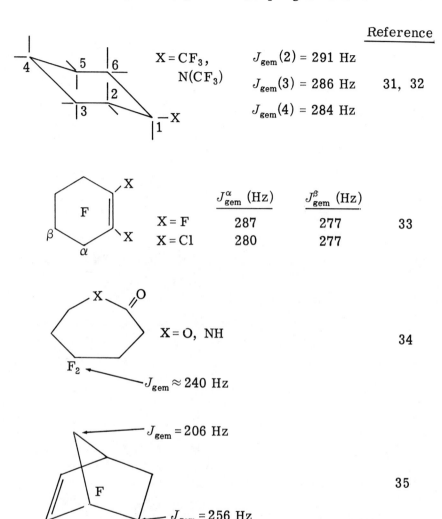

$$X = CF_3,$$
$$N(CF_3)$$

$$J_{gem}(2) = 291 \text{ Hz}$$
$$J_{gem}(3) = 286 \text{ Hz} \qquad 31, \ 32$$
$$J_{gem}(4) = 284 \text{ Hz}$$

| | $J^\alpha_{gem}$ (Hz) | $J^\beta_{gem}$ (Hz) | |
|---|---|---|---|
| X = F | 287 | 277 | 33 |
| X = Cl | 280 | 277 | |

$$X = O, \ NH$$

34

$$J_{gem} \approx 240 \text{ Hz}$$

$$J_{gem} = 206 \text{ Hz}$$

35

$$J_{gem} = 256 \text{ Hz}$$

Figure 19-10.  Geminal F-F couplings in cyclic compounds.

tronegativity of the X substituent is increased; (3) $J_{gem}$ increases as the angle $\alpha$ of $C{\overset{F}{\underset{F}{\diagup}}}\alpha$ decreases; and (4) the effect of an $sp^2$ carbon atom adjacent to a $>CF_2$ group seems to decrease the corresponding value of $J_{gem}$.

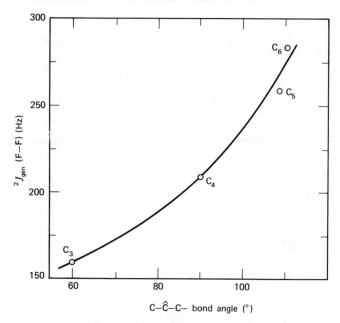

Figure 19-11.  Dependence of $J_{gem}$ on the internal $-C-\hat{C}-C-$ angle of some cyclic systems. [35]

## Vicinal F-F Coupling Constant, $^3J$

The early information on vicinal F-F couplings was obtained from the study of fluorinated alkanes, but it was incomplete because of problems of internal rotation and confusion about the relative signs of the couplings. One of the first important studies on $^3J$ in the fluoroalkane series is the one that gives an approximately linear relationship between the average vicinal F-F coupling $\langle ^3J \rangle$, and the substituent electronegativities[36]:

$$\langle ^3J \rangle = -91.4 + 6.15 \sum E ,$$

where $\sum E$ is the sum of the Huggins electronegativities of the first atom of all the substituents in the $>CF \cdot CF<$ fragment. Recently only selected and related ethane compounds were considered and a good linear dependence of $\langle ^3J \rangle$ with respect to $\sum E$

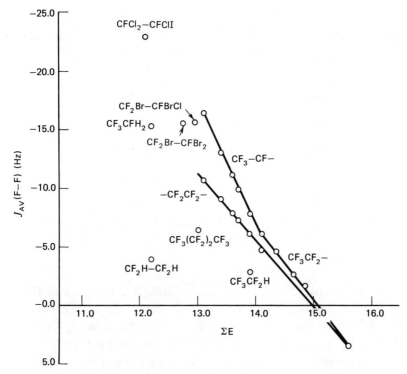

Figure 19-12. Rotationally averaged coupling, $J_{AV}$, plotted against the sum of the Huggins electronegativities $\sum E$ of the four substituent atoms in the $>$CF $\cdot$ CF$<$ fragment. [17]

in each of the three series of halogenated ethanes, $CF_3CF_2$-, $CF_3CF<$, and -$CF_2 \cdot CF_2$- was obtained (Fig. 19-12). [17] This fact leads to the conclusion that, for each series of compounds considered, the variation of $^3J$ can be related primarily to substituent electronegativities. The hydrofluoroethanes and other closely related molecules, as shown in Fig. 19-12, do not correlate at all with the halogenated ethanes; in addition, the $\langle ^3J \rangle$ values of the -$CF_2$-$CF_2$- fragment are clearly different from the corresponding values of molecules containing the same substituents. These observations suggest that other factors, such as an angularly dependent electronegativity effect and/or dihedral angle changes may also affect the value of $^3J$. The *trans*

| X-$CF_2 \cdot CF_2$-Y | | $^3J_{trans}$(F-F) Hz | $^3J_{gauche}$(F-F) Hz | Reference |
|---|---|---|---|---|
| I | I | – 18.5 | – 6.6 | 38 |
| I | Br | – 9.9 | – 8.5 | 17 |
| I | Cl | – 4.7 | – 9.2 | 17 |
| Br | Br | – 2.3 | – 9.7 | 37 |
| Br | Cl | + 5.6 | – 11.9 | 17 |
| Cl | Cl | + 6.6 | – 10.3 | 37 |

Figure 19-13.  *trans* and *gauche* F-F couplings for some tetrafluoro-
ethanes.

and *gauche* couplings, $^3J_t$ and $^3J_g$, of single rotamers of fluoro-
alkanes have been correlated with the substituent electronegativity.[17,37,38] In the fragment -$CF_2 \cdot CF_2$-, $^3J_t$ and $^3J_g$ were
evaluated by assuming that the value of a given coupling in one
rotamer is equal to the similarly oriented coupling in the other
rotamer of the same molecule (Fig. 19-13).  Despite this approximation it is clear that, in the -$CF_2 \cdot CF_2$- fragment, $^3J_t$
varies enormously (– 18.5 to + 6.6 Hz), becoming more positive
as the electronegativity of the substituent increases. $^3J_g$, on
the contrary, is almost unaffected and the direction of the small
change observed (– 6.6 to – 11 Hz) is opposite to that of $^3J_t$.
This finding is in agreement with the results obtained for some
hydrofluoroethanes (Fig. 19-14).[39] In these compounds $^3J_t$ varies enormously (– 30 to 0 Hz) with successive fluorine substitution; in contrast $^3J_g$ is relatively unaffected (– 13.7 to – 5.4
Hz) and shows no consistent change with substitution.

One characteristic of the value of $^3J$ in fluoroalkanes is
the anomalously low value of this coupling in perfluoroethyl
groups.  This low value could be attributed to the following two
causes: (*a*) the effect of electronegativity, and (*b*) the averaging
of *gauche* and *trans* couplings of opposite signs.

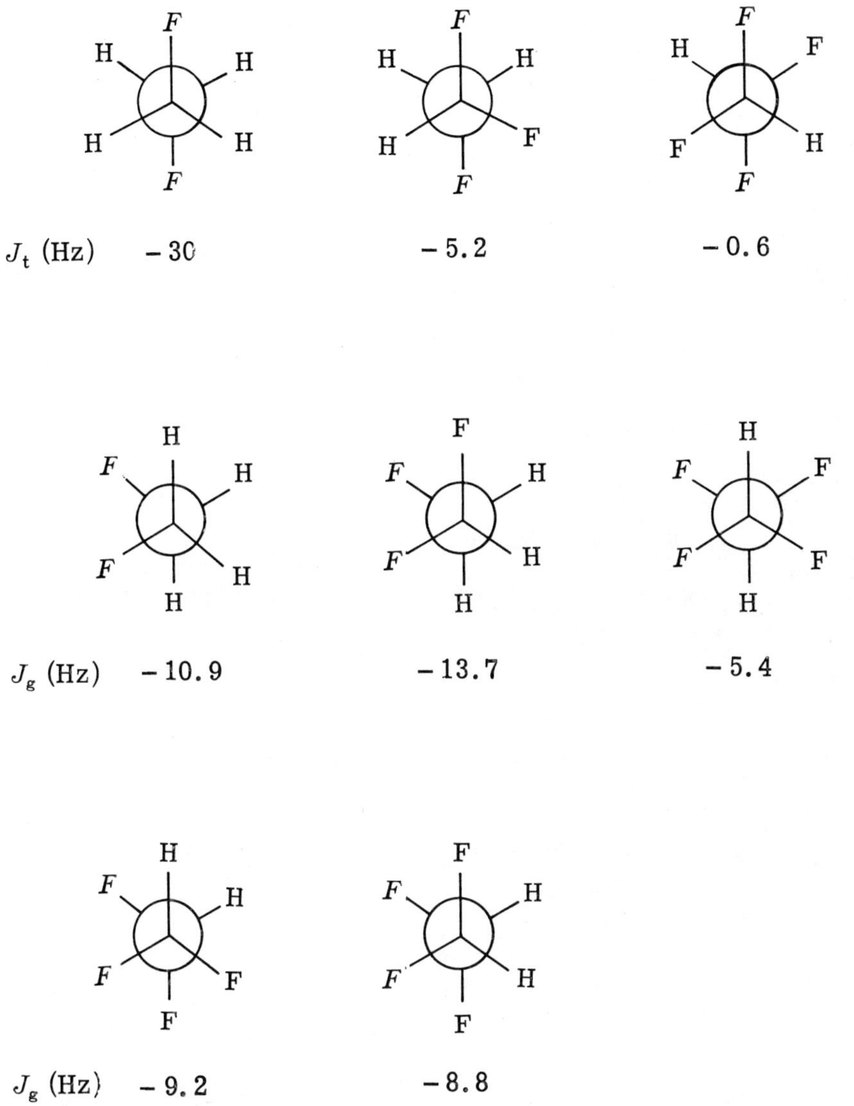

Figure 19-14. Vicinal F-F rotamer couplings of 1, 2-difluoro-, 1, 1, 2-trifluoro-, and 1, 1, 2, 2-tetrafluoroethane.[39]

| | $J_{cis}$ (Hz) | | | $J_{trans}$ (Hz) | | |
|---|---|---|---|---|---|---|
| K | $J_{AA}$ | $J_{BB}$ | $J_{BX}$ | $J_{AB}$ | $J_{AX}$ | Reference |
| 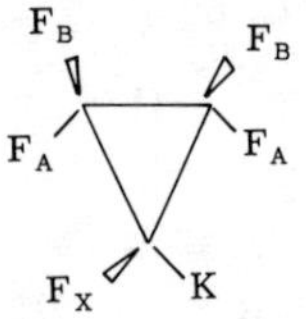 Cl | +4.87 | +10.73 | +8.27 | +0.88 | −5.82 | |
| Br | +1.91 | +10.41 | +6.59 | +2.42 | −10.13 | 20 |
| I | −1.39 | +9.78 | +4.72 | +4.34 | −15.44 | |
| 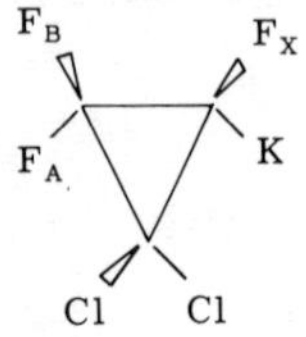 Cl | | | −4.1 | | −1.3 | |
| Br | | | −5.4 | | −5.4 | 19 |
| I | | | −6.8 | | −10.0 | |
| 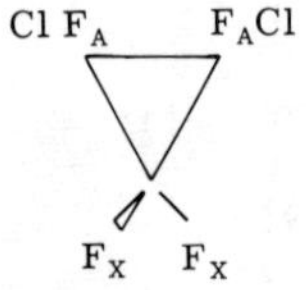 cis | | | +5.62 (AX) | | −5.25 | |
| trans | | | ~0 (AX) | −12.2 (AA) | ~−5 | 6 |
| | | | 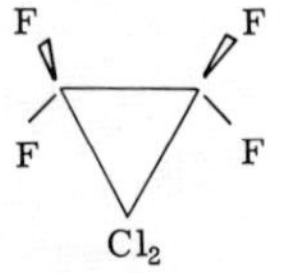 +1.44 | | +4.10 | 6 |

(Structure: $F_B$, $F_B$ top; $F_A$, $F_A$ middle; $F_X$, K bottom — $J_{cis}$ +5.5 for the $Cl F_A$ / $F_A Cl$ / $F_X$ $F_X$ structure)

Figure 19-15. Vicinal F–F couplings in fluorocyclopropanes.

Interest in $^3J$ has been aroused by measurements made on cyclic compounds. Fluorinated cyclopropanes[6, 19, 20, 40] have been studied by several workers and some $^3J$ values obtained are reported in Fig. 19-15. Both *cis* and *trans* couplings do depend upon the electronegativities of the substituents.

It is interesting to note that in the case of monosubstituted perfluorocyclopropanes $^3J$ is shown to be affected not only by the electronegativity of the X substituent but also by the orientation of X with respect to the interacting fluorine nuclei.[20] In the cyclobutene series *cis* and *trans* vicinal couplings of several compounds are known.[21, 22, 24, 41, 42] The value for $^3J_{cis}$ is fairly

| X | Y | $^3J_{cis}$ (Hz) | $^3J_{trans}$ (Hz) | Reference |
|---|---|---|---|---|
| F | F | – 12.27 | + 26.50 | 21 |
| Cl | Cl | – 12.88 | + 25.14 | 22 |
| I | I | – 15.35 | + 24.47 | 41 |
| I | F | – 14.3 | + 25.7 | 41 |
| I | Cl | – 13.8 | + 24.7 | 41 |
| H | F | – 12.5 | + 27.7 | 41 |
| H | Cl | – 12.2 | + 26.0 | 41 |
| H | I | – 14.0 | + 27.5 | 41 |
| H | H | – 12.92 | + 30.45 | 22 |

Figure 19-16.  Vicinal F-F couplings in fluorocyclobutenes.

constant in the range – 12 to – 17 Hz and the value for $^3J_{trans}$ falls in the range + 30 to + 21 Hz.  Furthermore in all cases $^3J_{cis}$ and $^3J_{trans}$ are of opposite sign to each other.

In Fig. 19-16 a few $^3J$ values of the cyclobutene series are listed.  As shown it is difficult to find a clear relationship between the $^3J$ values and the electronegativities of substituents; however, if we don't consider those molecules that have hydrogen as substituent, both $^3J_{cis}$ and $^3J_{trans}$ appear to increase with an increase of the X and Y substituent electronegativities.  The results in the cyclobutane compounds[22,25,43] are complicated by the puckered structure of the four-membered ring.  A careful study of some fluorinated cyclobutanes has been performed and a relationship between $^3J$ and a value for the dihedral angle between the interacting nuclei has been proposed.[43]  From this study it was noted that $J_{cis}$ (equatorial-axial, F-F coupling with

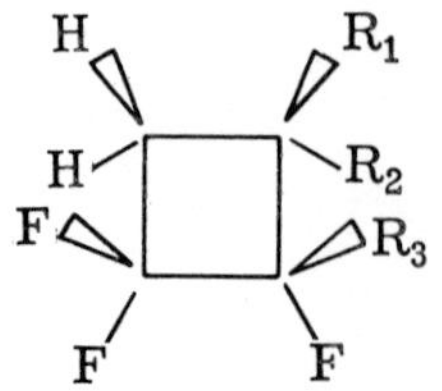

| $R_1$ | $R_2$ | $R_3$ | $J_{cis}$ (Hz) | $J_{trans}$ (Hz) | Reference |
|---|---|---|---|---|---|
| Cl | H | I | − 8.92 | − 5.56 | 25 |
| Cl | Cl | I | − 8.75 | − 5.05 | 25 |
| Cl | Cl | Cl | − 9.46 | + 2.82 | 25 |
| CH = CH$_2$ | CH$_3$ | Cl | − 8.60 | + 3.78 | 43 |
| CH$_3$ | CH = CH$_2$ | Cl | − 8.61 | − 0.82 | 43 |
| CH = CH$_2$ | H | Cl | − 7.81 | + 5.94 | 43 |
| H | CH = CH$_2$ | Cl | − 6.5 | − 1.25 | 43 |
| CN | H | Cl | − 6.75 | + 6.46 | 43 |
| H | CN | Cl | − 6.37 | + 0.76 | 43 |
| C$_6$H$_5$ | H | Cl | − 7.81 | + 6.97 | 43 |
| H | C$_6$H$_5$ | Cl | − 9.15 | − 0.20 | 43 |
| C(CH$_3$) : CH$_2$ | H | Cl | − 7.54 | + 7.23 | 43 |
| H | C(CH$_3$) : CH$_2$ | Cl | − 8.07 | − 0.42 | 43 |

Figure 19-17. Vicinal F–F couplings in fluorocyclobutanes.

dihedral angle near 20°) is consistently negative (− 7 to − 9 Hz), whereas $J_{trans}$ (equatorial-equatorial F-F coupling with dihedral angle near 100° and axial-axial, F-F coupling with dihedral angle near 140°) can be of either sign: in the former case it is positive and around 5 to 7 Hz, and for the latter it is mainly negative and around zero. A list of $^3J$ values for the cyclobutane series is shown in Fig. 19-17. In Fig. 19-18 the

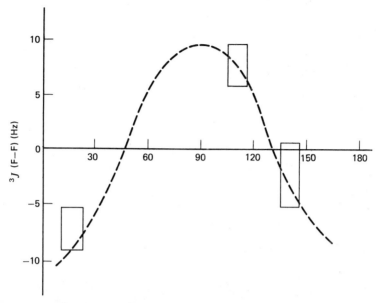

Figure 19-18.  Variation of vicinal F-F coupling constant with dihedral angle in fluorocyclobutanes. [43]

corresponding angular dependence of $^3J$ is reported. [43] A similar angular dependence for $^3J$ was proposed from a study of some fluoro sugars. [44] In this case maximum values of $^3J$ are observed for dihedral angles of 0° and 120°, whereas minima are at 60° and 180°. [44] The *cis* and *trans* vicinal couplings for the fragment -CF$_2$CF$_2$CF$_2$- of cyclopentenes have been obtained recently (Fig. 19-19). [45]

The values of $J_{cis}$ lie between − 1.9 and 0.7 Hz, whereas the values of $J_{trans}$ are from − 8.5 to − 4.4 Hz. $J_{trans}$ ($J_{cis}$ to a lesser extent) appears to increase; it becomes less negative as the electronegativity of the nearest substituent increases.

For other cyclic compounds such as cyclohexane, [31,32] cyclopentadienes, [46,47] and 1,3-cyclohexadienes [48] some F-F couplings are known, but no comprehensive work has been carried out. Figure 19-20 summarizes the ranges of $^3J$ values in fluoroalkanes and in several cyclic compounds. In parentheses

Compound

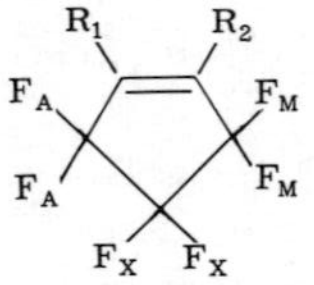

F–F Coupling Constant (Hz)[a]

| Compound | $^2J(F_A-F_A)$ | $^2J(F_M-F_M)$ | $^3J(F_M-F_X)$ trans | $^3J(F_M-F_X)$ cis | $^3J(F_A-F_X)$ trans | $^3J(F_A-F_X)$ cis | $^4J(F_A-F_m)$ trans | $^4J(F_A-F_m)$ cis | $^3J(F_B-F_A)$ | $^4J(F_B-F_X)$ | $^4J(F_B-F_M)$ |
|---|---|---|---|---|---|---|---|---|---|---|---|
| $R_1 = R_2 = I$[b] | +255.30 | +255.30 | −8.44 | −1.37 | −8.44 | −1.37 | +1.24 | +4.12 | — | — | — |
| $R_1 = R_2 = Cl$[b] | +254.33 | +254.33 | −5.75 | −1.03 | −5.75 | −1.03 | +1.32 | +3.01 | — | — | — |
| $R_1 = Cl, R_2 = I$ | +255.77 | +254.70 | −6.89 | −0.71 | −6.42 | −1.44 | +1.28 | +3.56 | — | — | — |
| $R_1 = F_B, R_2 = I$ | +260.46 | +250.88 | −5.55 | +0.66 | −4.45 | −1.85 | +1.68 | +3.07 | −15.71 | +4.91 | +10.64 |

[a]The value of $^2J(F_X-F_X)$ was kept constant at +240.0 Hz in the iterations.
[b]$F_M = F_A$.

Figure 19-19.  Coupling constants of fluorinated cyclopentenes.[45]

308

| $-CF_2-CF_2-$ | $^3J_{trans}$ (Hz)[a] | $^3J_{cis}$ (Hz)[a] | $^3J(sp^2)$ (Hz)[a] |
|---|---|---|---|
| Ethanes | − 30 to + 7 (~ + 20) | − 14 to − 7[b] (~ − 5) | |
| Cyclopropanes | + 5 to − 16 (~ − 1) | + 11 to − 7 (~ + 11) | |
| Cyclopropenes | | | (± )40 to 45[d] (± 43. 6)[c] |
| Cyclobutanes | + 6 to − 6 (~ + 3) | − 4 to − 12 (< − 9) | |
| Cyclobutenes | + 30 to + 21 (+ 26. 5)[c] | − 12 to − 17 (− 12. 3)[c] | (±)6 to 9[d] (± 6. 9)[c] |
| Cyclopentenes | − 5 to − 9 (~ − 4) | + 1 to − 2 (~ 0) | (±)14 to 16[d] (~ ± 15) |
| Cyclohexanes | (±) 1 to 5[d] | ~ ± 14[d] | |
| Cyclopentadienes | | | (±)7 ÷ 12[d] |
| Cyclohexadienes | | | (±)15[d] |

[a]In parentheses are the possible $^3J$ values for the corresponding completely fluorinated compounds.
[b]*gauche* coupling constants.
[c]Data for the totally fluorinated compound.
[d]The sign of the coupling is not known.

Figure 19-20.  Ranges of vicinal F-F coupling constants in fluoro-alkanes and in several cyclic compounds.

the guessed $^3J$ values for the corresponding completely fluorinated compounds are given.  The variations of sign and magnitude are remarkable and unfortunately no coherent pattern appears to emerge.  This presumably arises from an interplay of the effects of variation of dihedral angle, ring angle, and substituent electronegativity.

## Four-Bond F-F Coupling Constant, $^4J$

No comprehensive work has been carried out on this coupling constant. One important study on fluorocyclopropanes (Fig. 19-21) has shown the great stereospecificity of $^4J$.[49] Such results seem to be confirmed by the study of some fluoro sugars[50] and of some fluorochloroacetones.[51] The $^4J$ values of the $-CF_2CF_2CF_2-$ fragment in some cyclopentenes[45] are shown in Fig. 19-19. The *cis* and *trans* $^4J$ values are positive and fall into two distinct ranges. However the variations are rather small throughout the series (3.0 to 4.2 Hz for $^4J_{cis}$ and 1.2 to 1.7 Hz for $^4J_{trans}$). Other typical $^4J$ values of cyclic compounds are given in Fig. 19-22. It is difficult to compare and to rationalize on the same basis the $^4J$ values of cyclic and alicyclic compounds. More data are clearly necessary.

## Long-Range F-F Coupling Constants

This coupling is relative to interacting fluorine nuclei separated by at least five bonds. In the fluorine field this coupling can be "anomalously" large when the interacting nuclei are physically close in proximity.

Some typical examples of large F-F couplings, often cited as evidence for a through-space mechanism for F-F coupling, are shown in Fig. 19-23.

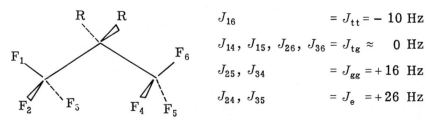

$$J_{16} = J_{tt} = -10 \text{ Hz}$$
$$J_{14}, J_{15}, J_{26}, J_{36} = J_{tg} \approx 0 \text{ Hz}$$
$$J_{25}, J_{34} = J_{gg} = +16 \text{ Hz}$$
$$J_{24}, J_{35} = J_{e} = +26 \text{ Hz}$$

Figure 19-21.   Four-bond F-F coupling constants in cyclopropanes.[49]

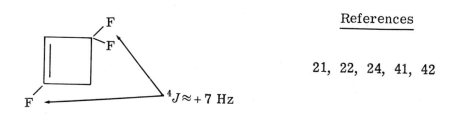

21, 22, 24, 41, 42

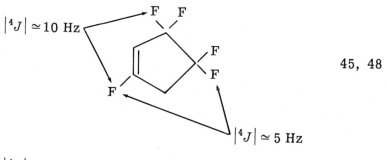

45, 48

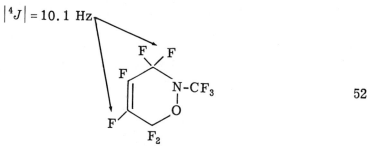

52

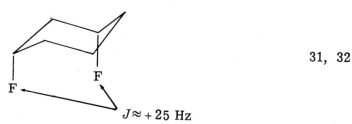

Figure 19-22. Four-bond F-F couplings in some cyclic compounds.

311

$J_{CF_3-CF_3}$ 
$\begin{cases} cis & \sim 0 & \text{Hz} \\ trans & 10 \text{ to } 12 & \text{Hz} \end{cases}$

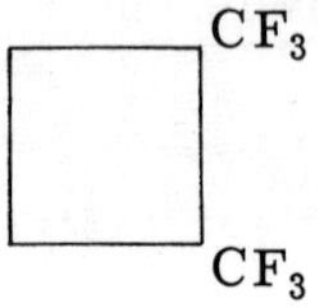

53

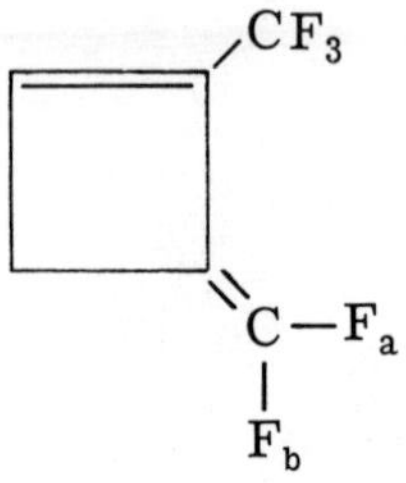

$J_{F_a-CF_3} = 8.5$ Hz

$J_{F_b-CF_3} \approx 0$ Hz

54

$CF_3 - C = C - CF_3$    $J_{CF_3-CF_3}$ 
$\begin{cases} cis & 1 \text{ to } 2 & \text{Hz} \\ trans & 10 \text{ to } 15 & \text{Hz} \end{cases}$

55

$J = 75$ Hz

$J \approx 0$ Hz

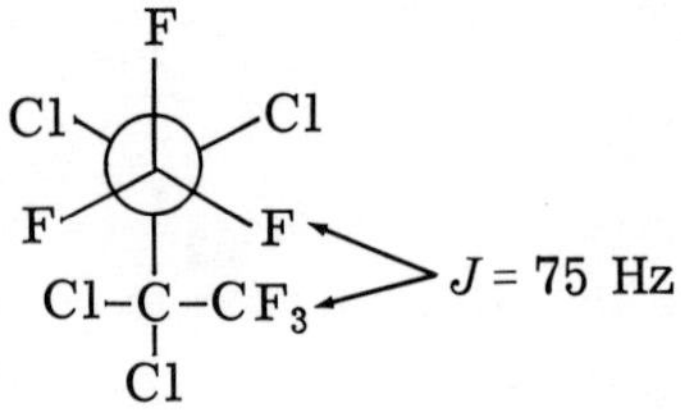

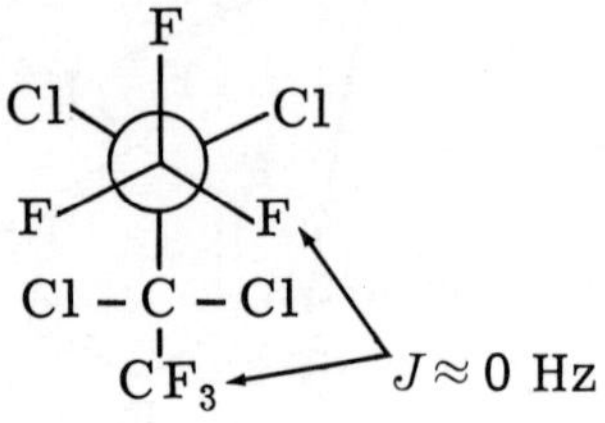

56

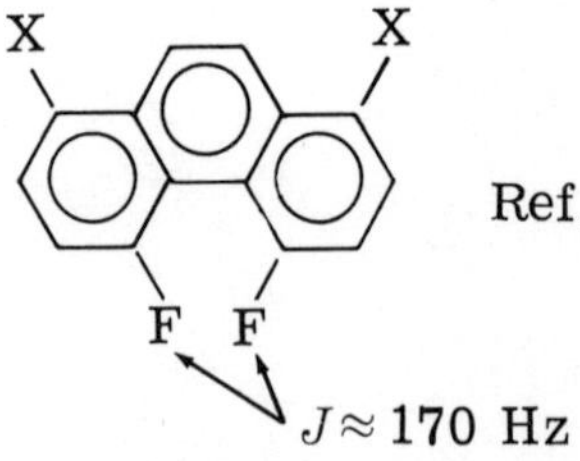

X      X

Ref. 57

$J \approx 170$ Hz      $J = 44$ Hz

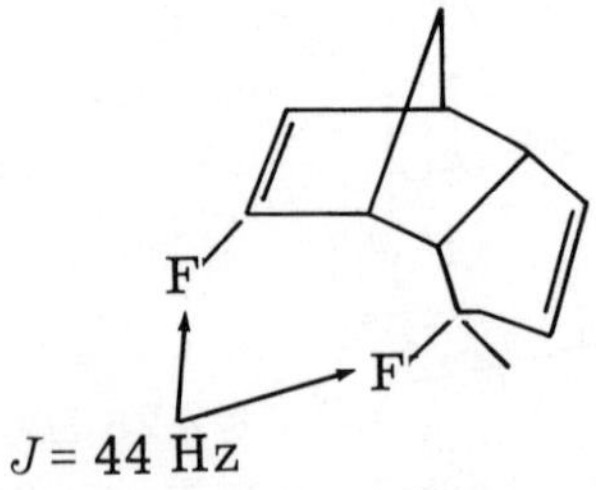

58

Figure 19-23.  "Long-range" F-F coupling constants.

## ACKNOWLEDGMENT

I am grateful to Dr. R. K. Harris of the University of East Anglia for communicating results on F-F coupling constants.

## References

1. J. W. Emsley and L. Phillips, "Fluorine Chemical Shifts," in *Progress in NMR Spectroscopy*, J. W. Emsley, J. Feeney, and L. H. Sutcliffe, Eds., Vol. 7, Pergamon Press, Oxford, 1971.

2. E. F. Mooney, *An Introduction to $^{19}F$ NMR Spectroscopy*, Heyden/Sadler, 1970.

3. C. H. Dungan and J. R. Van Wazer, *Compilation of Reported $^{19}F$ NMR Chemical Shifts*, Wiley-Interscience, New York, 1970.

4. N. Boden, J. W. Emsley, J. Feeney, and L. H. Sutcliffe, *Mol. Phys.*, **8**, 133 (1964).

5. J. Feeney, L. H. Sutcliffe, and S. M. Walker, *Mol. Phys.*, **11**, 117 (1966).

6. L. Cavalli, *Org. Magn. Resonance*, **2**, 233 (1970).

7. J. Feeney, L. H. Sutcliffe, and S. M. Walker, *Mol. Phys.*, **11**, 129 (1966).

8. J. W. Emsley, *Mol. Phys.*, **9**, 381 (1965).

9. J. Homer and D. Callaghan, *J. Chem. Soc. B*, 247 (1969).

10. L. Phillips and V. Wray, *J. Chem. Soc. B*, 2068 (1971).

11. L. Phillips and V. Wray, *J. Chem. Soc. Perkin*, **II**, 220 (1972).

12. L. Phillips and V. Wray, *J. Chem. Soc. Perkin*, **II**, 223 (1972).

13. J. Homer and L. F. Thomas, *Trans. Faraday Soc.*, **59**, 2431 (1963).

14. L. D. Hall and J. R. Manville, *Can. J. Chem.*, **47**, 1, 19 (1969).

15. J. Burdon and I. W. Parsons, *Tetrahedron*, **27**, 4553 (1971).

16. R. R. Dean and J. Lee, *Trans. Faraday Soc.*, **65**, 1 (1969).

17. L. Cavalli, *J. Magn. Resonance*, **6**, 298 (1972).

18. V. J. Robinson, Ph. D. Thesis, University of East Anglia, 1972.

19. K. L. Williamson and B. A. Braman, *J. Am. Chem. Soc.*, **89**, 6183 (1963).

20. M. G. Barlow, R. Fields, and F. P. Temme, *Chem. Commun.*, 1971 (1968).

21. R. K. Harris and R. Ditchfield, *Spectrochim. Acta*, **24A**, 2089 (1968).

22. R. K. Harris and V. J. Robinson, *J. Magn. Resonance*, **1**, 362 (1969).

23. R. K. Harris and V. J. Gazzard, *Org. Magn. Resonance*, **3**, 495 (1971).

24. R. A. Newmark, G. R. Apai, and R. O. Michael, *J. Magn. Resonance*, **1**, 418 (1969).

25. J. D. Park, R. O. Michael, and R. A. Newmark, *J. Am. Chem. Soc.*, **91**, 5933 (1969).

26. W. D. Phillips, *J. Chem. Phys.*, **25**, 949 (1956).

27. J. B. Lambert and J. D. Roberts, *J. Am. Chem. Soc.*, **87**, 3884 (1965).

28. R. Fields, M. Green, T. Harrison, R. N. Haszeldine, A. Jones, and A. B. P. Lever, *J. Chem. Soc. A*, 49 (1970).

29. P. Piccardi, M. Modena, and L. Cavalli, *J. Chem. Soc. C*, 3959 (1971).

30. P. Piccardi and M. Modena, *Chem. Commun.*, 1041 (1971).

31. K. W. Jolley, L. H. Sutcliffe, and S. M. Walker, *Trans. Faraday Soc.*, **64**, 269 (1968).

32. V. S. Plashkin, *J. Org. Chem. (USSR)*, **6**, 1010 (1970).

33. J. E. Anderson and J. D. Roberts, *J. Am. Chem. Soc.*, **92**, 97 (1970).

34. E. A. Noe and J. D. Roberts, *J. Am. Chem. Soc.*, **93**, 7261 (1971).

35. J. Homer and D. Callaghan, *J. Chem. Soc. B*, 2430 (1971).

36. R. J. Abraham and L. Cavalli, *Mol. Phys.*, **9**, 67 (1965).

37.  R. K. Harris and N. Sheppard, *Trans. Faraday Soc.*, **59**, 606 (1963).
38.  L. Cavalli, *Chim. Ind.* (*Milan*), **51**, 464 (1969).
39.  R. J. Abraham and R. H. Kemp, *J. Chem. Soc. B*, 1240 (1971).
40.  P. B. Sargeant, *J. Org. Chem.*, **35**, 678 (1970).
41.  L. Cavalli, unpublished results.
42.  L. Cavalli, *J. Chem. Soc. B*, 1616 (1970).
43.  R. R. Ernst, *Mol. Phys.*, **16**, 241 (1969).
44.  L. D. Hall and R. N. Johnson, *Chem. Commun.*, 463 (1970).
45.  L. Cavalli and R. K. Harris, *J. Magn. Resonance*, **10**, 355 (1973).
46.  R. E. Banks, M. Bridge, R. N. Haszeldine, D. W. Roberts, and N. I. Tucker, *J. Chem. Soc. C*, 2531 (1970).
47.  G. Camaggi and F. Gozzo, *J. Chem. Soc. C*, 925 (1971).
48.  J. Feeney, L. H. Sutcliffe, and S. M. Walker, *Trans. Faraday Soc.*, **62**, 2969 (1966).
49.  R. J. Abraham, *J. Chem. Soc. B*, 1922 (1969).
50.  L. D. Hall, R. N. Johnson, A. B. Foster, and J. H. Westwood, *Can. J. Chem.*, **49**, 236 (1971).
51.  L. H. Sutcliffe and B. Taylor, *Spectrochim. Acta*, **28A**, 619 (1972).
52.  J. Lee and K. G. Orrel, *Trans. Faraday Soc.*, **61**, 2342 (1965).
53.  S. Ng and C. H. Sederholm, *J. Chem. Phys.*, **40**, 2090 (1964).
54.  R. E. Banks, W. R. Deem, R. N. Haszeldine, and D. R. Taylor, *J. Chem. Soc. C*, 2051 (1966).
55.  R. D. Chambers and A. J. Palmer, *Tetrahedron Lett.*, 2799 (1968).
56.  F. J. Weigert and J. D. Roberts, *J. Am. Chem. Soc.*, **90**, 3577 (1968).
57.  K. L. Servis and K. N. Fang, *J. Am. Chem. Soc.*, **90**, 24 (1968).
58.  R. E. Banks, M. Bridge, R. Fields, and R. N. Haszeldine, *J. Chem. Soc. C*, 1282 (1971).

CHAPTER 20

# [19]F NMR STUDIES OF FLUOROCARBON POLYMERS

E. G. Brame, Jr.

*E. I. du Pont de Nemours & Company*
*Elastomer Chemicals Department*
*Experimental Station*
*Wilmington, Delaware*

## INTRODUCTION

High-resolution [19]F NMR has been applied only on a limited basis to polymers for structure study. The chief reason for this limited use of [19]F NMR has been the relatively few fluorine-containing polymers available for study in comparison with hydrocarbon polymers. However, for those that are available for such studies this technique has proved to be extremely valuable for the elucidation of polymer microstructure. For homopolymers, various kinds of tacticity confirmations and head-to-head versus head-to-tail configurations have been studied. For copolymers, in addition to the tacticity kinds of studies, sequencing of the comonomers has been studied by [19]F NMR.

In order to perform high-resolution NMR studies on fluorine-containing polymers, it is important that a proper solvent is chosen to dissolve the polymer so that useful high-resolution information can be obtained. The solvent chosen for use with fluorine-containing polymers is often much more difficult to select than it is for hydrocarbon polymers because the com-

317

mon chlorine-containing solvents are unacceptable. Nevertheless, certain common solvents such as acetone are applicable.
As a result, chemical shifts, spin-spin coupling constants, and
intensity relationship data must be obtained at the same concentration in a solvent that will dissolve them for microstructure studies in order to compare the data with those for other
kinds of fluorine-containing polymers. It is the chemical shift
and intensity relationships that often seem to play an extremely
important role, and probably the most important role, in NMR
studies on fluorine-containing polymers. The reason is that
the chemical shifts are much greater in $^{19}$F NMR spectra than
in $^{1}$H NMR spectra and can be utilized more fully for microstructure analysis of fluoropolymers than hydrocarbon polymers. Frequently, the preparation of the fluorine-containing
polymer in terms of varying composition can be used to aid in
the structure assignments of lines in the $^{19}$F NMR spectra because of observing the intensity relationships of lines. Discussed in this chapter are the following fluorine-containing
polymers and copolymers: polyvinyl fluoride, polyvinylidene
fluoride, and polytrifluorochloroethylene; chlorotrifluoroethylene–isobutylene, tetrafluoroethylene–isobutylene, and vinylidene fluoride–hexafluoropropylene.


## HOMOPOLYMERS

Of the three sources[1-3] relating to studies on polyvinyl
fluoride the work by Wilson and Santee[2] as well as the more recent work by Weigert[3] give the most significant results in the
studies to date on polyvinyl fluoride. Figure 20-1 shows the
$^{19}$F NMR spectrum obtained by Weigert on polyvinyl fluoride
with proton noise decoupling. The polymer was dissolved in
dimethyl sulfoxide and the sample solution heated to 110 °C for
the analysis that was performed at 94.1 MHz. It is seen from
this analysis that ten clearly distinct lines are observable in
the spectrum. According to the earlier work by Wilson and
Santee the lower field set of lines is attributed to head-to-tail
and tail-to-tail structure of the vinyl fluoride. The weaker set

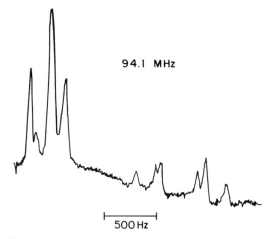

Figure 20-1.   ¹⁹F NMR spectrum of polyvinyl fluoride with proton noise decoupling in dimethyl sulfoxide at 110 °C (courtesy of *Organic Magnetic Resonance*).

of lines at higher fields and to the right in the spectrum is assigned to head-to-head structure.  It was not until Weigert performed the particular experiment of proton noise decoupling at the observed frequency of 94.1 MHz that the lines in the spectrum of this homopolymer were narrowed sufficiently to show the ten observable lines.  Through certain analogies with carbon-13 NMR spectra, in addition to the assignments previously performed by Wilson and Santee, Weigert was able to identify the different structures giving rise to each of the lines in this pattern.  Figure 20-2 shows the triad sequence analysis on this homopolymer for the head-to-tail and tail-to-tail structures. Figure 20-3 shows the triad sequence analysis for the head-to-head structures of this homopolymer.  The chemical shift values are in ppm relative to F-11 as zero.  The upfield values are negative relative to the F-11 zero value.

The work by Wilson[4] on polyvinylidene fluoride illustrates a study made by Wilson and amplified later by Wilson and Santee on this homopolymer.  In both these articles, Wilson developed a series of chemical shift values for various structure groups involving the sequencing of head-to-head, tail-to-tail,

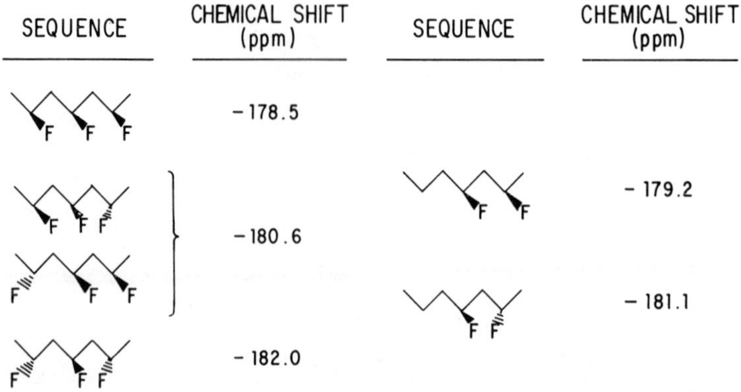

| SEQUENCE | CHEMICAL SHIFT (ppm) | SEQUENCE | CHEMICAL SHIFT (ppm) |
|---|---|---|---|
| (structure) F F F | −178.5 | | |
| (structure) F F F | −180.6 | (structure) F F | −179.2 |
| (structure) F F F | | (structure) F F | −181.1 |
| (structure) F F F | −182.0 | | |

Figure 20-2. Triad sequence structure analysis of polyvinyl fluoride; head-to-tail and tail-to-tail structures; chemical shift data referenced to $CCl_3F$ with negative values high field (courtesy of *Organic Magnetic Resonance*).

| SEQUENCE | CHEMICAL SHIFT (ppm) | SEQUENCE | CHEMICAL SHIFT (ppm) |
|---|---|---|---|
| (structure) F F F | −189.2 | (structure) F F F | −195.3 |
| (structure) F F | −191.1 | (structure) F F | −196.1 |
| (structure) F F F | −191.6 | (structure) F F F | −198.5 |

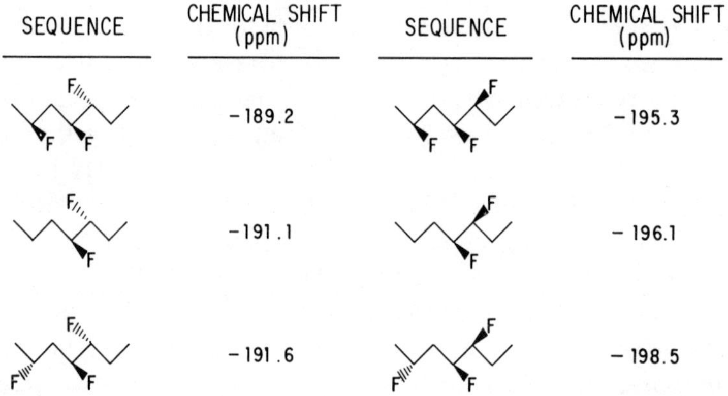

Figure 20-3. Triad sequence structure analysis of polyvinyl fluoride; head-to-head structures; chemical shift data referenced to $CCl_3F$ with negative values high field (courtesy of *Organic Magnetic Resonance*).

320

and head-to-tail for the vinylidene fluoride monomer. As a result, he was able to establish certain rules that hold for any homopolymer or copolymer involving the vinylidene fluoride monomer. For example, he noted that by substituting a $CH_2$ group for a $CF_2$ group the shift of the adjacent $CF_2$ group would be to low fields by about 10 ppm. On the other hand if a $CH_2$ group is substituted for a $CF_2$ group next to an adjacent $CF_2$ group the chemical shift of the $CF_2$ group is shifted to higher fields by about 2.5 ppm.

The homopolymer trifluorochloroethylene was studied by Tiers and Bovey.[5] One of the interesting points involving this homopolymer is that a special solvent, namely, 3, 3'-$bis$(trifluoromethyl)biphenyl was used in its examination by $^{19}$F NMR. The work reported by Tiers and Bovey on this homopolymer illustrates the point that often not until a model compound of five carbons or more in length is measured can the spectra obtained be interpreted. It is noted that the model compound spectra must be obtained under essentially the same conditions as the homopolymer spectrum in order to make any of the comparisons valid. Otherwise, changes in the chemical shift value occur and as a result comparisons in chemical shifts between the model compound spectra and the homopolymer spectrum would not be possible.

## COPOLYMERS

Studies have been done using $^{19}$F NMR on the copolymers chlorotrifluoroethylene–isobutylene,[6] tetrafluoroethylene–isobutylene,[7] and vinylidene fluoride–hexafluoropropylene.[8]

Figure 20-4 shows two spectra obtained by Ishigure[6] at 56.4 MHz and 94.1 MHz on the copolymer of chlorotrifluoroethylene and isobutylene. The question that was asked concerning the interpretation of the $^{19}$F NMR spectra for this copolymer was which one of two possible structures was the structure for this copolymer at about 50 mole %. Figure 20-5 shows these two possible structures for the copolymer. The chief difference between these two structures is whether the CFCl or a $CF_2$

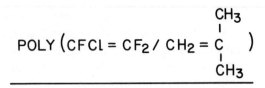

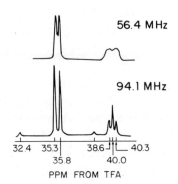

Figure 20-4.  $^{19}$F NMR spectra at 56.4 and 94.1 MHz on the copolymer of chlorotrifluoroethylene and isobutylene; ppm values upfield from TFA (trifluoroacetic acid) (courtesy of *Macromolecules*). [6]

POLY $\left( CFCl = CF_2 \middle/ CH_2 = C \substack{CH_3 \\ | \\ | \\ CH_3} \right)$

$$- CF_2CFCl - CH_2 \underset{\underset{CH_3}{|}}{\overset{\overset{CH_3}{|}}{C}} - CF_2CFCl - CH_2 \underset{\underset{CH_3}{|}}{\overset{\overset{CH_3}{|}}{C}} - CF_2CFCl -$$

STRUCTURE 1

$$- CFClCF_2 - CH_2 \underset{\underset{CH_3}{|}}{\overset{\overset{CH_3}{|}}{C}} - CFClCF_2 - CH_2 \underset{\underset{CH_3}{|}}{\overset{\overset{CH_3}{|}}{C}} - CFClCF_2$$

STRUCTURE 2

Figure 20-5.  Two different structure sequences for the copolymer of chlorotrifluoroethylene and isobutylene (50 : 50 mole %).

322

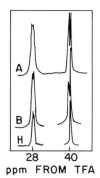

28          40
ppm FROM TFA

Figure 20-6.  $^{19}$F NMR spectra at 56.4 MHz on three different compositions of tetrafluoroethylene-isobutylene copolymers; chemical shift data upfield from TFA (trifluoroacetic acid).  A, 15 mole % TFE; B, 37 mole % TFE; H, 50 mole% TFE (assumed alternating copolymer) (courtesy of *Macromolecules*).

group is adjacent to the CH$_2$ group.  If structure 1 is correct, (with the sequence structure of the CFCl adjacent to the CH$_2$), then the CF$_2$ group adjacent to the CFCl would exhibit an AB type pattern.  In structure 2, however, since the CF$_2$ is adjacent to the CH$_2$, a greater degree of freedom is allowed for the two fluorines of the CF$_2$ group.  Therefore, these fluorines do not exhibit an AB type pattern.  Their chemical shift difference

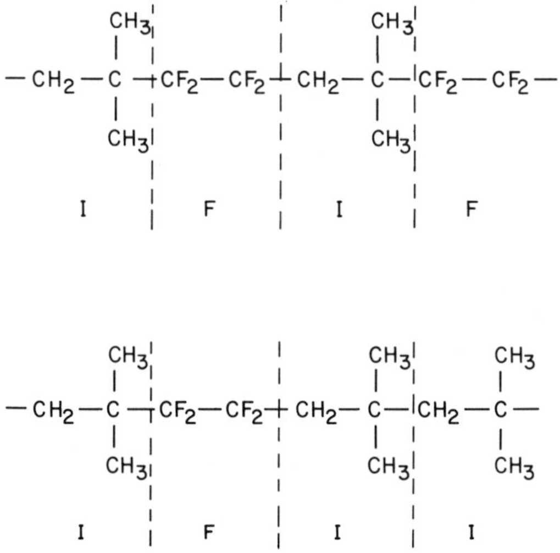

Figure 20-7.  Two different structure sequences for the copolymer of tetrafluoroethylene (F) and isobutylene (I); IFIF, IFII.

would be very small. From the $^{19}F$ NMR spectrum in Fig. 20-4 we see that at 94.1 MHz an AB type pattern is observed for the fluorines of the $CF_2$ group. Also, we see that the fluorine of the CFCl group shows a triplet pattern. In order for the fluorine of the CFCl group to show a triplet pattern it must be adjacent to the $CH_2$ group. Thus because the $CF_2$ group shows an AB pattern and the CFCl group shows a triplet pattern, structure 1 of Fig. 20-4 is the structure for this copolymer at 50 mole % copolymerization.

Next we consider the study performed by Ishigure on the copolymer of tetrafluoroethylene and isobutylene.[7] Figure 20-6 shows the $^{19}F$ NMR spectra obtained at 56.4 MHz on these copolymers of several compositions. Figures 20-7 and 20-8 show four sequence structures of TFE (F) incorporated into the copolymer with isobutylene (I). In Fig. 20-7 the sequence of a $CF_2$ group adjacent to the $CH_2$ group is likely to show coupling between them, and therefore this $CF_2$ group would most likely show a broader line in the spectrum than the $CF_2$ group adjacent to the branched carbon group. We see from the spectrum at 56.4 MHz in Fig. 20-6 that the low-field resonance at about 28 ppm from the trifluoroacetic acid reference is broader than the

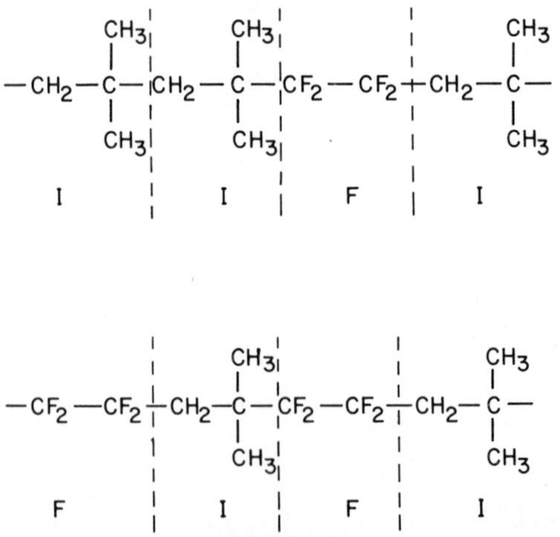

Figure 20-8. Two different structure sequences for the copolymer of tetrafluoroethylene (F) and isobutylene (I); IIFI, FIFI.

higher-field one at 40 ppm from the trifluoroacetic acid reference.  Thus it is probable that the 28 ppm line can be assigned
to the $CF_2$ group adjacent to the $CH_2$ group.  Now the question
is whether the effect of the third carbon group from this $CF_2$
group is influencing it.  From the spectrum it looks as if there
are at least two lines.  A more detailed analysis of that region
of the spectrum shows that there are indeed two lines that can
be attributed to this $CF_2$ group adjacent to the $CH_2$ group.  The
only interpretation that can be given to this observation is that
both structures shown in Fig. 20-7 are present in the copolymer.
If this is true for that particular $CF_2$ group then it is also true
for the $CF_2$ group adjacent to the branched methyls.  This agrees
with the observation of the resonance at 40 ppm.  Two lines are
observed at that position as well.  Since the coupling of the $CF_2$
with $CF_2$ is very weak and since there is essentially no coupling
of the $CF_2$ with the branched methyls, the resonance at 40 ppm
is seen to be much narrower than the one at 28 ppm.  Since the
pattern at 40 ppm is composed of two lines, this can be interpreted only in terms of two different structure sequences for
the $CF_2$ group adjacent to the branched methyls.  Thus as seen
in Fig. 20-8 the two different structures, namely, IIFI and
FIFI, can account for the two lines in this pattern at 40 ppm.
The question that arises next is which of these two structure
sequences can be assigned to the higher-field line and which to
the lower-field line.  The method that Ishigure selected to make
this assignment was to vary the composition of the copolymer
from 15 to 39 mole % TFE.  With increasing TFE content the
relative amounts of the structure sequence of FIFI to IIFI increase.  This increase is then related to the change in the intensities of the two lines.  The higher-field line of the two lines
is seen to increase relative to the lower-field line.  Therefore,
the higher-field line is assigned to the sequence structure of
FIFI and the lower-field line to the sequence structure IIFI.  At
94.1 MHz the $^{19}$F NMR spectrum of the copolymers shows the
difference between the two lines at about 40 ppm to be much
more resolved than the spectrum at 56.4 MHz.  Thus at 94.1
MHz the areas of the two lines can be measured in order to relate them to the composition of these two structures in the copolymer with varying monomer content.

The next fluorine-containing copolymer reported involves the copolymer of vinylidene fluoride and hexafluoropropylene. The NMR study on this material was first reported by Ferguson some years ago and more recently in a general discussion.[8] In Fig. 20-9 the 56.4 MHz $^{19}$F NMR spectrum is shown for the copolymer. Here the scale is referenced to trichlorofluoromethane (F-11). Below the spectrum in this figure is a listing

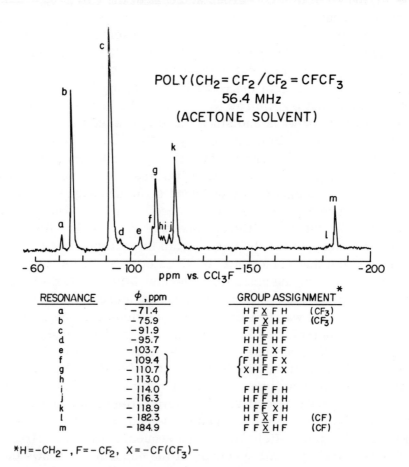

| RESONANCE | $\phi$, ppm | GROUP ASSIGNMENT* | |
|---|---|---|---|
| a | −71.4 | H F X F H | (CF$_3$) |
| b | −75.9 | F F X H F | (CF$_3$) |
| c | −91.9 | F H F H F | |
| d | −95.7 | H H F H F | |
| e | −103.7 | F H F X F | |
| f | −109.4 ⎫ | F H F F X | |
| g | −110.7 ⎬ | X H F F X | |
| h | −113.0 ⎭ | | |
| i | −114.0 | F H F F H | |
| j | −116.3 | H F F H H | |
| k | −118.9 | H F F X H | |
| l | −182.3 | H F X F H | (CF) |
| m | −184.9 | F F X H F | (CF) |

*H = −CH$_2$−, F = −CF$_2$, X = −CF(CF$_3$)−

Figure 20-9.   $^{19}$F NMR spectrum at 56.4 MHz on a copolymer of vinylidene fluoride and hexafluoropropylene along with triad group assignments for most of the observed resonance lines; ppm values upfield from CCl$_3$F (courtesy of *Kautschuk und Gummi Kunstoffe*).[8]

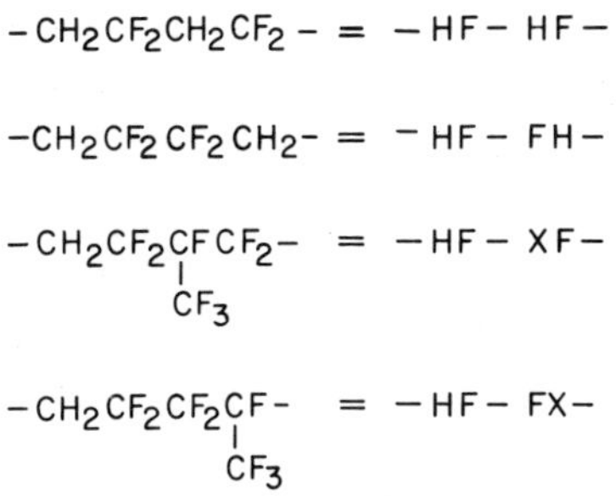

$-CH_2CF_2CH_2CF_2- = -HF-HF-$

$-CH_2CF_2CF_2CH_2- = -HF-FH-$

$-CH_2CF_2\underset{\underset{CF_3}{|}}{C}FCF_2- = -HF-XF-$

$-CH_2CF_2CF_2\underset{\underset{CF_3}{|}}{C}F- = -HF-FX-$

Figure 20-10. Dyad sequence structures for poly(vinylidene fluoride-hexafluoropropylene).

of the resonance lines observed at 56.4 MHz and the chemical shift values for each of these lines along with the probable group assignments for each. The group assignments are given by treating the data in terms of triad structures for the two monomers in the copolymer. The one assumption that is made in these assignments is that there are not any hexafluoropropylene–hexafluoropropylene sequences. This assumption appears to be reasonable considering the fact that hexafluoropropylene content is always less than 50 mole % and under the conditions of the polymerization there should be little, if any, of the HFP-HFP sequences formed. Figure 20-10 lists the different possible dyad structures for the copolymer and Fig. 20-11 lists

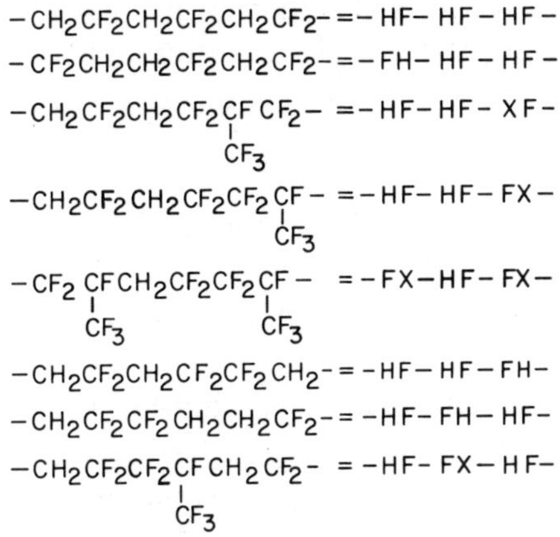

$-CH_2CF_2CH_2CF_2CH_2CF_2- = -HF-HF-HF-$

$-CF_2CH_2CH_2CF_2CH_2CF_2- = -FH-HF-HF-$

$-CH_2CF_2CH_2CF_2\underset{\underset{CF_3}{|}}{C}FCF_2- = -HF-HF-XF-$

$-CH_2CF_2CH_2CF_2CF_2\underset{\underset{CF_3}{|}}{C}F- = -HF-HF-FX-$

$-CF_2\underset{\underset{CF_3}{|}}{C}FCH_2CF_2CF_2\underset{\underset{CF_3}{|}}{C}F- = -FX-HF-FX-$

$-CH_2CF_2CH_2CF_2CF_2CH_2- = -HF-HF-FH-$

$-CH_2CF_2CF_2CH_2CH_2CF_2- = -HF-FH-HF-$

$-CH_2CF_2CF_2\underset{\underset{CF_3}{|}}{C}FCH_2CF_2- = -HF-FX-HF-$

Figure 20-11. Triad sequence structures for poly(vinylidene fluoride-hexafluoropropylene).

those triad structures that have been assigned to the different lines in the $CF_2$ region of the spectrum. The region between about $-91$ and about $-119$ ppm shows some resonances that have not been assigned with completeness as yet. All the assignments were made as a result of using the rules derived by Wilson and Santee and also by the use of relative intensities of the lines. In both the $CF_3$ and the CF regions of the spectrum a relatively weak line is observed as well as a very intense line. The assignments for these two lines in the two regions are based on the normal addition of $VF_2$ to HFP $(CF_2CFCF_3CH_2CF_2)$ relative to the inversion addition of $VF_2$ to HFP $(CF_2CFCF_3CF_2CH_2)$ in the copolymerization. The less probable inversion of $VF_2$ to HFP is assigned to the weak $CF_3$ line at about $-71.4$ ppm and the weak CF line at about $-182.3$ ppm. The lines marked c and d are assigned to the triad structures based on the work of Wilson and Santee; likewise, the lines marked i and j are based on the assignments made by Wilson and Santee for polyvinylidene fluoride. The remaining lines in the $CF_2$ region were assigned from the following criteria: (1) effects of $\alpha$ and $\beta$ carbon atom substituents, (2) relative intensity variations of lines with changes in monomer ratios, and (3) verification of the relative intensity variations with various group assignments.

From these spectra at 56.4 MHz and from the assignments made of the different lines observed it is apparent that $^{19}F$ NMR spectra can be useful in elucidating microstructures of copolymers. Following the microstructure analysis it then becomes possible to elucidate the mechanism(s) of the copolymerization.


## References

1. R. E. Naylor and S. W. Lasoski, *J. Polym. Sci.*, **44**, 1 (1960).
2. C. W. Wilson, III, and E. R. Santee, Jr., *J. Polym. Sci.* [C], **8**, 97 (1965).
3. F. J. Weigert, *Org. Magn. Resonance*, **3**, 373 (1971).
4. C. W. Wilson, III, *J. Polym. Sci.* [A], **1**, 1305 (1963).
5. G. V. D. Tiers and F. A. Bovey, *J. Polym. Sci.* [A], **1**, 833 (1963).

6.  K. Ishigure, Y. Tabata, and K. Oshima, *Macromolecules*, **3**, 27 (1970).
7.  K. Ishigure and Y. Tabata, *Macromolecules*, **3**, 450 (1970).
8.  R. C. Ferguson, *Kautschuk und Gummi Kunstoffe*, **11**, 723 (1965).

# CHAPTER 21

# $^{19}$F NMR STUDIES OF SOME MONO- AND BINUCLEAR AROMATIC CONJUGATED SYSTEMS

I. R. Ager and L. Phillips

*Organic Chemistry Department*
*Imperial College of Science*
*University of London*

For some time now we have been interested in trying to "understand" in qualitative terms the origins of substituent effects upon the shielding of $^{19}$F nuclei in aromatic molecules.[1-7] The discussion that follows is based upon recently published work.[5-7]

There have been several attempts to relate substituent effects upon $^{19}$F shielding in *para*-substituted aromatic molecules to other molecular parameters, such as $\pi$-electron density changes or reactivity parameters.[3] Possibly the analysis that has received widest recognition is that of Taft,[8] which relates substituent chemical shifts (S.C.S.) in these molecules to the reactivity parameters $\sigma_I$ and $\sigma_R$:

$$\Delta p = -5.83\sigma_I - 18.8\sigma_R + 0.8 . \tag{1}$$

This has been interpreted as demonstrating the dual dependence of *para*-substituent chemical shifts upon sigma-electron ($\sigma_I$) and pi-electron ($\sigma_R$) terms; an analogous equation relates *meta*-S.C.S., $\Delta m$, to $\sigma_I$ and has been taken as evidence that $\pi$-electron terms are unimportant.[8] Notwithstanding this treatment, several workers have related *para*-S.C.S. values to $\pi$-electron densities without reference to the importance of sigma-bonded interactions.[9, 10]

In order to study a system in which it is felt that only $\pi$-bonded interactions are of significance, we have examined a series of 10,4-substituted $4'$-fluoro-*trans*-stilbenes. The S.C.S. values for the various substituents in this series do not correlate at all with those for the corresponding series of 4-substituted fluorobenzenes, nor do they correlate with the parameter $\sigma_R$, as would be expected if this were a measure of $\pi$ interaction between a substituent and an aromatic ring. They do correlate very well indeed, however, with the original unmodified Hammett $\sigma_{para}$ parameter and if our basic assumption (that *para*-S.C.S. in this series arise solely from $\pi$ interactions) is correct, then it would indicate that $\sigma_{para}$ is a direct measure of the ability of a substituent to redistribute $\pi$ electrons alone and does not have a sigma-electron component as Taft and later workers imply.

Any LCAO-MO treatment of a delocalized $\pi$ system leads to the conclusion that the changes in electron density at the 1-carbon in 4-substituted fluorobenzenes (**1**) are linearly related to those on the $4'$-carbon in 4-substituted $4'$-fluoro-*trans*-stilbene (**2**). The value of the proportionality constant depends upon

**1**                                    **2**

the values of the coulomb integrals, $\alpha$, for the intervening carbon atoms and upon the bond integrals, $\beta$, for the C-C bonds; it does not depend, however, upon the values of $\alpha_X$ and $\beta_{C-X}$ for the substituents. Using simple Hückel MO theory, and not differentiating between olefinic and aromatic carbons, the value for the proportionality constant obtained is shown to be

$$\delta^{\pi}_{(C_1 \text{ benzene})} = 8.072 \; \delta^{\pi}_{(C'_4 \text{ stilbene})}, \qquad (2)$$

TABLE 21-1.   Calculated $\pi$ contributions to S.C.S. in 4-X
fluorobenzenes compared with experimentally
observed S.C.S. (ppm, positive shifts are to
low field)

| X | $\Delta_\pi$ (calc) | $\Delta$ (obs) | $\Delta$(obs) $- \Delta_\pi$(calc) |
|---|---|---|---|
| $NMe_2$ | $-12.92$ | $-14.22$ | $-1.30$ |
| $NH_2$ | $-10.57$ | $-14.14$ | $-3.57$ |
| OH | $-6.38$ | $-11.62$ | $-5.24$ |
| OMe | $-5.37$ | $-11.60$ | $-5.87$ |
| Me | $-2.93$ | $-5.47$ | $-2.48$ |
| H | $-0.00$ | $-0.00$ | $-0.00$ |
| F | $+1.94$ | $-6.90$ | $-8.84$ |
| Cl | $+5.01$ | $-3.30$ | $-8.31$ |
| Br | $+5.57$ | $-2.73$ | $-8.30$ |
| I | $+6.14$ | $-1.76$ | $-7.90$ |
| $^+N(Me)_3$ | $+12.03$ | $+1.03$ | $-11.00$ |
| $NO_2$ | $+19.21$ | $+8.79$ | $-10.42$ |

where $\delta^\pi_{(C_1 \text{ benzene})}$ and $\delta^\pi_{(C_4' \text{ stilbene})}$ are the appropriate $\pi$-electron
densities.

If the [19]F S.C.S. is proportional to the $\pi$-electron density
upon the carbon to which it is bonded,[3] the $\pi$ contributions to
*para*-S.C.S. in the benzene series should be in this same proportion to the observed S.C.S. in the stilbene series. Table
21-1 lists contributions calculated in this way together with experimentally observed S.C.S. for the benzenes and the differences between the two figures.

Apparently, all substituents are more shielding towards fluorine than may be accounted for by the proportionality to the
stilbene data. It would be interesting to compare the values of
$[\Delta(\text{obs}) - \Delta_\pi(\text{calc})]$ with S.C.S. for the series of 4-substituted
fluoro[2:2:2]bicyclooctanes (3). In this series, only through-

X—⟨benzene ring⟩—F

**3**

sigma bond or through-space interactions can contribute to the shielding effects of the substituents and the S.C.S. should be closely related to the values of $\Delta(obs) - \Delta_\pi(calc)$ if our treatment is of use. Very few data are available, but when X = F or COOMe the S.C.S. are large and negative; that is, both groups cause upfield shifts of the fluorine compared with X = H; this is in accord with the data of Table 21-1. A further comparison of some value may be made, however, with the ionization constants of an analogous series of [2:2:2]bicyclooctanecarboxylic acids (**4**). These have been utilized to define the Taft $\sigma_I$ parameters

X—⟨bicyclooctane⟩—$CO_2H$

**4**

for the substituent X as a measure of the inductive power of the group,[8] and the values of $\Delta(obs) - \Delta_\pi(calc)$ listed in Table 21-1 correlate with $\sigma_I$ to a high degree of precision ($\tau \sim 98\%$).

The conclusion may now be reached that $^{19}F$ *para*-S.C.S. in fluorobenzenes may be expressed in a two-parameter equation similar to that of Taft[1]:

$$\Delta p = -21.06\sigma_p + 15.75\sigma_I . \qquad (3)$$

For a wide range of substituents this gives better agreement with experiment than does equation 1 (rms for 12 substituents ±1.52 ppm compared with ±1.95 ppm); of greater importance, however, is the fact that the parameters used in equation 3 are experimentally derived ($\sigma_p$ from ionization constants of *p*-substituted benzoic acids and $\sigma_I$ from ionization constants of compounds of type **4**), whereas the $\sigma_R$ parameter used in equation 1 is not capable of independent measurement.

A similar analysis of S.C.S. in the series of *meta*-substituted fluorobenzenes (5) (using the observed S.C.S. in 4-substituted 3'-fluoro-*trans*-stilbenes) leads to the conclusion that $\pi$ interactions *alone* may account for the shielding changes; this is in direct opposition to the earlier suggestions of Taft et al., who state that $\pi$ terms are unimportant and the S.C.S. is dominated by sigma-bonded inductive interactions. It should be pointed out at this stage that there are some conceptual difficulties that arise from our treatment. If the S.C.S. in the stilbene series do indeed arise solely from $\pi$ interactions and are linearly dependent upon the $\pi$ electron density of the carbon to which they are bonded, we are led to the prediction that the halogens F, Cl, Br, and I are $\pi$-electron withdrawing from the *para* carbon in this series and in the *para*-substituted fluorobenzenes. Most modern MO treatments suggest that these groups are in fact $\pi$-electron releasing,[11] and a recent ab initio calculation for fluorobenzenes confirms this.[12] We are unable to comment upon this at the present time, but we have obtained some confirmation of the validity of our approach by successfully calculating S.C.S. values for a series of 4-substituted 4'-fluorodiphenyls and 4-substituted 4'-fluoro-*trans*, *trans*-diphenylbutadienes in which it is also reasonable to assume that only $\pi$ interactions are important.

   This work has led us to use <sup>19</sup>F chemical shift data to study the transmission of $\pi$-electronic interactions across the bridge in many series of 4-substituted 4'-fluoro bridged binuclear aromatic compounds (5). After correcting for solvent effects where

5

possible,[5,7] transmissive factors, $T$, have been defined in terms of the slopes of the observed linear correlation of <sup>19</sup>F S.C.S. (in *all* the series examined) with <sup>19</sup>F S.C.S. in the series of *trans*-stilbenes. These are quoted for various bridges Z in Table 21-2.

TABLE 21-2.  Transmissive factors, $T$, for substituent effects upon $^{19}F$ shielding in compounds of type 5

| Z | $T$ | | Z | $T$ |
|---|-----|---|---|-----|
| 1. ⚬ | 1.00 | 9. △ | | 0.21 |
| 2. ⚬ | 0.60 | 10. $>CH_2$ | | 0.53 |
| 3. — | 1.46 | 11. $-CH_2 \cdot CH_2 -$ | | 0.00 |
| 4. $>C=CH_2$ | 0.55 | 12. $>C<^H_{OH}$ | | 0.76 |
| 5. $>C=O$ | 1.03 | 13. $>C<^{CH_3}_{OH}$ | | 0.69 |
| 6. $>C=O \rightarrow BCl_3$ | 3.97 | 14. $-O-$ | | 1.64 |
| 7. $>C=O-H$ | 4.46 | 15. $-S-$ | | 2.35 |
| 8. $-NH-C{\small\nwarrow}^O$ | 0.32 | | | |

In discussing the values of $T$ in relation to the electronic structure of Z, a distinction has been drawn between bridges that "transmit" substituent effects, allowing transfer of charge from one ring to the other by means of conjugation or hyper-conjugation (groups Z = 1 to 11), and those for which a "relay" mechanism (which does not allow for charge transfer in this way) may operate (Z = 12 to 15). This aspect is discussed fully in reference 6.

## References

1.   J. W. Emsley and L. Phillips, *Mol. Phys.*, **11**, 437 (1966).
2.   J. W. Emsley and L. Phillips, *J. Chem. Soc. B*, **434** (1969).

3.  J. W. Emsley and L. Phillips, *Progress in Nuclear Magnetic Resonance Spectroscopy*, J. W. Emsley, J. Feeney, and L. H. Sutcliffe, Eds., Vol. 7, Pergamon Press, Oxford, 1971.

4.  L. Phillips and V. Wray, *J. Chem. Soc. Perkin*, **II**, 223 (1972).

5.  I. R. Ager and L. Phillips, *J. Chem. Soc. Perkin*, **II**, 1975 (1972).

6.  I. R. Ager, L. Phillips, T. J. Tewson, and V. Wray, *J. Chem. Soc. Perkin*, **II**, 1979 (1972).

7.  I. R. Ager, L. Phillips, and S. J. Roberts, *J. Chem. Soc. Perkin*, **II**, 1988 (1972).

8.  R. W. Taft, *J. Am. Chem. Soc.*, **79**, 1045 (1957).

9.  R. W. Taft, F. Prosser, L. Goodman, and G. T. Davis, *J. Chem. Phys.*, **38**, 380 (1963).

10. G. L. Caldow, *Mol. Phys.*, **11**, 71 (1966).

11. J. W. Emsley, *J. Chem. Soc. A*, 2018 (1968).

12. W. J. Hehre, L. Radom, and J. A. Pople, *J. Am. Chem. Soc.*, **94**, 1496 (1972).

CHAPTER 22

# A MULTINUCLEAR NMR STUDY OF AQUEOUS TETRAFLUOROBORATE ANION AND ITS REACTION WITH Al(H$_2$O)$_6^{3+}$

J. W. Akitt

*The School of Chemistry*
*The University of Leeds*

In this chapter we develop the theme that if in any system several nuclei are available for study, then far more useful information may be forthcoming if results are obtained from all of them than if reliance is placed on only one; the results reviewed here arise from studies involving eight different nuclei, none of which can provide a complete solution alone.

Aqueous salt solutions form an excellent subject for NMR investigation. The solvent can be studied using $^1$H and $^{17}$O spectroscopy, and most cations and many anions contain nuclei with attractive NMR properties, for example, the alkali metals, the group III metals, and the halide ions. Information can thus be obtained from several essentially independent sources and a full interpretation of the results should reconcile the whole of the available data.

Tetrafluoroborate solutions are even richer in nuclei than most since the anions contain three nuclei, $^{19}$F and $^{11}$B or $^{10}$B. This leads to two anion species, $^{11}$BF$_4^-$ and $^{10}$BF$_4^-$, which exhibit a large isotope shift in the $^{19}$F spectra. The $^{19}$F spectra

339

have been much studied[1-3] and it is known that both $^1J(B-F)$ and
the $^{19}F$ chemical shift $\delta_F$ depend upon the nature of the cation,
the concentration of salt, and the solvent. For example, $^1J(B-F)$
increases from 1.1 to 6.0 Hz in aqueous $NaBF_4$ solutions as the
concentration is increased and $\delta_F$ moves some 5 ppm upfield.
The value of $^1J$ is unusually small for a directly bonded coupling
and is believed to arise because of the near cancellation of large
contributions to $^1J$ which are opposite in sign. Certainly $^1J$
changes sign in some solvent systems.[3]

The changes in $^1J$ and $\delta_F$ have been attributed variously to
contact ion pairing,[1] hydrogen bonding between anion and solvent,[2] or interactions involving solvent-separated ion pairs.[3]
However, $^1H$ studies[4] show that $BF_4^-$ is a structure breaker like
$ClO_4^-$, the $^1H$ solid-state NMR of $NH_4BF_4$ is consistent only
with weak or nonexistent hydrogen bonding,[5] and water is lost
easily from the hydrates.[6] In addition the $^{11}B$ spectra consist
of well-resolved quintets with narrow lines and long relaxation
times, a fact difficult to reconcile with ion-pair formation,
which should reduce the symmetry of the ion and lead to quadrupole relaxation. On the other hand, the $^{23}Na$ and $^7Li$ nuclei
are very sensitive to the presence of $BF_4^-$ and move upfield with
increasing concentration, by as much as 10 ppm in the case of
$^{23}Na$. The $^{23}Na$ line widths are also broadened in the presence
of $BF_4^-$. Such data point to a strong interaction involving the
cation but not the anion, and appear to support a rather more
complex mechanism than just simple ion pairing.

Since the effect on the $BF_4^-$ parameters of the singly and
doubly charged cations has now been studied in some detail, it
seems likely that an ion such as $Al^{3+}$ might reveal further information on the nature of the interaction, either because of its
higher charge or because it is protected from direct contact
with the anion by a tight sheath of complexed water molecules.
The fluorine spectrum of aqueous $NaBF_4$ or $LiBF_4$ consists of
the two resonances due to $^{10}BF_4^-$ and $^{11}BF_4^-$ with a small low-
field quartet due to $^{11}BF_3OH^-$. This latter may not be detectable in a fresh solution and becomes evident only after a day
or so. When an aluminum salt solution is added to fresh $NaBF_4$
solution, three new resonances are observed whose relative in-

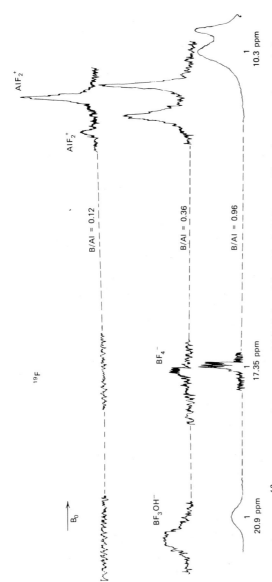

Figure 22–1. $^{19}$F spectra of aqueous $Al(NO_3)_3$ with added $LiBF_4$. The three regions of the spectra were run at different gain and power levels. The shifts reported are to low field of external $C_6F_6$.

341

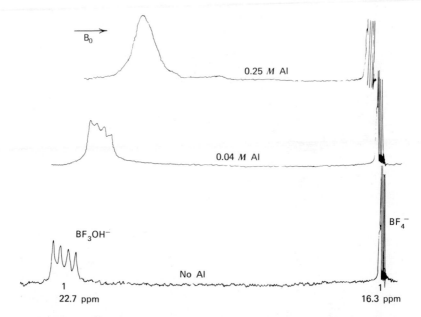

Figure 22-2.  $^{19}F$ spectra of 3 $M$ LiBF$_4$ previously heated to form an appreciable concentration of BF$_3$OH$^-$.  The shifts reported are to low field of external C$_6$F$_6$.

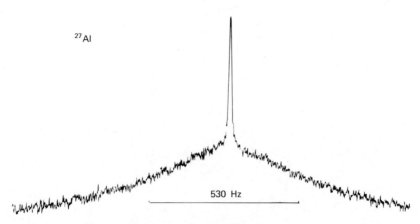

Figure 22-3.  $^{27}Al$ spectra of aqueous Al(NO$_3$)$_3$ with added LiBF$_4$.  The sharp singlet is due to Al(H$_2$O)$_6^{3+}$.

342

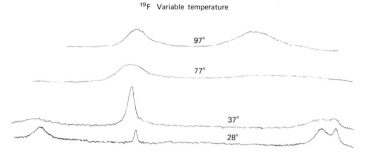

$^{19}F$  Variable temperature

97°

77°

37°

28°

Figure 22-4.  $^{19}F$ spectra of aqueous $Al(NO_3)_3$ with added $LiBF_4$ at elevated temperatures.

tensities depend upon the solution composition (Fig. 22-1). The low-field singlet arises from $BF_3OH^-$ whose formation is accelerated by the aluminum and whose chemical shift is altered and multiplicity destroyed because of exchange with other species in the system (Fig. 22-2).  The high-field pair of resonances falls close to those due to the aluminofluoro complexes $AlF^{2+}$ and $AlF_2^+$.[7]  The $^{27}Al$ spectrum is found to contain the broad line known to be given by these complexes, and the proportion of complexed aluminum is correct for the proportion of fluorine in their $^{19}F$ resonances (Fig. 22-3).  The $^{19}F$ spectra are in addition very temperature dependent and indicate exchange between $BF_3OH^-$, $AlF^{2+}$, and $AlF_2^+$ but not greatly involving the $BF_4^-$ ion which apparently serves mainly as the source of fluoride ion (Fig. 22-4).  A feature of the $^{19}F$ spectra is the extreme variability of the resonance positions in different solutions, this being apparently normal for many fluoro ions.[7]  The $BF_4^-$ environment, however, is relatively little affected in this system.

The $^{11}B$ spectra are also very informative (Fig. 22-5).  The quintet of $BF_4^-$ and the resonance of $BF_3OH^-$ are observed in their normal positions together with a broad resonance to low field, whose position and width depend upon solution composition.  This plainly arises from nonfluorinated species.  The chemical shift does not correspond to any known borate resonance but lies between those due to $B(OH)_3$ and $BO_3^{3-}$ and presumably originates from exchanging $BO_{3-n}(OH)_n^{(3-n)-}$ species. The absence of $BF_2(OH)_2^-$ or $BF(OH)_3^-$ is remarkable, and al-

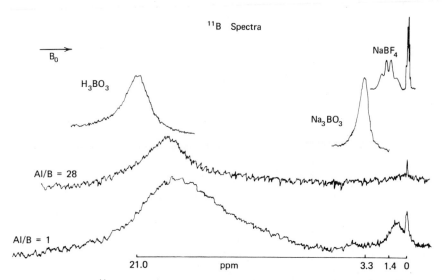

Figure 22-5.  $^{11}$B spectra of aqueous $Al(NO_3)_3$ with added $LiBF_4$ (two lower traces).  The three short, upper traces show for comparison the $^{11}$B spectra of aqueous $H_3BO_3$, $Na_3BO_3$, and $NaBF_4$ containing both $BF_4^-$ (quintet) and $BF_3OH^-$ (quartet).  The shifts reported are to low field of the $BF_4^-$ signal.

though such species are no doubt formed during the reaction, either their concentration must be small or their existence transitory.  They may contribute to the exchange on $BF_3OH^-$ but if they were present in large quantity they should introduce a low-field shift in its resonance rather than the high-field shift observed.

These spectra thus indicate a fairly complex set of equilibria, involving both fast and slow exchange, which can be formulated as follows:

$$H_2O + BF_4^- \xrightleftharpoons{slow} BF_3OH^- + F^- + H^+$$

$$n\ F^- + Al(H_2O)_6^{3+} \xrightleftharpoons{fast} Al(H_2O)_{6-n}F_n^{(3-n)+}$$

$$3\ Al(H_2O)_5F^{2+} + BF_3OH^- \xrightleftharpoons{fast} [BF_2(OH)_2^-,\ BF(OH)_3^-]$$

$$\xrightleftharpoons{fast} BO(OH)_2^- + 3\ Al(H_2O)_4F_2^+ + 3H^+ + H_2O\ .$$

## References

1.  K. F. Kuhlman and D. M. Grant, *J. Phys. Chem.*, **68**, 3208 (1964).
2.  R. Haque and L. W. Reeves, *J. Phys. Chem.*, **70**, 2753 (1966).
3.  R. J. Gillespie, J. S. Hartman, and M. Parekh, *Can. J. Chem.*, **46**, 1601 (1968).
4.  J. Davies, S. Ormondroyd, and M. C. R. Symons, *Trans. Faraday Soc.*, **67**, 3465 (1971).
5.  D. Pendred and R. E. Richards, *Trans. Faraday Soc.*, **51**, 468 (1955).
6.  D. W. A. Sharp, *Advan. Fluorine Chem.*, **1**, 68 (1961).
7.  N. A. Matwiyoff and W. E. Wageman, *Inorg. Chem.*, **9**, 1031 (1970).

# CHAPTER 23

# PULSE FOURIER TRANSFORM NMR WITH METAL NUCLEI

Gary E. Maciel

*Department of Chemistry*
*Colorado State University*
*Fort Collins*

## INTRODUCTION

Although a significant fraction of the early NMR experiments carried out in the decade of the 1950s was with the nuclei of metal elements, metal elements did not share proportionately in the great increase in popularity of NMR during the 1960s. A literature search on this subject reveals a substantial number of investigations that were carried out prior to 1960, [1] frequently by wide-line or rapid passage techniques, and a fairly large number of rather recent studies based upon the internuclear double resonance (INDOR) approach. [2] In some cases, a distorted assessment of the potential utility of current metal-nucleus NMR approaches in chemical studies can be reached if the apparent limitations of early techniques are given undue emphasis. There is very little information available on the application of Fourier transform (FT) methods to the nuclei of metal elements. Yet it appears that the same kinds of advancements in equipment and techniques that have made pulse FT routine for $^{13}C$ NMR are also applicable to a considerable number of metal nuclei. Preliminary experimental surveys confirm this. [3-7]

This paper is concerned with exploring the prospects for FT NMR applications with metal nuclei in diamagnetic substances, and the kinds of experimental techniques and limitations that are likely to be important. Included are some preliminary results from our laboratory which indicate the promise of the metal-element FT NMR approach. As most of our data are on $^{113}$Cd, $^{199}$Hg, and $^{207}$Pb, major emphasis is given to these nuclides.

## FUNDAMENTAL AND PRACTICAL CONSIDERATIONS

A variety of factors contribute to the suitability of a magnetic nuclide for pulse FT NMR applications. Some of these factors for the metal nuclides of main interest to us are summarized in Table 23-1. This table also includes, for comparison purposes, the characteristics of $^1$H, $^{29}$Si, and $^{13}$C (to which the data are normalized). The most obvious factors are the natural abundance and the intrinsic sensitivity for NMR detection. Although the experimentalist has no control over the latter (for a given field strength and customary detection systems), there are in special cases valid reasons to consider nuclide enrichment to offset an unfavorable natural abundance. This approach is attractive if the following two conditions are satisfied: (1) the natural abundance is sufficiently low so that enrichment provides a large increase in signal-to-noise ratio, and (2) enriched samples of the nuclide are available commercially at a cost commensurate with the research dividends. The most likely candidates in Table 23-1 for enrichment are $^{43}$Ca and $^{67}$Zn, especially the former. Both of these nuclides are of special interest in biological applications, and both are available in highly enriched states. If the history of $^{13}$C NMR can be taken as an indication of future developments, the awareness of a demand for a stable nuclide ultimately results in substantial price reductions by commercial suppliers.

In addition to the above-mentioned factors, there are such matters as relaxation times, quadrupole effects, and the nuclear Overhauser effect (NOE), [8] which are not well characterized for most cases of interest. Quadrupole relaxation mechanisms, for nuclei with $I > \frac{1}{2}$, can give rise to excessive broadening that can hinder detection in a high-resolution sense; or in some favorable cases, they may reduce relaxation times to a point where rapid repetitions become practicable. In addition, if the broadening effect is not excessive, then a study of relaxation times, like chemical shift studies, has the potential of providing information that is diagnostic of structure and dynamics. For example, if a quadrupolar nucleus is involved in a rapid chemical equilibrium between two sites A and B, then the resultant $T_2$ of the system can be expressed as

$$\frac{1}{T_2} = p_A\left(\frac{1}{T_{2_A}}\right) + p_B\left(\frac{1}{T_{2_B}}\right),$$

where $T_{2_A}$ and $T_{2_B}$ are the relaxation times of the quadrupolar nucleus in the species A and B, respectively, and the $p$'s are the fractions of the pertinent nuclei in the two states. If the symmetries of the local electronic environments of these two sites are far different, then the quadrupolar relaxation mechanisms are exerted to much different extents, and $T_{2_A}$ and $T_{2_B}$ are much different. In such a case, detailed $T_2$ studies provide valuable information on the equilibrium. However, if one of the individual $T_2$'s is too short, then the resulting weighted-average line width may be too large for convenient detection or interpretation.

It is interesting to note that, with the exception of $^{25}$Mg, all the quadrupolar nuclides listed in Table 23-1 either have a "relative natural abundance sensitivity" that is much higher than that of $^{13}$C, or else they are good candidates for isotope enrichment. Thus the line broadening effects that one expects for quadrupolar nuclides may not, in many cases, eliminate them from FT NMR studies.

Nuclear Overhauser effects (NOE) associated with proton decoupling are generally considered to be advantageous for nuclei with positive nuclear magnetic moments, except for the uncertainties associated with measured line intensities. However, for nuclei with negative nuclear magnetic moments, for

TABLE 23-1.  Pertinent properties of some metal nuclides[a]

| Nucleus | Natural Abundance (%) | Nuclear Spin[b] | NMR Frequency at 21 kg (MHz) | Relative NMR Sensitivity per Nucleus at Constant Field[c] | Relative Natural Abundance Sensitivity at Constant Field[d] |
|---|---|---|---|---|---|
| $^1$H | 100 | $\frac{1}{2}$ (+) | 90.0 | 63 | 5800 |
| $^{13}$C | 1.1 | $\frac{1}{2}$ (+) | 22.4 | 1 | 1 |
| $^{23}$Na | 100 | $\frac{3}{2}$ (+) | 23.8 | 5.8 | 526 |
| $^{25}$Mg | 10.1 | $\frac{5}{2}$ (−) | 5.5 | 0.17 | 1.5 |
| $^{27}$Al | 100 | $\frac{5}{2}$ (+) | 23.5 | 13 | 1200 |
| $^{29}$Si | 4.7 | $\frac{1}{2}$ (−) | 17.9 | 0.49 | 2.1 |
| $^{43}$Ca | 0.13 | $\frac{7}{2}$ (−) | 6.1 | 4.0 | 0.47 |
| $^{55}$Mn | 100 | $\frac{5}{2}$ (+) | 22.3 | 11.2 | 1100 |
| $^{59}$Co | 100 | $\frac{7}{2}$ (+) | 21.4 | 17.7 | 1610 |

| Nucleus | | Spin (sign) | | | |
|---|---|---|---|---|---|
| $^{63}$Cu | 69 | $\frac{3}{2}$ (+) | 23.8 | 5.8 | 365 |
| $^{67}$Zn | 4.1 | $\frac{5}{2}$ (+) | 5.6 | 0.18 | 0.67 |
| $^{75}$As | 100 | $\frac{3}{2}$ (+) | 15.4 | 1.58 | 144 |
| $^{113}$Cd | 12.3 | $\frac{1}{2}$ (−) | 20.0 | 0.69 | 7.7 |
| $^{119}$Sn | 8.7 | $\frac{1}{2}$ (−) | 33.6 | 3.3 | 26 |
| $^{195}$Pt | 33.7 | $\frac{1}{2}$ (+) | 19.4 | 0.64 | 19 |
| $^{199}$Hg | 16.9 | $\frac{1}{2}$ (+) | 16.1 | 0.37 | 5.6 |
| $^{205}$Tl | 70.5 | $\frac{1}{2}$ (+) | 51.9 | 12 | 770 |
| $^{207}$Pb | 21.1 | $\frac{1}{2}$ (+) | 18.8 | 0.57 | 11.0 |

[a] $^1$H, $^{13}$C, and $^{29}$Si included for comparison.
[b] Sign of the magnetic moment in parenthesis.
[c] Normalized to $^{13}$C.
[d] Product of natural abundance and relative NMR sensitivity per nucleus, normalized to carbon.

example, $^{15}$N, $^{29}$Si, $^{25}$Mg, $^{43}$Ca, $^{113}$Cd, and $^{119}$Sn, the NOE is negative, which can reduce the magnitude of an already hard-to-detect signal, or can even cause it to vanish.[9, 10] Techniques have been developed for reducing excessively long spin-lattice relaxation times, and for minimizing or eliminating the distortions due to NOE's; however, the applicability of each such technique remains to be determined for each nuclide and for each class of compounds. These techniques include the use of shift-less relaxation reagents (which provide new spin-lattice relaxation mechanisms to compete with those responsible for the NOE's)[11-13] and gated decoupler experiments (in which the proton irradiation power is kept on briefly for purposes of decoupling, but not long enough for the establishment of the NOE).[13, 14]

Two practical matters that one faces with NMR work on metal nuclides are how to determine and how to report chemical shifts. In determining chemical shifts one has basically the same dilemma to consider as previous workers have had to face for $^1$H and $^{13}$C: a homonuclear internal referencing technique requires a reference species that is stable and soluble in the desired medium and that provides a sharp signal that is easy to detect at low concentrations. An external referencing technique, homonuclear or heteronuclear, requires good bulk susceptibility corrections, which are often not easy to produce. Many of the metal systems that are of interest are quite labile, and what might otherwise serve as a useful internal reference may engage in a chemical reaction (ligand exchange) with either the solute species of interest or the solvent. Hence the internal reference approach is not useful in many, if not most, metal-nuclide studies. The bulk susceptibility corrections required for external referencing are at best inconvenient, although there are some NMR-based methods for determining accurate bulk susceptibilities[15-17] that may prove useful in many cases.

Perhaps the most generally useful referencing technique for metal nuclides will be a *heteronuclear* internal reference approach, in which an NMR signal of a (relatively) inert internal reference, say, $^1$H in TMS or $^2$H in perdeuterocyclohexane, is monitored or used as a lock signal. Of course, a proton signal would not be usable for experiments carried out under con-

ditions of broad-band proton decoupling. This kind of referencing approach is limited by the suitability of the internal reference signal (e.g., $^1H$ in TMS), but these limits are relatively
well established in many cases.[18] Furthermore, small errors,
for example, a few tenths of a ppm, due to solvent effects on the
heteronuclear reference are not likely to be serious in the light
of the wide ranges of chemical shifts often encountered with
metal nuclei.

Having decided upon an experimental approach for determining chemical shifts, one is faced with the question of how to
report them in the most useful manner. Ideally one would hope
to maintain the arbitrary, but currently accepted, convention of
choosing a primary reference (zero value of chemical shift)
near the high shielding end of the chemical shift range for each
nuclide, and then reporting shifts of less shielded species with
a positive sign. The primary reference should also have characteristics that render it a useful and reproducible standard for
purposes of calibrating the method of chemical shift determination of each laboratory. Thus the peak position of the primary
standard should not be excessively sensitive to inevitable experimental variations (e.g., small changes in temperature or
concentration).

## INSTRUMENTAL CONSIDERATIONS

In setting up a multinuclear NMR laboratory, one must decide on various aspects of instrumentation strategy. The most
common high-resolution technique currently in use for metal
nuclei is based upon the INDOR method. The INDOR method
has the advantage that it can be incorporated at relatively little
expense into any stable high-resolution spectrometer. It has
the further advantage, in the hands of an experienced spectros-
copist, of providing valuable information on spin-spin couplings.
However, INDOR is practical only if the "metal satellites" are
clearly identifiable in the spectrum of the high-sensitivity nucleus, usually $^1H$; in cases where there is no coupled high-
sensitivity nucleus, as with a simple solvated metal ion or a

metal halogen complex, INDOR is of course not applicable. The
FT approach is more applicable to routine spectral determinations, including samples of complex mixtures, for which complicated $^1$H spectra could preclude INDOR experiments. On the
basis of available data, it appears that the FT method provides
spectra that are *at least* as good in terms of line width and signal-to-noise ratio as the very best INDOR spectra reported for
the most favorable cases. The FT approach offers the further
advantage of a shorter time scale, permitting the convenient
study of a variety of time-dependent phenomena, for example,
reaction kinetics, chemically induced dynamic nuclear polarization, and spin-lattice relaxation.

If a pulse FT technique is chosen, then one must decide
whether the various nuclei will be covered by varying the magnetic field or by employing various resonance frequencies. A
variable field approach has the advantage of employing a small
number of highly tuned transmitter and receiver networks for
work with several nuclei.[3] In that sense, it may be less expen-
sive. However, field variation introduces complications into
the field/frequency lock system, and requires additional flexibility in decoupling equipment if decoupling is desired. Hence
for work with organometallic species or with metal complexes
that contain ligands with hydrogen atoms, $^1$H decoupling is usually desirable, and the variable frequency approach is to be
preferred.

The pulse FT system at Colorado State University is a
Bruker HFX-90-Digilab FTS/NMR-3 combination. In work with
metal nuclides the pulse frequency is derived from a frequency
synthesizer, the signal from which drives a gated pulse amplifier that is controlled by the data system. A wide-band power
amplifier ENI Model 3100L, 0.25–105 MHz, ~100 W pulse
power) has been employed, as has been a tuned amplifier (Digi-
lab 400-2, 200 W, tunable over the range 15–20 or 20–25 MHz).
A wide-band preamplifier (United Development Corporation) is
employed directly after the receiver coil (four of which cover
the nuclei studied). A second frequency synthesizer provides
a signal which, when mixed with that of the first synthesizer,

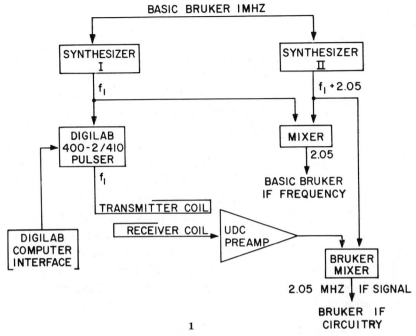

Figure 23-1. Block diagram of spectrometer configuration for multi-nuclear FT experiments.

generates a difference frequency ($\sim 2.05$ MHz) that is used as the reference for the IF section of the spectrometer's receiver system.  Aside from these variations, the basic spectrometer is utilized without modification.  Both single coil and cross-coil configurations have been employed with approximately equal advantage.  Converting the spectrometer for observation of one nuclide from another requires a minimal amount of retuning of matching circuits and, in some cases, changing an insert.  Even the insert change can be avoided if a broad-band insert is constructed.[19]  The configuration is shown in the form of a block diagram in Fig. 23-1.  Other spectrometer configurations have obtained similar results.[3-7]  With this system, we have observed FT NMR signals of all but two of the nuclides listed in Table 23-1.  Some typical spectra are shown in Figs. 23-2 through 23-7.

## METAL NMR STUDIES

During the past twenty years there has been an appreciable amount of NMR data on metal nuclei accumulated. Various stages of instrument sophistication and degrees of resolution are represented in the literature of this subject. The most commonly studied heavy metal nuclei have been $^{59}$Co and $^{119}$Sn, the first of these largely by "direct" methods (because of the extremely favorable natural abundance and NMR sensitivities of these nuclides)[20] and the latter mainly by INDOR techniques.[21] In addition there has been an appreciable amount of high-resolution NMR work on light metals, especially $^{23}$Na.[22] We make no attempt here to survey this entire body of knowledge, but rather summarize some of the main features that have emerged for heavy metals and present a few details in the areas that have occupied our attention in preliminary FT studies.

The general status of heavy metal-nuclide NMR can be summarized by the following general characteristics:

1. The ranges of chemical shifts are very large for the heavy metals, up to tens of thousands of ppm.

2. Solvent and temperature variations can have large effects on observed chemical shifts, betraying chemically significant interactions and transformations that can be studied conveniently by metal-nuclide NMR methods.

3. With the possible exceptions of $^{59}$Co and $^{119}$Sn, rather little detail is known about the relationships between structure and chemical shifts or relaxation times. In spite of a considerable amount of important work, none of these nuclides has been well characterized, by the standards of modern NMR research.

The theoretical interpretations that have been made on the chemical shifts of elements with $d$-orbital bonding have generally, although not exclusively, assumed that the chemical shifts are dominated by local paramagnetic contributions.[1b,23] These interpretations have been based largely upon attempts to understand the chemical shifts of $^{59}$Co and $^{31}$P, and have included treatments at about the level of the Karplus-Das,[24] or Jameson-Gutowsky[23a] formulations. It would seem that such approaches, which are not really suitable for computational work, may be more successful for understanding qualitative aspects of the chemical shifts of metal elements than the analogous theories have been for $^{13}C$ or $^{19}F$. This is because of the fundamentally richer variation of local electronic distributions characteristic of elements in which $d$ orbitals are in the valence shell. Such approximate theories are better able to demonstrate a greater sensitivity to major alterations in valence-shell electronic distributions than to the relatively minor perturbations associated with structural variations involving fluorine or even carbon. On the other hand, prospects for a computational theory of metal nuclide chemical shifts are not nearly as bright at the present time as for the $^{13}C$ or $^{19}F$ cases, [26-28] because of the difficulty of developing computationally useful molecular orbital approaches for atoms with $d$ orbitals.

The wide range of $^{199}Hg$ chemical shifts was first demonstrated by Dharmatti, Kanekar, and Mathur.[29] Using wide-line techniques, they demonstrated a chemical shift difference of more than 2100 ppm between $HgCl_4^{2-}$ and $HgI_4^{2-}$ (the more shielded), with aqueous $Hg(NO_3)_2$ approximately in the middle. Using similar experimental techniques, Schneider and Buckingham[30] carried out a study that included $(CH_3)_2Hg$, for which the shielding was found to be about 1200 ppm lower than that of $HgCl_4^{2-}$ and very sensitive to solvent variation. The results of similar experiments on four dialkyl mercury compounds had been reported by Dessy, Flautt, Jaffe, and Reynolds, [31] who found a chemical shift difference of 640 ppm between dimethylmercury and diisopropylmercury. The INDOR method has been applied to $^{199}Hg$ by several researchers, [1,2] providing $^{199}Hg$ shifts on a variety of organomercury species. This work is also characterized by a wide chemical shift range, and by large solvent effects. Some of the results are included in Tables 23-2 and 23-3.

Recently, the pulse FT method has been used to observe $^{199}Hg$ NMR.[4] Spectra of good quality were generally obtained in reasonable times (few minutes) on samples with concentrations ranging from 0.2 to 2.0 $M$. Line widths varied from about 3 to about 109 Hz. More than thirty chemical shifts, representing

TABLE 23-2.  Some selected $^{199}$Hg chemical shifts[a]

| Compound | Chemical Shift (ppm) | Linewidth (Hz) | Reference | Compound | Chemical Shift (ppm) | Linewidth (Hz) | Reference |
|---|---|---|---|---|---|---|---|
| $(CH_3)_2Hg$ | 2437 | | 1c | $CH_3HgCl$ (0.5 $M$, DMSO)[b] | 1548 | 7 | 4 |
| $(n\text{-}C_3H_7)_2Hg$ | 2197 | | 31 | $CH_3HgBr$ (0.5 $M$, DMSO) | 1430 | 21 | 4 |
| $(C_2H_5)_2Hg$ | 2157 | | 31 | $CH_3HgI$ (0.5 $M$, DMSO) | 1242 | 37 | 4 |
| | 2106 | | 30 | | | | |
| $(C_6H_5CH_2CH_2)_2Hg$ | 2086 | 9 | 4 | $CH_3HgOCOCH_3$ (0.5 $M$, $CH_3CO_2H$) | 1234 | 42 | 4 |
| $(i\text{-}C_3H_7)_2Hg$ | 1796 | | 31 | $HgCl_4^{2-}$ (sat'd, $H_2O$) | 1144 | | 30 |
| $(CH_2{=}CH)_2Hg$ | 1788 | | 2i | $Hg_2(NO_2)_2$(sat'd, $H_2O$) | 744 | | 29 |
| $(C_6H_5)_2Hg$ (0.2 $M$, THF) | 1596 | 6 | 4 | $HgBr_4^{2-}$ (sat'd, $H_2O$) | 553 | | 29 |
| $ClCH{=}CH)_2Hg$ | 1595 | | 1c | $Hg(NO_3)_2$(2 $M$, 0.84 $M$ $HNO_3$) | 23 | 4 | 4 |
| $(C_6F_5)_2Hg$ (sat'd, acetone) | 1505 | | 2f | $Hg(OCOCH_3)_2$ (1.05 $M$, $CH_3CO_2H$) | 0 | 5 | 4 |
| $(CF_2{=}CF)_2Hg$ | 1479 | | 2j | $HgI_4^{2-}$ (0.5 $M$, $H_2O$) | −920 | | 29 |
| $Hg(CN)_2$ (2 $M$, pyridine) | 1319 | 7 | 4 | | | | |

[a]Chemical shifts with respect to a 0.5 $M$ solution of $Hg(OCOCH_3)_2$ in 1.05 $M$ $CH_3CO_2H$. Increasing positive values correspond to decreasing shielding.
[b]DMSO: dimethyl sulfoxide.

358

TABLE 23-3. $^{199}$Hg chemical shifts of some aryl mercury compounds[a]

| Compounds | Chemical Shift (ppm) | Reference | Compounds | Chemical Shift (ppm) | Reference |
|---|---|---|---|---|---|
| $p$-CH$_3$C$_6$H$_4$HgCl | 1283 | 1c | (C$_6$H$_5$)$_2$Hg (1 M, CH$_2$Cl$_2$) | 1694 | 2i |
| $p$-FC$_6$H$_4$HgCl (sat'd, DMSO)[b] | 1263 | 1c | C$_6$H$_5$HgC$_6$F$_5$ (1 M, CH$_2$Cl$_2$) | 1606 | 2i |
| $p$-BrC$_6$H$_4$HgCl (sat'd, DMSO) | 1249 | 1c | (C$_6$F$_5$)$_2$Hg (sat'd, acetone) | 1505 | 2f |
| $p$-ClC$_6$H$_4$HgCl (sat'd, DMSO) | 1243 | 1c | C$_6$H$_5$HgCl (sat'd, DMSO) | 1255 | 1c |
| $p$-CH$_3$OC$_6$H$_4$HgCl (sat'd, DMSO) | 1228 | 1c | C$_6$F$_5$HgOCOCH$_3$ (sat'd, DMSO) | 1004 | 2f |
| $p$-CF$_3$C$_6$H$_4$HgCl (sat'd, DMSO) | 1177 | 1c | C$_6$H$_5$HgOCOCH$_3$ (sat'd, DMSO) | 994 | 2f |

[a] Chemical shifts with respect to a 0.5 M solution of Hg(OCOCH$_3$)$_2$ in 1.05 M CH$_3$CO$_2$H. Increasing positive values correspond to decreasing shielding.
[b] DMSO: dimethyl sulfoxide.

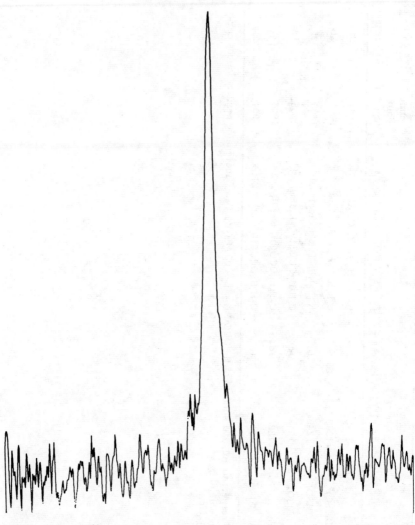

Figure 23-2. [199]Hg pulse FT spectrum of a 1 $M$ solution of $CH_3HgBr$ in pyridine. Result of 1000 40-$\mu$sec pulses under conditions of broadband proton decoupling. Total accumulation time: 409.6 sec. Field locked to [19]F signal of an external capillary of $C_6F_6$ at 84.67088 MHz. Peak position: 16.105336 MHz; spectrum width 625 Hz. Taken from reference 4a.

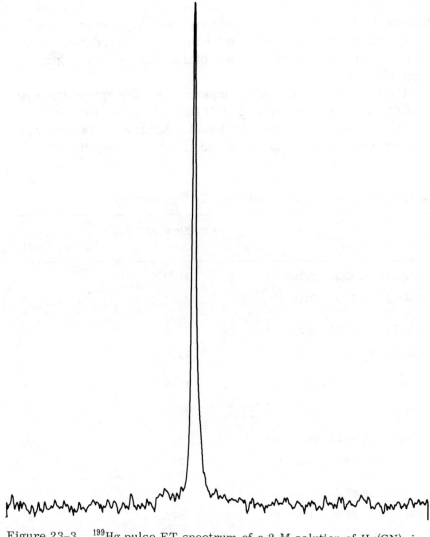

Figure 23-3. $^{199}$Hg pulse FT spectrum of a 2 $M$ solution of Hg(CN)$_2$ in pyridine. Result of 1000 40-$\mu$sec pulses. Total accumulation time: 800 sec. Field locked to $^{19}$F signal of an external capillary of C$_6$F$_6$ at 84.67088 MHz. Peak position, 16.102543 MHz; spectrum width, 625 Hz. Taken from reference 4a.

various combinations of solutes and solvents, were obtained.
Figures 23-2 and 23-3 show some typical $^{199}$Hg spectra using
the system described above and 10-mm tubes. If concentrations
are taken into account other kinds of samples generally give
comparable spectra.

Using a 180°-$\tau$-90° pulse sequence, a $^{199}$Hg spin-lattice relaxation time on a sample of 1.4 $M$ HgCl$_2$ in ethanol was obtained. A value of 1.4 sec was found. Although quantitative
$T_1$ measurements were not made on other samples, the relaxa-

TABLE 23-4. Some high-resolution FT $^{199}$Hg NMR data

| Sample | Chemical Shift[a] (ppm) | Line Width |
|--------|-------------------------|------------|
| CH$_3$HgCl  (0.52 $M$, pyridine) | 1609.0 | 12 |
| CH$_3$HgCl  (0.51 $M$, DMSO)[b] | 1547.5 | 12 |
| CH$_3$HgBr  (0.50 $M$, pyridine) | 1494.5 | 23 |
| CH$_3$HgBr  (0.49, DMSO) | 1429.9 | 21 |
| Hg(CN)$_2$  (2.0 $M$, pyridine) | 1318.9 | 8 |
| Hg(CN)$_2$  (0.41 $M$, pyridine) | 1315.0 | 10 |
| CH$_3$HgI  (0.50 $M$, pyridine) | 1291.5 | 34 |
| CH$_3$HgI  (0.49 $M$, DMSO) | 1242.0 | 37 |
| HgCl$_2$  (0.25 $M$, ethanol) | 895.2 | 3 |
| HgCl$_2$  (1.39 $M$, ethanol) | 877.2 | 3 |
| Hg(NO$_3$)$_2$  (2.0 $M$, 0.84 $M$ HNO$_3$) | 8.9 | 4 |
| Hg(NO$_3$)$_2$  (4.0 $M$, 70% HNO$_3$) | −195.4 | 4 |

[a]Chemical shifts with respect to a 0.5 $M$ solution of Hg(OCOCH$_3$)$_2$ in 1.05 $M$
CH$_3$CO$_2$H. Increasing positive values correspond to decreasing shielding. Taken from reference 4a.
[b]DMSO: dimethyl sulfoxide.

TABLE 23-5. Selected $^{207}$Pb chemical shifts

| Compound | | Chemical Shift[a] (ppm) | Reference |
|---|---|---|---|
| Pb(CH$_3$)$_4$ | (3.6 $M$, toluene) | 2961 | 5a |
| (CH$_3$)$_3$PbCl | (sat'd, CHCl$_3$) | 2111 | 1c |
| Pb$_2$(CH$_3$)$_6$ | (sat'd, C$_6$H$_6$) | 2680 | 2a |
| Pb$_2$(C$_6$H$_5$)$_6$ | (0.3 $M$, CS$_2$) | 2888 | 5a |
| C$_6$H$_5$Pb(CH$_3$)$_3$ | (sat'd, CDCl$_3$) | 2923 | 1c |
| (C$_6$H$_5$)$_2$Pb(CH$_3$)$_2$ | (sat'd, CDCl$_3$) | 2881 | 1c |
| PbI$_4^-$ | (0.81 $M$, conc. HI) | 3942 | 5b |

[a]Chemical shifts in ppm with respect to a 1.0 $M$ aqueous solution of Pb(NO$_3$)$_2$; higher values correspond to lower shielding.


TABLE 23-6. $^{207}$Pb chemical shifts of X–C$_6$H$_4$Pb(CH$_3$)$_3$ compounds[a]

| meta-X | Chemical Shift | para-X | Chemical Shift |
|---|---|---|---|
| H | 0 | H | 0 |
| CF$_3$ | 12 | CF$_3$ | 15 |
| Cl | 11 | Cl | 15 |
| F | 14 | F | 10 |
| OCH$_3$ | 10 | OCH$_3$ | 3 |
| CH$_3$ | 4 | CH$_3$ | 3 |

[a]Chemical shifts in ppm with respect to C$_6$H$_5$Pb(CH$_3$)$_3$; higher values correspond to higher shieldings. Data taken from reference 1c.

tion properties of this sample appears qualitatively to be typical of the substances studied to date.

Moyes and Wells[1c] have studied the effect of cyanide complexation on the [199]Hg chemical shift of methylmercury cyanide, using INDOR. Borzo and Maciel have obtained [199]Hg data on several species in a variety of solvents.[4] Some of the data are given in Table 23-4. These and the earlier results show the great potential of [199]Hg NMR approaches for investigating the structure and dynamics of mercury complexations.

Early studies of [207]Pb NMR by broad-line techniques were reported by Piette and Weaver[32] and by Rocard, Bloom, and Robinson.[33] These studies, employing both solid and liquid samples, demonstrated a [207]Pb chemical shift range of about 8000 ppm (excluding Pb metal). A similar study by Schneider and Buckingham[30] demonstrated that the [207]Pb chemical shift of tetramethyllead is roughly in the middle of the chemical shift range. The INDOR method has also been applied to [207]Pb NMR.[1c, 2] The sensitivity of [207]Pb shielding in organolead compounds to structure variation is high, as one has come to expect for heavy metals. Some of the published data are summarized in Tables 23-5 and 23-6.

The FT approach has been applied to the [207]Pb NMR of several systems[5, 34]; Table 23-7 summarizes some of the data. Figures 23-4 and 23-5 show some typical [207]Pb FT NMR spectra. Considered together with the very large structural selectivity of [207]Pb chemical shifts obtained from the earlier wide-line and INDOR work, the data of Table 23-7 demonstrate that even rather minor structural variations give rise to large chemical shifts, and demonstrate the promise of [207]Pb chemical shift measurements for studies of structure and dynamics in lead compounds and for analytical applications. In addition, very large effects of solvent and temperature variation (e.g., about 0.5 ppm decrease in shielding per degree increase in temperature for 2 $M$ triisobutyllead acetate in acetic acid) on the [207]Pb chemical shifts of lead compounds have been observed in this work,[35b] implying that significant changes in species identity and/or environment accompany such variations. Some

TABLE 23-7. Typical high-resolution FT $^{207}$Pb NMR data[a]

| Compound | Solvent (Molar Conc.) | Line Width (Hz) | Chemical Shift (ppm)[b] | $T_1$ (sec) |
|---|---|---|---|---|
| Pb(CH$_3$)$_4$ | Toluene (3.7) | 2 | 2961.2 | 0.6 |
| Pb(C$_2$H$_5$)$_4$ | Neat | 6 | 3031.8 | 1.2 |
| Pb(CH$_3$)$_3$OCOCH$_3$ | Acetic acid (0.8) | 6 | 3369.2 | |
| Pb(C$_2$H$_5$)$_3$OCOCH$_3$ | Acetic acid (2.0) | 8 | 3373.2 | |
| Pb($n$–C$_3$H$_7$)$_3$OCOCH$_3$ | Acetic acid (1.0) | 12 | 3390.0 | |
| Pb($n$–C$_4$H$_9$)$_3$OCOCH$_3$ | Acetic acid (2.0) | 12 | 3385.2 | |
| Pb($i$–C$_4$H$_9$)$_3$OCOCH$_3$ | Acetic acid (2.0) | 9 | 3393.2 | 0.08 |
| Pb(C$_6$H$_5$)$_3$C$_2$H$_5$ | Carbon tetrachloride (0.5) | 11 | 2846.7 | |
| Pb(OCOCH$_3$)$_2$ | Water (1.1) | 100 | 1624.2 | |
| Pb(OCOCH$_3$)$_2$ | Acetic acid (2.1) | 100 | 1442.2 | |
| Pb(NO$_3$)$_2$ | Water (1.0) | 12 | 0.0 | 1.1 |

[a]Results taken from reference 5.
[b]Chemical shifts given with respect to a 1.0 $M$ aqueous solution of Pb(NO$_3$)$_2$; higher values correspond to lower shielding.

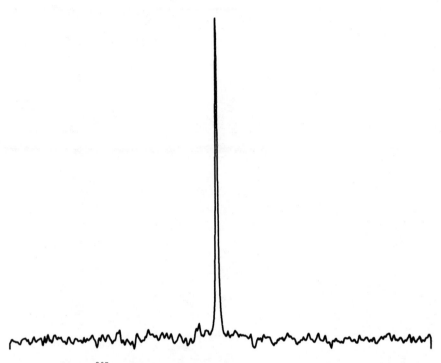

Figure 23-4.   $^{207}$Pb pulse FT spectrum of a 3.7 $M$ solution of $(CH_3)_4Pb$ in toluene under conditions of broad-band proton decoupling. Result of one 90° pulse. Field locked to $^{19}$F signal of an external capillary of $C_6F_6$ at 84.67088 MHz. Peak position, 18.828633 MHz; spectrum width, 750 Hz. Taken from reference 5a.

details of these observations are given in Tables 23-8 and 23-9. Systematic studies of these types of phenomena should elucidate the detailed nature of structural alterations and the dynamics and thermodynamics of complexation equilibria in these and other systems. Moyes and Wells[1c] have employed the $^{207}$Pb-$\{^1H\}$ INDOR technique to study the dissociation and complexation equilibria of trimethyllead chloride in aqueous solution, demonstrating the utility of metal-nuclide NMR for such investigations.

Spin-lattice relaxation times have been determined by a 180°-$\tau$-90° method for some of the samples indicated in Table 23-7.[35a] These $T_1$ values range typically from about 0.1 to 2

Figure 23-5. $^{207}$Pb pulse FT spectrum of a 3.7 $M$ solution of $(CH_3)_4Pb$ in toluene with no proton decoupling. Result of 100 pulses with 0.4 sec between pulses. Total accumulation time: 40 sec. Field locked to $^{19}$F signal of an external capillary of $C_6F_6$ at 84.67088 MHz. Peak position, 18.828633 MHz; spectrum width, 750 Hz. Taken from reference 5a.

TABLE 23-8. Temperature dependence of $^{207}$Pb chemical shifts of $(i$-butyl)$_3$ PbOCOCH$_3$ (2 $M$ in CH$_3$CO$_2$H)[a]

| Temperature (°K) | Chemical Shift (ppm) |
|---|---|
| 320 | 2533 |
| 310 | 2539 |
| 300 | 2543 |
| 290 | 2548 |
| 280 | 2554 |
| 270 | 2559 |

[a]Taken from reference 5b. Chemical shifts with respect to a 1.0 $M$ aqueous Pb(NO$_3$)$_2$ solution.

TABLE 23-9.  Solvent effect on $^{207}$Pb chemical shifts of Pb(CH$_3$)$_4$ and Pb(CH$_2$CH$_3$)$_4$ (each 0.36 $M$)[a]

| Solvent (Including 10% Toluene) | Chemical Shift[b] (ppm) | | |
|---|---|---|---|
| | Pb(CH$_3$)$_4$ | Pb(C$_2$H$_5$)$_4$ | Difference |
| Carbon tetrachloride | 2960.9 | 3032.3 | −71.4 |
| Acetone | 2962.7 | 3029.7 | −67.0 |
| Carbon disulfide | 2954.1 | 3032.7 | −78.6 |

[a]Data taken from reference 5b.
[b]Chemical shifts given with respect to a 1.0 $M$ aqueous solution of Pb(NO$_3$)$_2$; higher values correspond to lower shielding.

sec.  These values are remarkably short, in view of the generally larger $T_1$'s for $^{13}$C and $^{29}$Si, especially the latter in analogous compounds.[10,11] With no possibility of quadrupole relaxation mechanisms, this implies substantial contributions to spin-lattice relaxation from mechanisms such as chemical exchange, chemical shift anisotropy, and spin rotation. However, it should be kept in mind that the $T_1$ values reported in Table 23-7 are properties of *typical samples*, not scrupulously purified materials.

The $^{113}$Cd nuclide is one of the nuclides that has been largely neglected from the point of view of nuclear magnetic resonance.  Preliminary pulse FT experiments on this nuclide have been reported using essentially the same techniques that have been described above.[6] Figures 23-6 and 23-7 show some typical $^{113}$Cd spectra.

The spin-lattice relaxation times for a $(n\text{-}C_4H_9)_2$Cd sample and a saturated aqueous solution of CdCl$_2$ give values of about 0.2 and 16 sec, respectively.[6] This range appears to be typical of the samples studied so far and indicates that common $^{13}$C NMR approaches will generally be suitable for $^{113}$Cd studies. Although a quantitative nuclear Overhauser effect was not determined for $(n\text{-}C_4H_9)_2$Cd in this study, there is clearly not a signal inversion.

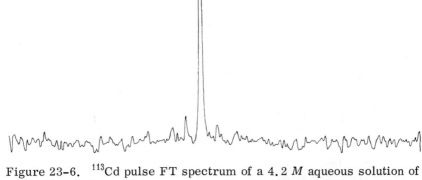

Figure 23-6. $^{113}$Cd pulse FT spectrum of a 4.2 $M$ aqueous solution of Cd(ClO$_4$)$_2$. Result of 25 90° pulses. Total accumulation time: 500 sec. Field locked to $^{19}$F signal of an external capillary of C$_6$F$_6$ at 84.67088 MHz. Peak position, 19.960986 MHz; spectrum width, 625 Hz. Taken from reference 6.

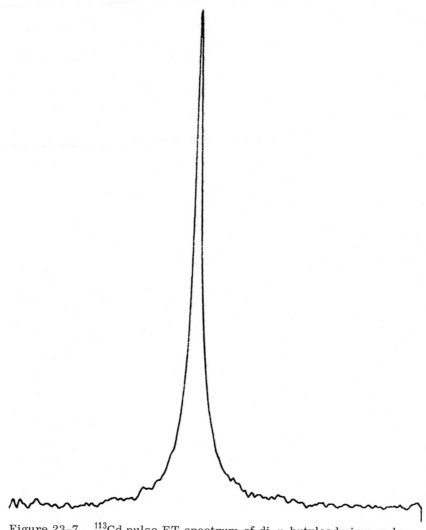

Figure 23-7. $^{113}$Cd pulse FT spectrum of di-*n*-butylcadmium under conditions of broad-band proton decoupling. Result of 1000 pulses. Total accumulation time: 100 sec. Field locked to $^{19}$F signal of a capillary of $C_6F_6$ at 84.67088 MHz. Peak position, 19.970900 MHz; spectrum width, 1250 Hz.

TABLE 23-10.  Typical high-resolution $^{113}$Cd NMR data

| Sample | Chemical Shift[a] (ppm) | Line Width (Hz) |
|---|---|---|
| $(CH_3CH_2CH_2CH_2)_2Cd$[b] | 489.11 | 15 |
| 3.8 $M$ aqueous $CdCl_2$ | 124.75 | 4 |
| 0.1 $M$ aqueous $CdCl_2$ | 57.03 | 5 |
| 0.1 $M$ aqueous $Cd(ClO_4)_2$ | 0 | 3 |
| 0.1 $M$ aqueous $Cd(NO_3)_2$ | $-3.72$ | 4 |
| 4.2 $M$ aqueous $Cd(ClO_4)_2$ | $-8.11$ | 5 |
| 4.5 $M$ aqueous $Cd(NO_3)_2$ | $-49.41$ | 4 |

[a]With respect to 0.1 $M$ aqueous solution of $Cd(ClO_4)_2$; increasing values correspond to decreasing shielding.  All spectra recorded at 298 °K unless otherwise indicated.  Taken from reference 6.
[b]228 °K.

Chemical shift values[4b, 6] for a few species are reported in Table 23-10.  These results display the wide range of chemical shift values one has come to expect of heavy metal nuclides. Of special interest are the results obtained from some aqueous solutions of cadmium salts.  It is seen that the $Cd(ClO_4)_2$ results at 0.1 $M$ and 4.2 $M$ are nearly the same (only a 8.11 ppm difference), whereas the analogous $Cd(NO_3)_2$ results differ by 45.69 ppm, and the $CdCl_2$ chemical shifts differ by 67.72 ppm. Together with the fact that the chemical shifts of 0.1 $M$ aqueous solution of $Cd(NO_3)_2$ and $Cd(ClO_4)_2$ and a 4.2 $M$ solution of $Cd(ClO_4)_2$ in water are within 8 ppm of each other, whereas an aqueous 0.1 $M$ $CdCl_2$ solution gives a chemical shift that differs from these by about 57 ppm, this leads to the conclusion that the cadmium exists primarily as hydrated $Cd^{2+}$ in the former three solutions mentioned, and is more strongly involved in complexation with chloride in the 0.1 $M$ $CdCl_2$ solution.  These results are in qualitative agreement with prevailing views of the structure of cadmium salts in solution.

The high sensitivity of $^{113}$Cd chemical shifts of cadmium salts in solution to changes in the concentration and/or identity of the salt provides a means of studying the equilibrium and dynamics of species interconversion in such systems. Kostelnik and Bothner-By have given a preliminary report of such studies, based on cw experiments.[35]

From preliminary data on the quadrupolar nuclei, $^{25}$Mg, $^{43}$Ca and $^{67}$Zn, it appears that relaxation data may be extremely useful in complexation studies.[22d,f; 36, 37] Chemical shifts are rather small, especially for the former two.

## References

1. The following reviews contain references to most of the early work.

   (a) P. C. Lauterbur, in *Determination of Organic Structures by Physical Methods*, F. C. Nachod and W. D. Phillips, Eds., Vol. 2, Academic Press, New York, 1962.

   (b) J. W. Emsley, J. Feeney, and L. H. Sutcliffe, *High Resolution Nuclear Magnetic Resonance Spectroscopy*, Vol. 2, Pergamon Press, Oxford, 1966, Chap. 12.

   (c) P. R. Wells, in *Determination of Organic Structures by Physical Methods*, F. C. Nachod and J. J. Zuckerman, Eds., Vol. 4, Academic Press, New York, 1971.

2. The following references are representative of recent work in this area.

   (a) R. J. H. Clark, A. G. Davies, R. J. Puddlephatt, and W. McFarlane, *J. Am. Chem. Soc.*, **91**, 1334 (1969).

   (b) P. D. Godfrey, M. L. Heffernan, and D. F. Kerr, *Aust. J. Chem.*, **17**, 701 (1964).

   (c) K. A. McLauchlan, D. H. Whiffen, and L. W. Reeves, *Mol. Phys.*, **10**, 131 (1966).

   (d) V. S. Petrosyan, B. P. Bespalov, and O. A. Reutov, *Izv. Akad. Nauk SSSR, Ser. Khim.*, **1967**, 2127.

(e)  F. A. L. Anet and J. L. Sudmeier, *J. Magn. Resonance*, **1**, 124 (1969).

(f)  W. McFarlane, *J. Chem. Soc.*, **1968**, 2280.

(g)  R. R. Dean and J. C. Green, *J. Chem. Soc.*, **1968**, 3047.

(h)  P. J. Banney, D. C. McWilliam, and P. R. Wells, *J. Magn. Resonance*, **2**, 235 (1970).

(i)  A. P. Tupčiauskas, N. M. Sergeyev, Y. A. Ustynyuk, and A. N. Kashin, *J. Magn. Resonance*, **7**, 124 (1972).

(j)  R. B. Johannesen and R. W. Duerst, *J. Magn. Resonance*, **5**, 355 (1971).

(k)  A. P. Tupčiauskas, N. M. Sergeyev, and Y. A. Ustynyuk, *Org. Magn. Resonance*, **3**, 655 (1971).

(l)  W. McFarlane and R. J. Wood, *J. Organomet. Chem.*, **40**, C17 (1972).

(m)  L. Verdonck and G. P. Van Der Kelen, *J. Organomet. Chem.*, **40**, 139 (1972).

(n)  P. J. Smith, R. F. M. White, and L. Smith, *J. Organomet. Chem.*, **40**, 341 (1972).

(o)  K. Hildenbrand and H. Dreeskamp, *Z. Phys. Chem.*, **69**, 171 (1970).

3.  D. D. Traficante and J. A. Simms, *Rev. Sci. Instr.*, **43**, 1122 (1972).

4.  (a)  G. E. Maciel and M. Borzo, *J. Magn. Res.*, **10**, 388 (1973).

    (b)  M. Borzo and G. E. Maciel, unpublished results.

5.  (a)  G. E. Maciel and J. Dallas, *J. Am. Chem. Soc.*, **95**, 3039 (1973).

    (b)  J. Dallas and G. E. Maciel, unpublished results.

6.  G. E. Maciel and M. Borzo, *Chem. Commun.*, 394 (1973).

7.  P. D. Ellis, H. C. Walsh, C. S. Peters, and R. Codrington, *J. Magn. Res.*, **11**, 426 (1973).

8.  J. H. Noggle and R. E. Schirmer, *The Nuclear Overhauser Effect—Chemical Applications*, Academic Press, New York, 1971.

9.  R. Scholl, G. E. Maciel, and W. K. Musker, *J. Am. Chem. Soc.*, **94**, 6376 (1972).

10. G. C. Levy, *J. Am. Chem. Soc.*, **94**, 4793 (1972).

11. S. Barcza and N. Engstrom, *J. Am. Chem. Soc.*, **94**, 1792 (1972).

12. R. Freeman, K. G. R. Pachler, and G. N. LaMar, *J. Chem. Phys.*, **55**, 4586 (1971).

13. G. C. Levy, see Chapter 17 of this volume.

14. R. Freeman, H. D. W. Hill, and R. Kaptein, *J. Magn. Resonance*, **7**, 327 (1972).

15. R. F. Spanier, T. Vladimiroff, and E. R. Malinowski, *J. Chem. Phys.*, **45**, 4355 (1966).

16. E. R. Malinowski and A. R. Pierpaoli, *J. Magn. Resonance*, **1**, 509 (1969).

17. J. K. Becconsall, G. D. Daves, Jr., and W. R. Anderson, Jr., *J. Am. Chem. Soc.*, **92**, 430 (1970).

18. M. Bacon and G. E. Maciel, *J. Am. Chem. Soc.*, **95**, 2413 (1973).

19. D. D. Traficante, abstract of 13th Experimental NMR Conference, Asilomar, California, April, 1972.

20. In addition to the references cited in Ref. 1, the following are representative of more recent work.

   (a) H. W. Spiess and R. K. Sheline, *J. Chem. Phys.*, **53**, 3036 (1970).

   (b) S. Fujiwara, F. Yajima, and A. Yamasaki, *J. Magn. Resonance*, **1**, 203 (1969).

21. In addition to Refs. 2*l*, *m*, and *n*, the reader is referred to the earlier work cited in Ref. 1.

22. (a) G. W. Canters, *J. Am. Chem. Soc.*, **94**, 5230 (1972).

   (b) D. M. Chen and L. W. Reeves, *J. Am. Chem. Soc.*, **94**, 4384 (1972).

   (c) E. Shchori, J. Jagur-Grodzinski, Z. Luz, and M. Shporer, *J. Am. Chem. Soc.*, **93**, 7133 (1971).

   (d) R. Bryant, *J. Am. Chem. Soc.*, **91**, 1870 (1969).

   (e) E. M. Arnett, H. C. Ro, and R. J. Minaz, *J. Phys. Chem.*, **76**, 2474 (1972).

   (f) R. Bryant, *J. Magn. Resonance*, **6**, 159 (1972).

23. (a) C. J. Jameson and H. S. Gutowsky, *J. Chem. Phys.*, **40**, 1714 (1964).

(b)  J. H. Letcher and J. R. Van Wazer, *J. Chem. Phys.*, **44**, 815 (1966).

24.  M. Karplus and T. P. Das, *J. Chem. Phys.*, **34**, 1683 (1961).

25.  M. Karplus and J. A. Pople, *J. Chem. Phys.*, **38**, 2803 (1963).

26.  R. Ditchfield, D. P. Miller, and J. A. Pople, *J. Chem. Phys.*, **54**, 4861 (1970).

27.  R. Ditchfield, *J. Chem. Phys.*, **56**, 5688 (1972).

28.  P. D. Ellis, G. E. Maciel, and J. W. McIver, Jr., *J. Am. Chem. Soc.*, **94**, 4069 (1972).

29.  S. S. Dharmatti, C. R. Kanekar, and S. C. Mathur, *Proc. U. N. Intern. Conf. Peaceful Uses At. Energy*, 2nd, **28**, 644 (1958).

30.  W. G. Schneider and A. D. Buckingham, *Discussions Faraday Soc.*, **34**, 147 (1962).

31.  R. E. Dessy, T. J. Flautt, H. H. Jaffe, and G. F. Reynolds, *J. Chem. Phys.*, **30**, 1422 (1959).

32.  L. H. Piette and H. E. Weaver, *J. Chem. Phys.*, **28**, 735 (1958).

33.  J. M. Rocard, M. Bloom, and L. B. Robinson, *Can. J. Phys.*, **28**, 735 (1959).

34.  O. Lutz and G. Stricker, *Physics Lett.*, **35A**, 397 (1971).

35.  R. J. Kostelnik and A. A. Bothner-By, Abstract, 162nd National A.C.S. Meeting, Washington, D.C., September, 1971.

36.  (a)  O. Lutz, A. Schwenk and A. Uhl, *Z. Naturforsch.*, **28a**, 1534 (1973).

     (b)  B. W. Epperlein, H. Krüger, O. Lutz and A. Schwenk, *Phys. Letters*, **45A**, 255 (1973).

37.  L. Simeral and G. E. Maciel, unpublished results.

CHAPTER 24

# NMR STUDIES OF NONHYDROGEN NUCLEI
## IN ORIENTED MOLECULES

P. Diehl

*Department of Physics*
*University of Basel*

## INTRODUCTION

Although the field of NMR in oriented molecules, [1,2] which was initiated in 1963, [3] immediately concentrated on [1]H spectra, the basic papers by Saupe and Englert[4] pointed out the possibilities of obtaining valuable information from [2]D spectra. In 1965 the first [19]F spectra of oriented molecules were recorded. [5] The expansion to other nuclei followed more slowly and only isolated examples were published: 1965, [2]D[22]; 1967, [13]C[6]; 1968, [35]Cl[7]; 1970, [14]N[8]; 1971, [23]Na[30]; and 1972, [31]P[9].

In this chapter we first discuss the differences between normal NMR spectroscopy and the spectroscopy of oriented molecules, and point out the various possibilities and difficulties of the latter. Then we treat the individual nuclei beginning with spin $I = \frac{1}{2}$ in the order of their importance: [19]F, [13]C, and [31]P. Further on the discussion is expanded to $I > \frac{1}{2}$ nuclei and concentrates on [2]D, [14]N, and finally, [35]Cl, as well as [23]Na.

## THE POSSIBILITIES OF NMR IN ORIENTED MOLECULES

If we consider that practically all the NMR parameters, that is, the chemical shifts and indirect, direct, and quadrupole couplings, change with the orientation of the molecules with respect to the applied external magnetic field, it becomes obvious that a wealth of information must be lost in normal NMR spectroscopy. Only traces of the shift and indirect-coupling tensors are observed, and the direct and quadrupole couplings are averaged to zero and affect the spectra only via relaxation processes. In the NMR of oriented molecules, on the other hand, all the anisotropic parameters are observable in principle, and the following very impressive table of measurement possibilities can be made.

1. Relative distances and absolute angles (molecular shape)
2. Intramolecular motion (rotation and vibration)
3. Molecular orientation
4. Anisotropy of chemical shift
5. Anisotropy of indirect coupling
6. Quadrupole coupling constant
7. Observation of indirect coupling in magnetically equivalent but not fully equivalent groups
8. Observation of indirect coupling by removal of deceptive simplicity

After inspection of this table one wonders at first why everybody is not immediately entering this extremely promising field. There are various reasons why not.

The experimental problem is usually not too difficult; the samples must be studied dissolved in liquid crystals, preferably at room temperature and at concentrations of approximately 20 mole %. The sample temperature must be kept extremely constant in time and over the sample volume, because a temperature change affects the orientation as much as $1/1000$ per $0.1\,^{\circ}C$.

The first part of the spectral analysis, that is, the iterative determination of the spectral parameters, may be time-consuming but does not constitute a principal barrier. However, the second part, that is, the evaluation of the structural and motional information and of the various anisotropies, is much more problematical for the following reasons:

1. Quite generally, the number of unknowns is large so that it is often difficult to find overdetermined, or even just determined, systems. Consequently, assumptions must be made, and information from different spectroscopies often must be considered simultaneously.

2. The precision in structure determination is affected by molecular vibration so that the NMR average distances, $\langle r^{-3} \rangle^{-1/3}$, may differ from microwave data, for example.

3. The anisotropies of the indirect couplings have the same directional dependence as the direct couplings. For this reason the observed "direct coupling" may contain an unknown amount of anisotropic indirect coupling which is difficult to separate out.

4. Measured anisotropies in chemical shifts are affected by unknown and uncontrollable solvent effects. This is particularly embarrassing for small anisotropies and small molecular orientations.

5. Measurements of quadrupole couplings are based on simultaneous determinations of the molecular orientation, which in turn depends upon an absolute internuclear distance which cannot be obtained from NMR studies.

It can be concluded that the table of difficulties is almost as impressive as the table of possibilities. Whereas for [1]H NMR it is mainly the points about vibration and shift anisotropy that are important, all the other factors also affect nonproton NMR. In this chapter we therefore devote a lot of attention to problems and difficulties.

## DISCUSSION OF THE VARIOUS NON-H SPECTROSCOPIES

### $^{19}$F Spectroscopy

*The Anisotropy of the Indirect F-F Coupling.* One of the most obvious tests of the $r^{-3}$ dependence of the direct couplings is their ratio in hexafluorobenzene, which should be $D_{ortho} : D_{meta} : D_{para} = 1 : \frac{1}{3} \cdot \sqrt{3} : \frac{1}{8} = 1 : 0.1925 : 0.1250$. Whereas this ratio is found as predicted in normal benzene, it deviates considerably in hexafluorobenzene $(1 : 0.1872 : 0.1338)$.[10] This discrepancy has been attributed to an anisotropic indirect coupling $(\Delta J = D_{exp} - D_{calc})$ which may arise from the orbital term of the $p$ electrons involved. Unfortunately, the anisotropies of these indirect couplings cannot be determined separately. For the four unknowns (orientation, and three anisotropies) there are only three measurable couplings. In principle, one way out of this difficulty is the assumption that there are molecules with identical anisotropies, the experimental data of which may be combined. However, this assumption is difficult to prove.

In systems of lower symmetry there is an additional possibility which consists of a combination of data for the same molecule at various orientations. Experimentally, this variation may be obtained by a change of temperature, concentration, or liquid crystal. Of course it must be assumed here that the changes do not affect the molecular shape or anisotropy.

Summarizing the situation, it must be stated that in each method of coupling anisotropy determination we have some uncertainties and "if's." This conclusion is illustrated by the considerable controversy that exists on the magnitude of the relevant tensor elements in fluorobenzenes. The errors are large in all cases; still there seems to be a certain trend. Usually $\Delta J_{ortho}$ is small, $\Delta J_{meta}$ is zero or negative, and $\Delta J_{para}$ is positive.[11] As pointed out, the situation is more hopeful in systems of lower symmetry. A typical example is 1, 1-difluoroethylene,[12] which has been studied at various concentrations and temperatures with the following results:

$$J_{xx} = -720 \pm 181 \text{ Hz}$$

$$J_{yy} = +339 \pm 181 \text{ Hz}$$

$$J_{zz} = +478 \pm 115 \text{ Hz},$$

with:

$$J_{FF} = \tfrac{1}{3}(J_{xx} + J_{yy} + J_{zz}) = 32.5 \text{ Hz}.$$

The theoretical prediction[13] is

$$J_{\mu\nu}(\text{Hz}) = \begin{vmatrix} -93 & 2 & 0 \\ 2 & 69 & 0 \\ 0 & 0 & 87 \end{vmatrix}.$$

Obviously the situation here is not clear yet. The observation that the experimental results also depend upon the liquid crystal used as solvent certainly adds to the trouble.

This pessimistic point of view does not claim that such measurements are impossible, but many more data of higher precision must be collected and the various influences carefully studied.

*The Anisotropy of Chemical Shifts.* The measurement of the chemical shift anisotropy contributions (i.e., combination of tensor elements) in F spectroscopy is straightforward and solvent effects are not very important. However, the determination of the individual chemical shift tensor elements, which are of considerable theoretical interest, is again difficult because usually data from several molecules have to be combined. As a typical illustration the results for the anisotropy in C-F for fluorobenzenes are presented in Table 24-1. The most reliable data seem to be those of the second column as they avoid ortho effects and non-F substituents. (The ortho effect increases the shielding perpendicular to the C-F bond and decreases it along the bond.)

A comparison of the shift anisotropy data with theory, which predicts a proportionality of the elements with C-F bond ionicity, double bond character, and the degree of $sp$ hybridization, does not seem very promising. The theory predicts[15]:

TABLE 24-1.  Principal values of the anisotropic part of the fluorine shift tensor in fluorobenzenes ($\hat{\sigma}' = \hat{\sigma} - \frac{1}{3} \operatorname{Tr} \hat{\sigma}$) and anisotropic part of the ortho effect increments ($\Delta\sigma' = \Delta\sigma - \frac{1}{3} \operatorname{Tr}\Delta\sigma$)

|  | 11-$p$-substituted Fluorobenzenes[14] | Combination of $C_6H_5F$ and 1,3-$C_6H_4F_2$[15] | Combination of $C_6H_5F$ and $C_6F_6$ |
|---|---|---|---|
| $\sigma_{xx}'$ (pmm) | $-5$ | $+30 \pm 20$ | $+75$ |
| $\sigma_{yy}'$ (ppm) | $-35$ | $-100 \pm 30$ | $-180$ |
| $\sigma_{zz}'$ (ppm) | $+40$ | $+70 \pm 10$ | $+105$ |

Anisotropy of ortho effects:

$$\Delta\sigma_{xx}' \qquad -55 \pm 30 \text{ ppm}$$
$$\Delta\sigma_{yy}' \qquad +35 \pm 30$$
$$\Delta\sigma_{zz}' \qquad +20 \pm 10$$

$$\sigma_{xx}' = +140 \text{ ppm}$$
$$\sigma_{yy}' = -130 \text{ ppm}$$
$$\sigma_{zz}' = -\ 10 \text{ ppm}.$$

## $^{13}$C Spectroscopy

In this field there have been very few measurements. The earliest was a determination of the $^{13}$C-shift anisotropy in $H_3^{13}CI$ by means of $H - \{^{13}C\}$ heteronuclear tickling.[6] The result of $-30 \pm 3$ ppm for $\Delta\sigma$ disagrees with a later one of $-75 \pm 20$ ppm.[16]

There are only two papers in which the $^{13}$C resonances of larger molecules have been studied directly. The first is on natural abundance 1,3,5-trichlorobenzene for which 86,000 pulses of 40 $\mu$sec duration and a repetition rate of 0.4 sec were collected so that the total measuring time was 6 hr.[17] The authors at-

tribute the contradictions in internuclear distance, which are
of the order of 0.02–0.03 Å, to isotope effects in the orientation or to anisotropy of the C-H couplings. However, because
vibration corrections have not been applied, these conclusions
are not valid.

In the second paper the spectrum of $^{13}C$-enriched benzene
was analyzed.[18] In 71 hr measuring time 374,000 pulses of
4 $\mu$sec were collected at a repetition rate of 0.68 sec. The
spectrum obtained is of very high quality and the $^{13}C$ anisotropy
of $\Delta\sigma = +190 \pm 4$ ppm determined compares favorably with data
from a polycrystalline sample ($\Delta\sigma = +180$ ppm).

On the whole, in spite of the considerable experimental difficulties, $^{13}C$ spectroscopy of oriented molecules seems to be
very promising. However, it should be noted that particularly
for the small bond length C-H, vibration corrections become
important if NMR data are compared with results from different spectroscopic measurements.[19]


## $^{31}P$ Spectroscopy


Only preliminary measurements have been made in our
group at Basel.[9] The main problems seem to be poor solu-
bility and sensitivity. On the other hand, the anisotropies of
chemical shift and indirect coupling will be of considerable interest.


## $^{2}D$ Spectroscopy


As pointed out in the introduction there are, for nuclei with
spin $I > \frac{1}{2}$, two possible types of measurements in oriented molecules. The observed doublet splitting $\Delta$ for deuterons is a
measure of the degree of orientation as well as of the quadrupole coupling.

$$\Delta = \left(\frac{3}{2}\right)\left(\frac{eQV_{zz}}{h}\right)\left(\frac{1}{S_{ij}}\right)$$

$$S_{ij} = -\left(\frac{4\pi^2 r_{ij}{}^3}{h\gamma_i\gamma_j}\right) \cdot D_{ij} \ .$$

Consequently, either on the basis of an assumed absolute distance $(r_{ij})$ the quadrupole coupling $(eQV_{zz}/h)$ is determined via a measured orientation $(S_{ij})$, or from a known quadrupole coupling the orientation is obtained directly. Both methods are of considerable interest. Whereas earlier the quadrupole coupling constants of molecules in liquids had to be determined from the quadrupole relaxation, with all the corresponding well-known problems of the correlation time interfering, the coupling is now measurable directly. Its precision is actually determined predominantly by the uncertainty of the assumed absolute distance.

As an example, for monodeuterobenzene we have obtained[20]

$$\left(\frac{eQV_{zz}}{h}\right) = 196.5 \pm 1.3 \text{ kHz}$$

for

$$(r_{HH})_{ortho} = 2.481 \pm 0.005 \text{ Å}.$$

Similarly, the variation of the coupling with increasing substitution of $NO_2$ groups in benzene has been observed[21]:

|  | $eQV_{zz}/h$ |
|---|---|
| Nitrobenzene | $192 \pm 4$ kHz |
| 1, 3-Dinitrobenzene | $187 \pm 4$ kHz |
| 1, 3, 5-Trinitrobenzene | $181 \pm 4$ kHz . |

In the second type of measurement, that is, the determination of orientation, protons can also be used, with a known distance instead of a known quadrupole coupling as a basis. With deuterons, however, we have all the well-known advantages of the method: due to the reduction in the direct couplings by factors of 6.5 for H-D and 42 for D-D, the spectra are simple. Well-defined positions in larger molecules may be studied by $^2$D labeling. Finally, the "sensitivity" of deuterons for orientation is larger than that of protons; this may be seen in monodeuterobenzene where the largest observed direct H-H coupling of approximately 400 Hz is exceeded by far by the corresponding quadrupole splitting of 15,000 Hz.[20]

The first application of $^2$D resonance in an orientation study was performed relatively early on pure liquid crystals[22] in which the side-chain conformation was determined.

Later on, when for the first time lyotropic nematic liquid crystals (i.e., soaps) were used as solvents[23] instead of thermotropic ones, $^2$D resonance again helped to elucidate the nature and structure of the oriented phase. With dilution of, say, $D_2O$ in such a soap, at first the $^2$D spectra corresponded to a powder type continuous distribution of O-D orientations. With time in the magnetic field the spectra changed to a sharp doublet, known from thermotropic solvents, and corresponding to an average orientation. The observation that a slight rotation of the sample by a few degrees again drastically changed the spectrum can be explained by the orientation of the liquid crystal with a planar distribution of optical axis at right angles to the magnetic field. With constant sample rotation, on the other hand, a uniaxial orientation along the sample axis at right angles to the applied field can be obtained.

As these examples demonstrate, not only the orientation but also the dynamics of orientation and, of course, differences in orientations between various parts in complex molecules may, in principle, be studied. It is obvious that such a method can be a powerful tool in the study of biological systems. Still only very few measurements have been made. In one, a unique confirmation of the structure of collagen has been derived;[24] in another, an attempt was made to study chain mobility in membranes.[25] Although at present such studies are perhaps still qualitative in nature, they will doubtlessly become more quantitative in the future.

## $^{14}$N Spectroscopy

What has been stated for $^2$D resonance is also valid for $^{14}$N resonance. Again quadrupole couplings have been determined. In $CH_3CN^{26}$ the coupling in the liquid crystal solvent (3.6 MHz) was found to be much closer to the value for the solid (3.74 MHz) than for the gas (4.35 MHz).

$^{14}$N resonance has also been used to measure very weak orientations of the order of $S = 3 \times 10^{-4}$ which arise in dipolar liquids oriented by applied electric fields.[27]

In $CH_3NC$ a considerable shift anisotropy of $\sigma_\parallel - \sigma_\perp = 360 \pm 70$ ppm has been detected.[8]

## $^{35}$Cl Spectroscopy

With the increased quadrupole coupling of $^{35}$Cl (40 MHz) it would seem that the sensitivity to orientation is still better. This is not correct, however. Because the line width is proportional to the square of the quadrupole coupling and the splitting only linearly depends upon it, the sensitivity, which may be defined as splitting divided by line width, depends upon the inverse coupling. Consequently, deuterons are the ideal nuclei for such studies. Still, some measurements have been made with $^{35}$Cl. They concentrate on a structural study of oriented phases, in one case of the lyotropic type,[28] and in others of the system poly-$\gamma$-benzyl-$l$-glutamate in dichloromethane and 1,2-dichloroethane.[7,29] Finally, $^{23}$Na NMR has been applied to similar problems, that is, for a determination of structure of lyotropic lamellar phases as well as to studies of ion binding to membrane surfaces.[30,31]

## CONCLUSIONS

To conclude, we refer to the introduction. Undoubtedly NMR of oriented molecules is a field that provides a wealth of new parameters that are unobtainable, or as yet unobtainable by conventional methods. Many of these are, as we hope to have shown, particularly interesting for non-$^1$H NMR. On the other hand, the complexity of the spectroscopy has also increased considerably. Obviously many problems are still to be solved, and many extremely precise and time-consuming measurements will have to be made. Even so, we are convinced that this field is one of the very promising non-$^1$H domains of NMR.

## References

1. P. Diehl and C. L. Khetrapal, *NMR—Basic Principles and Progress*, Vol. 1, Springer-Verlag, Berlin, Heidelberg, New York, 1969.
2. P. Diehl and P. M. Henrichs, *Specialist Periodical Report—NMR*, The Chemical Society, London, 1972.
3. A. Saupe and G. Englert, *Phys. Rev. Lett.*, **11**, 462 (1963).
4. G. Englert and A. Saupe, *Z. Naturforsch.*, **19a**, 172 (1964).
5. L. C. Snyder, *J. Chem. Phys.*, **43**, 4041 (1965).
6. C. S. Yannoni et al., *J. Chem. Phys.*, **47**, 2508 (1967).
7. D. Gill et al., *J. Am. Chem. Soc.*, **90**, 6870 (1968).
8. C. S. Yannoni, *J. Chem. Phys.*, **52**, 2005 (1970).
9. P. Diehl and J. Vogt, unpublished results.
10. L. C. Snyder and E. W. Anderson, *J. Chem. Phys.*, **42**, 336 (1965).
11. U. Lienhard, Thesis, Basel, 1972.
12. J. Gerritsen and C. MacLean, *J. Magn. Resonance*, **5**, 44 (1971).
13. H. Nakatsuji et al., *Bull. Chem. Soc. Japan*, **44**, 2010 (1971).
14. C. T. Yim and D. F. R. Gilson, *J. Am. Chem. Soc.*, **91**, 4360 (1969).
15. J. Nehring and A. Saupe, *J. Chem. Phys.*, **52**, 1307 (1970).
16. I. Morishima et al., *Chem. Phys. Lett.*, **7**, 633 (1970).

17.  R. Price and C. Schumann, *Mol. Cryst. Liq. Cryst.*, **16**, 291 (1972).
18.  G. Englert, *Z. Naturforsch.*, **27a**, 715 (1972).
19.  P. Diehl and W. Niederberger, to be published in *J. Magn. Resonance*.
20.  P. Diehl and C. L. Khetrapal, *Can. J. Chem.*, **47**, 1411 (1969).
21.  I. Y. Wei and B. M. Fung, *J. Chem. Phys.*, **52**, 4917 (1970).
22.  J. C. Rowell et al., *J. Chem. Phys.*, **43**, 3442 (1965).
23.  K. D. Lawson and T. J. Flautt, *J. Am. Chem. Soc.*, **89**, 5489 (1967).
24.  G. E. Chapman et al., *Nature*, **225**, 639 (1970).
25.  E. Oldfield et al., *FEBS Lett.*, **16**, 102 (1971).
26.  M. J. Gerace and B. M. Fung, *J. Chem. Phys.*, **53**, 2984 (1970).
27.  C. W. Hilbers and C. MacLean, *Ber. Bunsenges.*, **75**, 277 (1971).
28.  G. Lindblom et al., *Chem. Phys. Lett.*, **8**, 489 (1971).
29.  B. M. Fung et al., *J. Phys. Chem.*, **74**, 83 (1970).
30.  N. O. Persson and A. Johanson, *Acta Chem. Scand.*, **25**, 2118 (1971).
31.  G. Lindblom, *Acta Chem. Scand.*, **25**, 2767 (1971).

# CHAPTER 25

# PHOSPHORUS-PHOSPHORUS AND PHOSPHORUS-METAL COUPLING CONSTANTS IN TRANSITION METAL PHOSPHINE COMPLEXES

John F. Nixon

*School of Molecular Sciences*
*University of Sussex*
*Brighton*

## INTRODUCTION

There is current interest in the factors affecting the magnitude and signs of phosphorus-phosphorus coupling constants $^2J(\text{P-M-P}')$ and phosphorus-metal coupling constants $^1J(\text{P-M})$ in transition metal phosphine complexes. [1,2]

No *systematic* study has been made on systems in which both of these parameters can be measured within the *same* complex. In this chapter we present results for some phosphine complexes of rhodium(I) ($I = \frac{1}{2}$ for $^{103}\text{Rh}$ in 100% natural abundance) in different stereochemistries and for disubstituted octahedral complexes of zero-valent tungsten ($I = \frac{1}{2}$ for $^{185}\text{W}$ in 14.28% natural abundance).

We have chosen a number of fluorophosphine ligands $R_n PF_{3-n}$ for this study since (*a*) the nature of R can be varied easily, and (*b*) the large magnitude of $^1J(\text{P-F})$ facilitates the accurate evaluation of $^2J(\text{P-M-P}')$. (See references 1 and 3 for a fuller discussion.)

## RESULTS AND DISCUSSION

The magnitude of the directly bonded coupling constant $^1J$(P-M) in a metal-phosphine complex is given by the expression[1]:

$$^1J(P\text{-}M) = \gamma_P \gamma_M \cdot \frac{\hbar}{2\pi} \cdot \frac{256\pi^2}{9} \cdot \beta^2 \frac{|S_P(0)|^2 |S_M(0)|^2}{^3\Delta E} \frac{a^2(1-a^2)}{n} \alpha_P^2 ,$$

where $\gamma_x$ is the magnetogyric ratio of nucleus x, $|S_x(0)|^2$ is the electron density in the $s$ orbital of atom x evaluated at the nucleus, $a^2$ is the $s$ character of the metal hybrid orbital, $\alpha_P^2$ is the $s$ character of the phosphorus lone pair orbital, $n$ is the number of ligands, and $^3\Delta E$ is an average triplet excitation energy.

It has been shown[1,4,5] that changes in the magnitude of $^1J$(P-M) induced in those terms that are associated with the metal nucleus are usually relatively small and variations in $^1J$(P-M) within a series of metal-phosphine complexes are largely dependent on changes in $|S_P(0)|^2$ and $\alpha_P^2$. Both these terms increase with increasing electronegativity of the substituent groups attached to phosphorus; for example, $^1J$(P-M) for a tertiary phosphite-metal complex is larger than for the corresponding tertiary phosphine compound.

In earlier studies we have discussed the factors affecting the magnitude and sign of the phosphorus-phosphorus coupling constant $^2J$(P-M-P′) in a number of disubstituted phosphine complexes of chromium, molybdenum, and tungsten, using a simple molecular orbital model.[1] We have successfully correlated the magnitude of $^2J$(P-M-P′) with the nature of the substituent on phosphorus, which affects both the $s$ character of the phosphorus hybrid orbital forming the $\sigma$ bond to the transition metal atom and the $|S_P(0)|^2$ term.[1,3,6]

Since phosphorus-phosphorus couplings are transmitted via the metal atom it might be expected that, to a first approximation, the magnitude of $^1J$(P-M) and $^2J$(P-M-P′) should be related. In order to test this hypothesis we have synthesised a number of new complexes of rhodium(I) and tungsten(O) containing fluorophosphine ligands $R_nPF_{3-n}$ in order to measure

$^1J$(M-P) and $^2J$(P-M-P$'$) and to see how these vary as a function of the substituent R.

The five structural types of complex studied, **1** to **5** below, each have two fluorophosphine ligands in mutually *cis* positions and represent examples of nuclear-spin systems of the type $[AX_n]_2$ (A = phosphorus, X = fluorine, n = 2 or 3). In the rhodium complexes the presence of the rhodium nucleus causes each line in the NMR spectrum to be split further into a 1 : 1 doublet pattern, whereas in the tungsten complexes each signal appears as a 7 : 86 : 7 triplet. The much lower natural abundance of $^{185}$W usually makes evaluation of $^1J$(W-P) rather more difficult than $^1J$(Rh-P). As mentioned earlier, analysis of the X part of the NMR spectrum of the $[AX_n]_2$ type leading to *accurate* evaluation of $J$(A-A$'$), $^2J$(P-M-P$'$) in this case, is facilitated in fluoro-phosphine transition metal complexes because in general $^1J$(P-F) = $J$(A-X) is at least 1000 Hz and therefore very much larger than $^2J$(P-M-P$'$) = $J$(A-A$'$).

$$RPF_2 \diagdown Rh \diagup Cl \diagdown Rh \diagup PF_2R$$
$$RPF_2 \diagup \quad \diagdown Cl \diagup \quad \diagdown PF_2R$$

**1**

$$Rh \diagup PF_2R$$
$$\diagdown PF_2R$$

**2**

$$(acac)\ Rh \diagup PF_2R$$
$$\diagdown PF_2R$$

**3**

$$(facac)\ Rh \diagup PF_2R$$
$$\diagdown PF_2R$$

**4**

$$(CO)_4\ W \diagup PF_2R$$
$$\diagdown PF_2R$$

**5**

(acac = $CH_3COCHCOCH_3$;          facac = $CF_3COCHCOCF_3$. )

A complicating feature of the systems **1**, **3**, and **4** is a result of a rather easy *inter*molecular ligand exchange process which can lead to broadened featureless spectra unless the complexes are carefully purified immediately before use. The ready ligand exchange in square-planar Rh(I) phosphine complexes is well known. We have recently described the *inter*-molecular ligand exchange of *both* types of phosphine in the complex *trans*-RhCl(PF$_3$)(PPh$_3$)$_2$.[7,8] Ligand exchange is entirely absent for complexes of types **2** and **5**; in these cases the formal attainment of the electronic configuration of the next inert gas may be important.

The NMR results for the *cis*-(RPF$_2$)$_2$W(CO)$_4$ complexes are summarized in Table 25-1 together with available literature data on analogous complexes of the ligands PH$_3$, PMe$_3$, and P(OPh)$_3$. It can be seen that the magnitudes of $^1J$(W-P) and $^2J$(P-W-P$'$) are considerably affected by the nature of the substituents attached to phosphorus and the attachment of more electronegative groups leads as expected to larger absolute values for both parameters. In line with other previous results[1,3,9] the complex of trifluorophosphine, PF$_3$, which contains the most electronegative substituents, has the largest values of $^1J$(W-P) and $^2J$(P-W-P$'$), whereas phosphines such as PH$_3$ and PMe$_3$ give values that are very much smaller. There is also a stepwise decrease in both parameters as R changes from F to CH$_2$Cl.

The more limited data presently available from our studies on the fluorophosphine rhodium complexes **1** to **4**, although in general agreement with the results found for the phosphine complexes of zero-valent tungsten [e.g., the PF$_3$ complexes always have the largest values for $^1J$(Rh-P) and $^2J$(P-Rh-P) within their respective series], show some variations from the trends expected on the basis of the simplified arguments presented earlier.

In particular, those complexes where R = Me$_2$N- afford anomalous results. Verkade et al.[10,11] have previously noted anomalies in the magnitude of $^2J$(P-M-P$'$) for a number of phosphine ligands containing R$_2$N- groups and have suggested

TABLE 25-1. Some $^1J$(W-P) and $^2J$(P-W-P$'$) coupling constants in zero-valent tungsten complexes of the type $cis$-$L_2W(CO)_4$

| Ligand L | $^1J$(W-P) (Hz) | $^2J$(P-W-P$'$) (Hz) |
|---|---|---|
| $PF_3$ | $465 \pm 15$[b] | $38 \pm 1$[a] |
| $PF_2CCl_3$ | $420 \pm 10$[b] | $37.6 \pm 0.8$[a] |
| $PF_2CCl_2H$ | $415 \pm 6$[b] | $37.0 \pm 0.5$[a] |
| $P(OPh)_3$ | $416.0$[e] | — |
| $P(OCH_2)_3CC_3H_7$ | — | $35 \pm 5$[c] |
| $PF_2CClH_2$ | $370 \pm 10$[b] | $33 \pm 1$[b] |
| $PF_2Ph$ | $375 \pm 40$[b] | $28 \pm 3$[b] |
| $PBu_3$ | $226.5$[e] | — |
| $PMe_3$ | $209.8$[c] | $25.0 \pm 0.1$[c] |
| $PH_3$ | $125$[d] | $13.4$[d] |

[a]R. M. Lynden-Bell, J. F. Nixon, J. Roberts, J. R. Swain, and W. McFarlane, *Inorg. Nucl. Chem. Lett.* **7**, 1187 (1971).
[b]This work.
[c]References 10 and 11.
[d]E. Moser and E. O. Fischer, *J. Organomet. Chem.*, **15**, 157 (1968).
[e]Reference 4.

that this may be due to the bulk of this type of ligand, which causes the P-M-P angle to be unusually large.

The apparently nonsystematic variation of $^1J$(Rh-P) with $^2J$(P-Rh-P$'$) is being studied further. A *selection* of values from complexes of types **1** to **4** is shown below. More recently we have extended our studies to include complexes containing trifluoromethylphosphines which are also amenable to accurate evaluation of $^2J$(P-M-P$'$).

| | Complex | $^1J(\text{Rh-P})(\text{Hz})$ | $^2J(\text{P-Rh-P}')(\text{Hz})$ |
|---|---|---|---|
| **1** | R = F | 341 | $63.5 \pm 0.5$ |
| | R = $CCl_3$ | 311 | $52 \pm 1$ |
| | R = $Me_2N$ | 302 | $49 \pm 2$ |
| | R = Ph | $277 \pm 10$ | $48 \pm 2$ |
| **2** | R = F | 380.4 | 147.5 |
| | R = $CCl_3$ | 337.3 | 100.8 |
| | R = $CCl_2H$ | 325.1 | 100.6 |
| | R = $Me_2N$ | 332.6 | 114.0 |
| | R = $CF_3$ | 322.8 | 105.1 |
| **3** | R = F | 334.2 | 159.0 |
| | R = $CCl_3$ | — | 106.0 |
| **4** | R = F | 341.4 | 145.5 |
| | R = $CCl_3$ | 302.0 | 95.5 |
| | R = $Me_2N$ | 300.6 | 108.0 |

## ACKNOWLEDGMENTS

It is a pleasure to acknowledge the contributions of my co-workers in this study.  The rhodium work was carried out by Dr. A. A. Pinkerton and Dr. J. R. Swain, and  the tungsten studies by Dr. T. R. Johnson and Dr. J. R. Swain.

## References

1.  J. F. Nixon and A. Pidcock, *Annual Reports on NMR Spectroscopy*, E. Mooney, Ed., Vol. 2, Academic Press, London, 1969, p. 345.
2.  E. G. Finer, R. K. Harris, and J. R. Woplin, *Progress in NMR Spectroscopy*, J. W. Emsley, J. Feeney, and L. H. Sutcliffe, Eds., Vol. 6, Pergamon Press, Oxford, 1971, p. 61.
3.  C. G. Barlow, J. F. Nixon, and J. R. Swain, *J. Chem. Soc. A*, 1082 (1969).

4. G. G. Mather and A. Pidcock, *J. Chem. Soc. A*, 1226 (1970).
5. A. Pidcock, R. E. Richards, and L. M. Venanzi, *J. Chem. Soc.*, 1707 (1966).
6. J. F. Nixon and J. R. Swain, *J. Chem. Soc. Dalton*, 1038 (1972).
7. D. A. Clement, J. F. Nixon, and M. D. Sexton, *Chem. Commun.*, 1509 (1969).
8. D. A. Clement and J. F. Nixon, *J. Chem. Soc. Dalton*, 2553 (1972).
9. R. L. Keiter and J. G. Verkade, *Inorg. Chem.*, **8**, 2115 (1969).
10. R. D. Bertrand, F. B. Ogilvie, and J. G. Verkade, *J. Am. Chem. Soc.*, **92**, 1908 (1970).
11. F. B. Ogilvie, J. M. Jenkins, and J. G. Verkade, *J. Am. Chem. Soc.*, **92**, 1916 (1970).

# SUBJECT INDEX

Number preceding colon denotes chapter. Numbers following colon indicate pages. Numbers in parentheses preceded by T indicate Tables and those preceded by F indicate Figures.